Springer Proceedings in Mathematics & Statistics

Volume 180

Springer Proceedings in Mathematics & Statistics

This book series features volumes composed of selected contributions from workshops and conferences in all areas of current research in mathematics and statistics, including operation research and optimization. In addition to an overall evaluation of the interest, scientific quality, and timeliness of each proposal at the hands of the publisher, individual contributions are all refereed to the high quality standards of leading journals in the field. Thus, this series provides the research community with well-edited, authoritative reports on developments in the most exciting areas of mathematical and statistical research today.

More information about this series at http://www.springer.com/series/10533

Lluís Alsedà i Soler · Jim M. Cushing
Saber Elaydi · Alberto A. Pinto
Editors

Difference Equations, Discrete Dynamical Systems and Applications

ICDEA, Barcelona, Spain, July 2012

Springer

Editors
Lluís Alsedà i Soler
Departament de Matemàtiques
Universitat Autònoma de Barcelona
Cerdanyola del Vallès, Barcelona
Spain

Jim M. Cushing
Department of Mathematics
University of Arizona
Tucson, AZ
USA

Saber Elaydi
Department of Mathematics
Trinity University
San Antonio, TX
USA

Alberto A. Pinto
Faculty of Sciences, Department of Mathematics,
University of Porto
Porto
Portugal

ISSN 2194-1009 ISSN 2194-1017 (electronic)
Springer Proceedings in Mathematics & Statistics
ISBN 978-3-662-52926-3 ISBN 978-3-662-52927-0 (eBook)
DOI 10.1007/978-3-662-52927-0

Library of Congress Control Number: 2016944486

Mathematics Subject Classification (2010): 37-XX, 39-XX, 58-XX

Printed on acid-free paper

This Springer imprint is published by Springer Nature
The registered company is Springer-Verlag GmbH Berlin Heidelberg

Group picture of the participants of the ICDEA 2012 on the main stairs of Casa de Convalescència (venue of the meeting)

Preface

This volume contains the proceedings of the 18th International Conference on Difference Equations and Applications held in Barcelona (Catalonia, Spain) from July 23rd to 27th, 2012. The conference was organized by the Departament de Matemàtiques of the Universitat Autònoma de Barcelona (UAB), under the auspices of the International Society of Difference Equations (ISDE).

The purpose of the conference was to bring together experts and novices in the theory and applications of difference equations and discrete dynamical systems. The main theme of the meeting was the interplay between difference equations and dynamical systems.

The plenary speakers were experts chosen from many different areas of difference equations, *broadly defined*, and discrete dynamical systems and their interplay with nonlinear science.

There were 129 presentations which included 14 plenary talks (including the Special talk of the winner of the "Best JDEA paper" prize), 39 contributed talks, and 87 special sessions talks. The main topics of the meeting were represented by the six special sessions which were organized during the conference. They cover the theory of difference equations and discrete dynamical systems and their applications to biology and economics and are described below.

- *Combinatorial and Topological Dynamics* (14 talks organized by S. Kolyada and L'. Snoha);
- *Complex Dynamics* (23 talks organized by B. Devaney, N. Fagella and X. Jarque);
- *Applications of Difference Equations to Biology* (10 talks organized by J. Cushing, S. Elaydi and J. Li);
- *Asymptotic Behavior and Periodicity of Difference Equations* (20 talks organized by I. Győri and M. Pituk);
- *Chaotic Linear Dynamics* (9 talks organized by J.P. Bès, P. Oprocha and A. Peris); and
- *Economic Dynamics and Control* (11 talks organized by A.A. Pinto and A. Yannacopoulos).

There were 157 participants (120 male and 37 female) from Austria, Belarus, Belgium, Bosnia-Herzegovina, Brazil, Canada, Chile, Czech Republic, Denmark, Finland, France, Germany, Greece, Hungary, India, Iran, Ireland, Israel, Italy, Jamaica, Japan, Latvia, Mexico, Oman, Poland, Portugal, Russia, Serbia, Slovakia, Spain, Sweden, Ukraine, United Kingdom, and USA.

We would like to acknowledge the financial support of the following institutions: Centro Internacional de Matemática, Institut de Matemàtiques de la Universitat de Barcelona, Centre de Recerca Matemàtica, Ministerio de Economía y Competitividad, Generalitat de Catalunya, Grup de Sistemes Dinàmics de la Universitat Autònoma de Barcelona. Also, we would like to thank the organizing committee that ensured a good organization and the success of the conference as well as the scientific committee that took care of the high scientific standards and the quality of the conference.

We warmly thank all the speakers and participants of the meeting for their contributions and helping to create a wonderful, friendly, and fruitful atmosphere.

All participants of ICDEA2012 were invited to submit a contribution to these proceedings, and all papers that were accepted had to pass through a refereeing process appropriate for a mathematical research journal.

Cerdanyola del Vallès, Spain
Tucson, USA
San Antonio, USA
Porto, Portugal

Lluís Alsedà i Soler
Jim M. Cushing
Saber Elaydi
Alberto A. Pinto

Organization Committees

Organizing Committee

Lluís Alsedà i Soler, Universitat Autònoma de Barcelona, Barcelona, Spain
Francisco Balibrea, Universidad de Murcia, Murcia, Spain
Zhaoyang Dong, Universitat Autònoma de Barcelona, Barcelona, Spain
Eduardo Liz, Universidad de Vigo, Vigo, Spain
Leopoldo Morales, Universitat Autònoma de Barcelona, Barcelona, Spain
Henrique Oliveira, Instituto Superior Técnico de Lisboa, Lisboa, Portugal
David Romero i Sànchez, Universitat Autònoma de Barcelona, Barcelona, Spain

Scientific Committee

Lluís Alsedà i Soler, Universitat Autònoma de Barcelona, Barcelona, Spain
Jim M. Cushing, University of Arizona, Tucson, Arizona, USA
Sergiĭ Kolyada, National Academy of Sciences, Kiev, Ukraine
Alberto A. Pinto, Universidade de Porto and International Centre for Mathematics, Porto, Portugal
Robert Sacker, University of Southern California, Los Angeles, California, USA
André Vanderbauwhede, Gant University, Gant, Belgium

Contents

On the Second Order Rational Difference Equation $x_{n+1} = \beta + \frac{1}{x_n x_{n-1}}$

Aija Anisimova

Abstract The author investigates the local and global stability character, the periodic nature, and the boundedness of solutions of the second-order rational difference equation

$$x_{n+1} = \beta + \frac{1}{x_n x_{n-1}}, \quad n = 0, 1, \ldots,$$

with parameter β and with arbitrary initial conditions such that the denominator is always positive. The main goal of the paper is to confirm Conjecture 8.1 and to solve Open Problem 8.2 stated by A.M. Amleh, E. Camouzis and G. Ladas in *On the Dynamics of a Rational Difference Equations I* (International Journal of Difference Equations, Volume 3, Number 1, 2008, pp.1–35).

Keywords Boundedness · Periodicity · Rational difference equations · Stability

AMS Subject Classifications 39A10 · 39A20 · 39A30

1 Introduction and Preliminaries

The author investigates the local and global stability character, the periodic nature, and the boundedness of solutions of the second-order rational difference equation in the form

$$x_{n+1} = \beta + \frac{1}{x_n x_{n-1}}, \quad n = 0, 1, \ldots, \tag{1}$$

with parameter β and with arbitrary initial conditions such that the denominator is always positive.

A. Anisimova (✉)
University of Latvia, Zellu 8, Riga, Latvia
e-mail: aija-anisimova@inbox.lv

L. Alsedà i Soler et al. (eds.), *Difference Equations, Discrete Dynamical Systems and Applications*, Springer Proceedings in Mathematics & Statistics 180, DOI 10.1007/978-3-662-52927-0_1

In the paper [1] have been given several open problems and conjectures about such equations:

Conjecture 8.1 ([1]) *Every positive solution of (1) has a finite limit.*

Open Problem 8.2 ([1]) Assume that β is a real number. Determine the set G of real initial values x_{-1}, x_0 for which the equation (1) is well defined for all $n \geq 0$, and investigate the character of solutions of (1) with $x_{-1}, x_0 \in G$.

In this paper the author investigates the local and global stability character of the difference equation (1), describes the periodic nature of the difference equation (1), investigates the boundedness of solutions of (1) and poses some ideas how to confirm Conjecture 8.1 and to solve Open Problem 8.2.

Equation (1) is a special case of the second-order quadratic rational difference equation

$$x_{n+1} = \frac{\alpha + \beta x_n x_{n-1} + \gamma x_{n-1}}{A + B x_n x_{n-1} + C x_{n-1}}, \quad n = 0, 1, 2, \ldots \tag{2}$$

with non-negative parameters and with arbitrary non-negative initial conditions such that the denominator is always positive.

Related non-linear second order rational difference equations have been investigated in [1, 2, 5–9].

Now we give well-known results, which will be useful in the investigation of (1).

For the next results, we consider the difference equation defined by

$$x_{n+1} = f(x_n, x_{n-1}), \quad n = 0, 1, \ldots. \tag{3}$$

Let I be some interval of real numbers and let

$$f : I \times I \to I$$

be a continuously differentiable function.

Then for every set of initial conditions $x_{-1}, x_0 \in I$, the difference equation (3) has a unique solution $\{x_n\}_{n=-1}^{\infty}$.

A point $\overline{x} \in I$ is called an equilibrium point of (3) if

$$\overline{x} = f(\overline{x}, \overline{x})$$

that is,

$$x_n = \overline{x}, \quad \forall n \geq 0$$

is a solution of (3), or equivalently, $\overline{x}$ is a fixed point of f.

Let $Dom(f)$ be a domain of function f of (1). The Forbidden set (denote with F) of function f is a set such that:

$$F := \{(c, d) \in R^2 : \quad \exists n \in N : (x_0, x_{-1}) = (c, d), (x_k, x_{k-1}) \in Dom(f) \quad \forall k = 0, 1, \ldots, n,$$

and $\quad (x_n, x_{n+1}) \notin Dom(f)\}$.

Let $p = \frac{\partial f}{\partial u}(\overline{x}, \overline{x})$ and $q = \frac{\partial f}{\partial v}(\overline{x}, \overline{x})$ denote the partial derivatives of $f(u, v)$ evaluated at the equilibrium $\overline{x}$ of (3). Then the equation

$$y_{n+1} = p y_n + q y_{n-1}, \quad n = 0, 1, \ldots \tag{4}$$

is called the linearized equation associated with (3) about the equilibrium point $\overline{x}$ and the equation

$$\lambda^2 - p\lambda - q = 0 \tag{5}$$

called the characteristic equation of (4) about $\overline{x}$.

Theorem 1 ([7])

1. *If both roots of quadratic equation (5) lie in the open unit disk $|\lambda| < 1$, then the equilibrium $\overline{x}$ of (3) is locally asymptotically stable.*
2. *If at least one of the roots of (5) has absolute value greater than one, then the equilibrium $\overline{x}$ of (3) is unstable.*
3. *A necessary and sufficient condition for both roots of (5) to lie in the open unit disk $|\lambda| < 1$, is*
$$|p| < 1 - q < 2. \tag{6}$$

In the next theorem we make use of the following notation associated with a function $f(z_1, z_2)$, which is monotonic in both arguments. For each pair of numbers (m, M) and for each $i \in \{1, 2\}$, define

$$M_i(m, M) = \begin{cases} M, & \text{if } f \text{ is increasing in } z_i \\ m, & \text{if } f \text{ is decreasing in } z_i \end{cases}$$

and

$$m_i(m, M) = M_i(M, m).$$

Theorem 2 ([4, 7]) *Let $[a, b]$ be a closed and bounded interval of real numbers and let $f \in C([a, b]^2, [a, b])$ satisfy the following conditions:*

1. *$f(z_1, z_2)$ is monotonic in each of its arguments.*
2. *If (m,M) is a solution of the system*

$$\begin{cases} M = f(M_1(m, M), M_2(m, M)) \\ m = f(m_1(m, M), m_2(m, M)) \end{cases}$$

then $M = m$.

Then the difference equation (3) has a unique equilibrium point $\overline{x} \in [a, b]$ and every solution of (3), with initial conditions in $[a, b]$, converges to $\overline{x}$.

A solution $\{x_n\}_{n=-1}^{\infty}$ of (3) is said to be periodic with period p if

$$x_{n+p} = x_n \quad for \quad all \quad n \geq -1. \tag{7}$$

A solution $\{x_n\}_{n=-1}^{\infty}$ of (3) is said to be periodic with prime period p, or a p-cycle if it is periodic with period p and p is the least positive integer for which (7) holds.

With the change of variables

$$x_n = \frac{1}{y_n\sqrt{\beta}} \tag{8}$$

difference equation (1) can be transformed to the difference equation in the form

$$y_{n+1} = \frac{\alpha}{1 + y_n y_{n-1}}, \qquad n = 0, 1, \ldots, \tag{9}$$

where

$$x_n = \frac{1}{y_n\sqrt{\beta}}, \quad x_{n-1} = \frac{1}{y_{n-1}\sqrt{\beta}}, \quad x_{n+1} = \frac{1}{y_{n+1}\sqrt{\beta}}$$

and by these equalities we obtain that

$$\frac{1}{y_{n+1}\sqrt{\beta}} = \beta + y_n\sqrt{\beta}y_{n-1}\sqrt{\beta} \Rightarrow$$

$$\Rightarrow y_{n+1}\sqrt{\beta} = \frac{1}{\beta(1 + y_n y_{n-1})} \Rightarrow y_{n+1} = \frac{\frac{1}{\beta\sqrt{\beta}}}{1 + y_n y_{n-1}},$$

where

$$\alpha = \frac{1}{\beta\sqrt{\beta}}. \tag{10}$$

Equality (10) and transformation to the difference equation (9) are true for all $\beta > 0$ (or $\alpha > 0$).

In paper [1] had been proved that:

1. Every solution of (9) is bounded by positive constants, precisely

$$\frac{\alpha}{1 + \alpha^2} \leq y_{n+1} = \frac{\alpha}{1 + y_n y_{n-1}} \leq \alpha, \quad \forall n \geq 1; \tag{11}$$

2. Assume that

$$0 < \alpha \leq 2. \tag{12}$$

Then the positive equilibrium of (9) is globally asymptotically stable.
These characteristics of (9) can be useful in the investigation of (1).

2 Boundedness

In this section we investigate the boundedness of (1).

Theorem 3 *Every positive solution of (1) is bounded from above and below by positive constants.*

Proof Obviously we can estimate equation (1) from below by

$$x_{n+1} = \beta + \frac{1}{x_n x_{n-1}} \geq \beta \quad \forall n \geq 1.$$

Set that $\beta \leq x_n$ and in equation (1) replacing it with a smaller value, the fraction increases and we get the estimation from above

$$x_{n+1} = \beta + \frac{1}{x_n x_{n-1}} \leq \beta + \frac{1}{\beta^2}, \quad \forall n \geq 1.$$

Finally we obtain

$$\beta \leq x_{n+1} = \beta + \frac{1}{x_n x_{n-1}} \leq \beta + \frac{1}{\beta^2}, \quad \forall n \geq 1. \quad \square \tag{13}$$

3 Local and Global Asymptotic Stability

In this section we investigate the local and global stability of the solutions of the difference equation (1) where the parameter β and the initial values x_{-1}, x_0 are arbitrary real numbers, such that denominator is not equal to zero.

Assume $x_{-1}, x_0 \in I$, where I is the set of all real initial values x_{-1}, x_0 for which the difference equation (1) is well defined for all $n \geq 0$.

The Equilibrium equation of (1)

$$\overline{x}^3 - \beta\overline{x}^2 - 1 = 0 \tag{14}$$

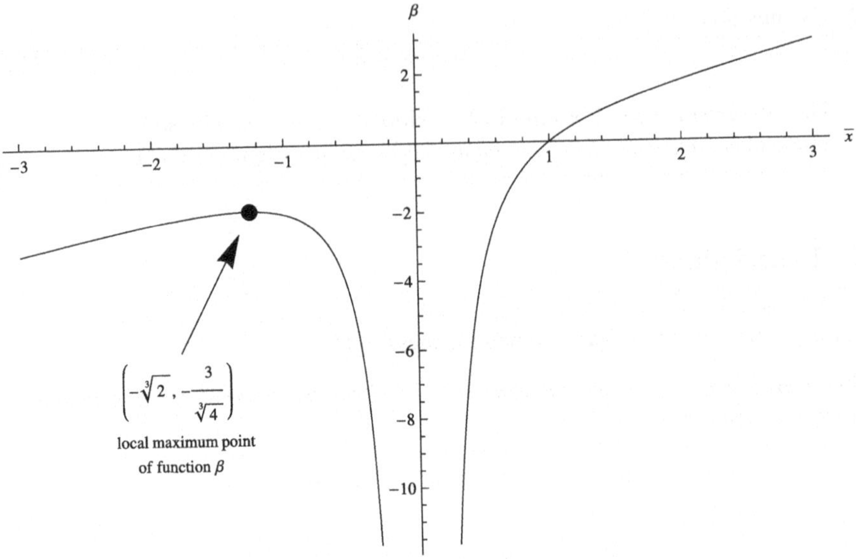

Fig. 1 Graph of function β

can be easily considered as a function of $\overline{x}$:

$$\beta = \frac{\overline{x}^3 - 1}{\overline{x}^2}. \tag{15}$$

Equilibrium equation (14) has three roots, the real roots can be obtained from the graph of the function (15) (see Fig. 1). From Fig. 1 we see that the parameter β is related to the equilibrium point and the function β has one local maximum point $(-\sqrt[3]{2} \approx -1.26; -\frac{3}{\sqrt[3]{4}} \approx -1.89)$, and behaviour of function β is as follows:

1. If $\beta > 0$ the function β is increasing and difference equation (1) has an unique real equilibrium point;
2. In the interval $-\frac{3}{\sqrt[3]{4}} < \beta < 0$ function is increasing. In this interval difference equation (1) has only one real equilibrium point;
3. In case $\beta < -\frac{3}{\sqrt[3]{4}}$ function β is decreasing and difference equation (1) has three real equilibrium points;
4. There exists vertical asymptote at zero ($\overline{x} = 0$).

We can make an assertion that the only local maximum point of function β is a very important point of reference when analysing character of solutions of difference equation (1), so further we will investigate behaviour of (1) depending on the different values of the parameter β.

Looking at many different examples and analysing the equilibrium equation for different values of the parameter β and investigating the function of β it can be concluded that the behaviour of solutions of (1) is very sensitive to different values of parameter β, and if we take arbitrary initial conditions such that the denominator is not equal to zero we can make the following hypotheses:

1. If $\beta > 0$, then the equilibrium equation has one real root and two complex conjugate roots. The solution of difference equation (1) converges to the unique real equilibrium point.
2. If $\beta = 0$, then the equilibrium equation has one real root and two complex conjugate roots. The solution of difference equation (1) is periodic with prime period 3.
3. If $-\frac{3}{\sqrt[3]{4}} < \beta < 0$, then the equilibrium equation has one real root and two complex conjugate roots, but solution of difference equation (1) is with oscillating character.
4. If $\beta = -\frac{3}{\sqrt[3]{4}}$, then the equilibrium equation has three real equilibrium points (two equal and one different). The solution of difference equation (1) converges to the double root $\overline{x}_2 = -\sqrt[3]{2}$ of the equilibrium equation.
5. If $\beta < -\frac{3}{\sqrt[3]{4}}$, then the equilibrium equation has three real roots. The solution of difference equation (1) converges to the root that is greatest by the absolute value.

In numerical calculations we have observed that in these four situations behaviour of solutions of difference equations (1) is similar with arbitrary initial values such that difference equation (1) is well defined.

In our investigation we have established the following results.

Theorem 4 *Assume $x_{-1}, x_0 \in I$, where I is the set of all real initial values x_{-1}, x_0 for which the difference equation (1) is well defined for all $n \geq 0$. If $\beta \in (-\infty; -\frac{3}{\sqrt[3]{4}}] \cup [\frac{1}{\sqrt[3]{4}}; \infty)$, then the solution of difference equation (1) is locally asymptomatically stable.*

Proof The proof is based on the theorem of linearized stability of second order difference equation (see Theorem 1). Let write the characteristic equation of (1) about $\overline{x}$:

$$f = \beta + \frac{1}{uv}$$

$$u := x_n, \quad v := x_{n-1}$$

$$f_u^{'} = (\beta + (uv)^{-1})_u^{'} = -\frac{v}{(uv)^2} \Rightarrow p = f_u^{'}(\overline{x}, \overline{x}) = -\frac{1}{\overline{x}^3}$$

$$f_v^{'} = (\beta + (uv)^{-1})_v^{'} = -\frac{u}{(uv)^2} \Rightarrow q = f_v^{'}(\overline{x}, \overline{x}) = -\frac{1}{\overline{x}^3}$$

$$p = q = -\frac{1}{\overline{x}^3}$$

$$z_{n+1} = pz_n + qz_{n-1}, \quad n = 0, 1, \ldots$$

$$z_{n+1} = -\frac{1}{\overline{x}^3} z_n - \frac{1}{\overline{x}^3} z_{n-1}. \tag{16}$$

Equation (16) is the linearized equation associated with (1) about the equilibrium point $\overline{x}$. The characteristic equation of (1) about $\overline{x}$ is in the form:

$$\lambda^2 + \frac{1}{\overline{x}^3}\lambda + \frac{1}{\overline{x}^3} = 0 \tag{17}$$

By Condition 1 of Theorem 1 if both roots of (17) lie in the open unit disk $|\lambda| < 1$ then the equilibrium $\overline{x}$ of (1) is locally asymptotically stable. Roots of (17) lie in the open unit disk if Condition 3 of Theorem 1 holds, that is,

$$\left|-\frac{1}{\overline{x}^3}\right| < 1 - \frac{1}{\overline{x}^3} < 2.$$

If $\overline{x} > 0$, we have

$$\begin{cases} \frac{1}{\overline{x}^3} < 1 - \frac{1}{\overline{x}^3} \\ \frac{1}{\overline{x}^3} > -1 + \frac{1}{\overline{x}^3} \\ 1 - \frac{1}{\overline{x}^3} < 2 \end{cases} \Rightarrow \overline{x} > \sqrt[3]{2}.$$

Since β can be expressed from (15) we obtain that

$$\beta = \overline{x} - \frac{1}{\overline{x}^2} > \sqrt[3]{2} - \frac{1}{\sqrt[3]{4}} = \frac{1}{\sqrt[3]{4}}.$$

We have established Condition 3 of Theorem 1 when $\beta > \frac{1}{\sqrt[3]{4}}$ and this means that in this case the equilibrium point $\overline{x}$ is locally asymptotically stable.
If $\overline{x} < 0$, we have

$$\begin{cases} \frac{1}{\overline{x}^3} < 1 - \frac{1}{\overline{x}^3} \\ \frac{1}{\overline{x}^3} > -1 + \frac{1}{\overline{x}^3} \\ 1 - \frac{1}{\overline{x}^3} < 2 \end{cases} \Rightarrow \overline{x} < -\sqrt[3]{2}$$

and by Eq. (15) we get that

$$\beta = \overline{x} - \frac{1}{\overline{x}^2} < -\sqrt[3]{2} - \frac{1}{\sqrt[3]{4}} = -\frac{3}{\sqrt[3]{4}}.$$

In this case we see that Condition 3 of Theorem 1 holds, that is, the equilibrium point $\overline{x}$ is locally asymptotically stable when $\beta < -\frac{3}{\sqrt[3]{4}}$. □

Theorem 5 *Assume $x_{-1}, x_0 \in I$, where I is the set of all real initial values x_{-1}, x_0 for which the difference equation (1) is well defined for all $n \geq 0$ and $\beta \in [\frac{1}{\sqrt[3]{4}}; \infty)$. Then every positive solution $\{x_n\}_{n=-1}^{\infty}$ of difference equation (1) has a finite limit.*

Proof With the change of variables $x_n = \frac{1}{y_n \sqrt{\beta}}$ difference equation can be transformed to the difference equation (1) in the form

$$y_{n+1} = \frac{\alpha}{1 + y_n y_{n-1}}, \quad n = 0, 1, \ldots,$$

where $x_n = \frac{1}{y_n \sqrt{\beta}}, \quad x_{n-1} = \frac{1}{y_{n-1} \sqrt{\beta}}, \quad x_{n+1} = \frac{1}{y_{n+1} \sqrt{\beta}}$ and $\alpha = \frac{1}{\beta \sqrt{\beta}}$.

In the paper [1] it has been shown that every positive solution of (9) converges to $\bar{y}$ for all values of $0 < \alpha \leq 2$. Since $\alpha = \frac{1}{\beta \sqrt{\beta}}$ then

$$0 < \frac{1}{\beta \sqrt{\beta}} \leq 2 \Rightarrow \infty > \beta \sqrt{\beta} \geq \frac{1}{2} \Rightarrow \beta^3 \geq \frac{1}{4} \Rightarrow \beta \geq \frac{1}{\sqrt[3]{4}}. \quad \square$$

4 Periodicity

In this section we discuss the periodicity of equation (1). We will show that difference equation (1) has no periodic solutions with period 2 and difference equation (1) has periodic solution with period 3 if and only if $\beta = 0$.

Assume $x_{-1}, x_0 \in I$, where I is the set of all real initial values x_{-1}, x_0 for which the difference equation (1) is well defined for all $n \geq 0$ then that next two results are true.

Theorem 6 *Difference equation (1) has no periodic solutions with period* 2.

Proof Assume that x_{-1}, x_0 are initial conditions such that difference equation (1) is well defined and solution of difference equation (1) is periodic with period 2 $(x_{-1} \neq x_0)$, that is,

$$\ldots x_{-1}, x_0, x_{-1}, x_0, \ldots.$$

Then it must be that

$$x_1 = \beta + \frac{1}{x_0 x_{-1}} = x_{-1}$$

$$x_2 = \beta + \frac{1}{x_1 x_0} = \beta + \frac{1}{x_{-1} x_0} = x_{-1} \Rightarrow x_0 = x_{-1}$$

which is a contradiction from which follows that equation (1) has no periodic solutions with period 2. $\square$

Theorem 7 *Difference equation (1) has periodic solution with period* 3 *if and only if* $\beta = 0$.

Proof Assume that x_{-1}, x_0 are well defined initial conditions and the solution of difference Eq. (1) is periodic with period 3, that is,

$$\ldots x_{-1}, x_0, x_1, x_{-1}, x_0, x_1, \ldots.$$

1. Assume $\beta \neq 0$. Then it must be that

$$x_1 = \beta + \frac{1}{x_0 x_{-1}}$$

$$x_2 = \beta + \frac{1}{x_1 x_0} = x_{-1}$$

$$x_3 = \beta + \frac{1}{x_2 x_1} = x_0$$

$$x_4 = \beta + \frac{1}{x_3 x_2} = x_1$$

From which follows that

$$x_2 = \beta + \frac{1}{x_1 x_0} = \beta + \frac{1}{(\beta + \frac{1}{x_0 x_{-1}})x_0} = x_{-1} \Rightarrow$$

$$\Rightarrow \frac{\beta(\beta x_0 x_{-1} + 1) + x_{-1}}{\beta x_0 x_{-1} + 1} = x_{-1} \Leftrightarrow \beta = 0,$$

which is a contradiction from which follows that if $\beta \neq 0$ then (1) has no periodic solutions with period 3.

2. Assume $\beta = 0$. Then we can write that

$$x_1 = \frac{1}{x_0 x_{-1}}$$

$$x_2 = \frac{1}{x_1 x_0} = \frac{x_0 x_{-1}}{x_0} = x_{-1}$$

$$x_3 = \frac{1}{x_2 x_1} = \frac{x_{-1} x_0}{x_{-1}} = x_0$$

$$x_4 = \frac{1}{x_3 x_2} = \frac{1}{x_0 x_{-1}} = x_1$$

$$\vdots$$

$$x_{n-1} = \frac{1}{x_{n-2}x_{n-3}}$$

$$x_n = \frac{1}{x_{n-1}x_{n-2}} = \frac{x_{n-2}x_{n-3}}{x_{n-2}} = x_{n-3}$$

$$x_{n+1} = \frac{1}{x_n x_{n-1}} = \frac{x_{n-2}x_{n-3}}{x_{n-3}} = x_{n-2}$$

$$x_{n+2} = \frac{1}{x_{n+1}x_n} = \frac{1}{x_{n-2}x_{n-3}} = x_{n-1}$$

$$x_{n+3} = \frac{1}{x_{n+2}x_{n+1}} = \frac{1}{x_{n-1}x_{n-2}} = x_{n-3}$$

$$\vdots$$

From which follows that if $\beta = 0$, then (1) has a periodic solution with period 3. □

After these two last results we obtain the following conclusions.

Corollary 1 *1. If $\beta \geq \frac{1}{\sqrt[3]{4}}$ and $\beta \leq -\frac{3}{\sqrt[3]{4}}$ then (1) has no periodic solutions with period $p > 1$, because in these cases can be obtained that the solution of (1) is locally asymptotically stable.*

2. *If $\beta = 0$, $x_{-1}, x_0 > 0$, $x_{-1} \neq x_0$ then (1) has a periodic solution with period 3 and solution is bounded in interval $[\frac{1}{x_0 x_{-1}}; \max\{x_{-1}, x_0\}]$.*
3. *If $\beta = 0$, $x_{-1} = x_0 > 0$, $x_{-1} = x_0 \neq 1$ then the solution of (1) is periodic with period 3 and bounded in interval $[\frac{1}{x_{-1}^2}; x_{-1}]$. If $x_{-1} = x_0 = 1$ then the solution of (1) is $x_n = \{1\}_{n \geq -1}$ and it is periodic with period 1.*
4. *If $\beta = 0$, $x_{-1}, x_0 < 0$, $x_{-1} \neq x_0$ then the solution of (1) is periodic with period 3 and bounded from below by $\min\{x_{-1}, x_0\}$ and from above by $\frac{1}{x_0 x_{-1}}$.*
5. *If $\beta = 0$, $x_{-1} = x_0 < 0$ then the solution of (1) is periodic with period 3 and bounded in interval $[x_{-1}, \frac{1}{x_{-1}^2}]$.*
6. *If $\beta = 0$, $x_{-1} > 0$, $x_0 < 0$ or $x_{-1} < 0$, $x_0 > 0$ then the solution of (1) is periodic with period 3 and bounded from below by $\min\{x_{-1}, x_0\}$ and from above by $\max\{x_{-1}, x_0\}$.*
7. *If $\beta = 0$, then difference equation (1) can be written in the form*

$$x_{n+1} = \frac{1}{x_n x_{n-1}}, \quad n = 0, 1, \ldots, \tag{18}$$

and the positive solution of difference equation (18) can be written in the form

$$x_n = e^{c_1 \sin\left(\frac{2\pi}{3}n\right)+c_2 \cos\left(\frac{2\pi}{3}n\right)}, \quad c_1, c_2 \in R \tag{19}$$

and it is periodic with prime period 3 for all positive well defined initial conditions.

5 Forbidden Set

Open Problem 8.2 ([1]) Assume that β is a real number. Determine the set I of real initial values x_{-1}, x_0 for which the difference equation (1) is well defined for all $n \geq 0$, and investigate the character of solutions of difference equation (1) with $x_{-1}, x_0 \in I$.

Let $Dom(f)$ be domain of function $f(x_n, x_{n-1}) = \beta + \frac{1}{x_n x_{n-1}}$ of difference equation (1).

If there exist initial values (x_0, x_{-1}) and such $n \in N$ for which difference equation (1) is well defined and in iteration $n+1$ difference equation (1) becomes equal to zero then initial values (x_0, x_{-1}) belong to forbidden set (F). This holds if

$$\beta = -\frac{1}{x_n x_{n-1}}, \quad n = 0, 1, \ldots, \quad x_n, x_{n-1} \neq 0 \Rightarrow \quad \beta x_n x_{n-1} + 1 = 0. \tag{20}$$

If initial values (x_0, x_{-1}) belong to the forbidden set then holds one of these conditions:

1. $\beta \in R$ and x_{-1} or x_0 is equal to zero;
2. $\beta \in R$ and $x_{-1} = x_0 = 0$;
3. $\beta < -1$ and $x_n x_{n-1} \in (0; 1)$;
4. $\beta = -1$ and $x_n x_{n-1} = 1$;
5. $-1 < \beta < 0$ and $x_n x_{n-1} \in (1; +\infty)$;
6. $\beta = 0$ and $x_n x_{n-1} = 0$;
7. $0 < \beta < 1$ and $x_n x_{n-1} \in (-\infty; -1)$;
8. $\beta = 1$ and $x_n x_{n-1} = -1$;
9. $\beta > 1$ and $x_n x_{n-1} \in (-1; 0)$.

Let $\beta \neq 0, \quad x_{-1}x_0 \neq 0$. Now we determine the forbidden set in each iteration.

1. $n = 0: \quad x_1 = 0 \Leftrightarrow \beta x_0 x_{-1} + 1 = 0 \Rightarrow x_0 = -\frac{1}{\beta x_{-1}}$
   In this step initial conditions $(x_{-1}, -\frac{1}{\beta x_{-1}}) \in F$.
2. $n = 1: \quad x_2 = 0 \Leftrightarrow \beta^2 x_0 x_{-1} + \beta + x_{-1} = 0 \Rightarrow x_0 = \frac{-\beta - x_{-1}}{\beta^2 x_{-1}}$
   $(x_{-1}, \frac{-\beta - x_{-1}}{\beta^2 x_{-1}}) \in F$.
3. $n = 2: \quad x_3 = 0 \Leftrightarrow \beta^3 x_0 x_{-1} + \beta^2 + \beta x_{-1} + x_0 x_{-1} = 0 \Rightarrow x_0 = \frac{-\beta^2 - \beta x_{-1}}{\beta^3 x_{-1} + x_{-1}}$
   $(x_{-1}, \frac{-\beta^2 - \beta x_{-1}}{\beta^3 x_{-1} + x_{-1}}) \in F$.

4. $n = 3: \quad x_4 = 0 \Leftrightarrow \beta^4 x_0 x_{-1} + \beta^3 + \beta^2 x_{-1} + 2\beta x_0 x_{-1} + 1 = 0 \Rightarrow x_0 = \frac{-\beta^3 - \beta^2 x_{-1} - 1}{\beta^4 x_{-1} + 2\beta x_{-1}}$
   $(x_{-1}, \frac{-\beta^3 - \beta^2 x_{-1} - 1}{\beta^4 x_{-1} + 2\beta x_{-1}}) \in F.$
5. $n = 4: \quad x_5 = 0 \Leftrightarrow \beta^5 x_0 x_{-1} + \beta^4 + \beta^3 x_{-1} + 3\beta^2 x_0 x_{-1} + 2\beta + x_{-1} = 0 \Rightarrow x_0 = \frac{-\beta^4 - \beta^3 x_{-1} - 2\beta - x_{-1}}{\beta^5 x_{-1} + 3\beta^2 x_{-1}}$
   $(x_{-1}, \frac{-\beta^4 - \beta^3 x_{-1} - 2\beta - x_{-1}}{\beta^5 x_{-1} + 3\beta^2 x_{-1}}) \in F.$
6. $n = 5: \quad x_6 = 0 \Leftrightarrow \beta^6 x_0 x_{-1} + \beta^5 + \beta^4 x_{-1} + 4\beta^3 x_0 x_{-1} + 3\beta^2 + 2\beta x_{-1} + x_0 x_{-1} = 0 \Rightarrow x_0 = \frac{-\beta^5 - \beta^4 x_{-1} - 3\beta^2 - 2\beta x_{-1}}{\beta^6 x_{-1} + 4\beta^3 x_{-1} + x_{-1}}$
   $(x_{-1}, \frac{-\beta^5 - \beta^4 x_{-1} - 3\beta^2 - 2\beta x_{-1}}{\beta^6 x_{-1} + 4\beta^3 x_{-1} + x_{-1}}) \in F.$
7. $n = 6: \quad x_7 = 0 \Leftrightarrow \beta^7 x_0 x_{-1} + \beta^6 + \beta^5 x_{-1} + 5\beta^4 x_0 x_{-1} + 4\beta^3 + 3\beta^2 x_{-1} + 3\beta x_0 x_{-1} + 1 = 0 \Rightarrow x_0 = \frac{-\beta^6 - \beta^5 x_{-1} - 4\beta^3 - 3\beta^2 x_{-1} - 1}{\beta^7 x_{-1} + 5\beta^4 x_{-1} + 3\beta x_{-1}}$
   $(x_{-1}, \frac{-\beta^6 - \beta^5 x_{-1} - 4\beta^3 - 3\beta^2 x_{-1} - 1}{\beta^7 x_{-1} + 5\beta^4 x_{-1} + 3\beta x_{-1}}) \in F.$
8. $n = 7: \quad x_8 = 0 \Leftrightarrow \beta^8 x_0 x_{-1} + \beta^7 + \beta^6 x_{-1} + 6\beta^5 x_0 x_{-1} + 5\beta^4 + 4\beta^3 x_{-1} + 6\beta^2 x_0 x_{-1} + 3\beta + x_{-1} = 0 \Rightarrow x_0 = \frac{-\beta^7 - \beta^6 x_{-1} - 5\beta^4 - 4\beta^3 x_{-1} - 3\beta - x_{-1}}{\beta^8 x_{-1} + 6\beta^5 x_{-1} + 6\beta^2 x_{-1}}$
   $(x_{-1}, \frac{-\beta^7 - \beta^6 x_{-1} - 5\beta^4 - 4\beta^3 x_{-1} - 3\beta - x_{-1}}{\beta^8 x_{-1} + 6\beta^5 x_{-1} + 6\beta^2 x_{-1}}) \in F.$
9. $n = 8: \quad x_9 = 0 \Leftrightarrow \beta^9 x_0 x_{-1} + \beta^8 + \beta^7 x_{-1} + 7\beta^6 x_0 x_{-1} + 6\beta^5 + 5\beta^4 x_{-1} + 10\beta^3 x_0 x_{-1} + 6\beta^2 + 3\beta x_{-1} + x_0 x_{-1} = 0 \Rightarrow x_0 = \frac{-\beta^8 - \beta^7 x_{-1} - 6\beta^5 - 5\beta^4 x_{-1} - 6\beta^2 - 3\beta x_{-1}}{\beta^9 x_{-1} + 7\beta^6 x_{-1} + 10\beta^3 x_{-1} + x_{-1}}$
   $(x_{-1}, \frac{-\beta^8 - \beta^7 x_{-1} - 6\beta^5 - 5\beta^4 x_{-1} - 6\beta^2 - 3\beta x_{-1}}{\beta^9 x_{-1} + 7\beta^6 x_{-1} + 10\beta^3 x_{-1} + x_{-1}}) \in F.$
10. $\cdots$

The general case (when $n = k: \quad x_{k+1} = 0$) is still in investigation process.

Example 1 If $\beta = -\frac{3}{\sqrt[3]{4}}$, then with the change of variables $x_{n-1} = -\sqrt[3]{2} y_{n-1}$ equation (1) can be written in the form without irrationality

$$y_{n+1} = \frac{3}{2} - \frac{1}{2 y_n y_{n-1}}, \quad n = 0, 1 \ldots. \tag{21}$$

$$\overline{y}_1 = -0.5, \quad \overline{y}_2 = \overline{y}_3 = 1$$

Let assume

$$y_{-1} = \frac{k}{k+1}, y_0 = \frac{k-1}{k}, \quad k \in N \Rightarrow y_k = \frac{1}{2}, y_{k+1} = 0.$$

As we can take k as big as we like initial points y_{-1}, y_0 will be close to the equilibrium point $\overline{y} = 1$, but after limited amount of iterations we get $y_{k+1} = 0$. Hence

$$(y_{-1}, y_0) = \left(\frac{k}{k+1}, \frac{k-1}{k}\right) \in F, \quad k \in N.$$

Acknowledgments These results were presented in 18th International Conference on Difference Equations and Applications (ICDEA 2012) ([3]).
This research is partially supported by the project of European Social Fund "Support for doctoral studies at the Latvian University" No. 2009/0138/1DP/1.1.2.1.2/09/IPIA/VIAA/004.

References

1. Amleh, A.M., Camouzis, E., Ladas, G.: On the dynamics of a rational difference equations I. Int. J. Differ. Equ. **3**(1), 1–35 (2008)
2. Amleh, A.M., Camouzis, E., Ladas, G.: On the dynamics of a rational difference equations II. Int. J. Differ. Equ. **3**, 195–225 (2008)
3. Anisimova, A.: On the Second Order Rational Difference Equation $x_{n+1} = \beta + \frac{1}{x_n x_{n-1}}$. In: Abstract 18th International Conference on Difference Equations Applied, 22-27 July, Barcelona, Spain (2012). http://www.gsd.uab.cat/icdea2012/Abstracts-pdf/COM-02.pdf
4. Camouzis, E., Ladas, G.: Advances in Discrete Mathematics and Aplications, Volume 5, Dynamics of Third - Order Rational Difference Equations with Open Problems and Conjectures. Chapman & Hall/CRC, Boca Raton (2008)
5. Drymonis, E., Kostrov, Y., Kudlak, Z.: On the difference equations with nonnegative periodic coefficients. Int. J. Differ. Equ. **7**(1), 1–16 (2012)
6. Hu, L., Li, W., Stevic, S.: Global asymptotic stability of a second order rational difference equation. Int. J. Differ. Equ. **14**(8), 779–797 (2008)
7. Kulenovic, M.R.S, Ladas, G.: Dynamics of Second Order Rational Difference Equations: With Open Problems and Conjectures. Chapman & Hall/CRC, Boca Raton (2001)
8. Kulenovic, M.R.S., Ladas, G., Martins, L.F., Rodrigues, I.W.: The dynamics of $x_{n+1} = \frac{\alpha+\beta x_n}{A+Bx_n+Cx_{n-1}}$: facts and conjectures. Comput. Math. Appl. **45**(6–9), 1087–1099 (2003)
9. Stevic, S.: On the difference equation $x_{n+1} = \alpha + \frac{x_{n-1}}{x_n}$. Comput. Math. Appl. **56**, 1159–1171 (2008)

Moment Vanishing of Piecewise Solutions of Linear ODEs

Dmitry Batenkov and Gal Binyamini

Abstract We consider the "moment vanishing problem" for a general class of piecewise-analytic functions which satisfy on each continuity interval a linear ODE with polynomial coefficients. This problem, which essentially asks how many zero first moments can such a (nonzero) function have, turns out to be related to several difficult questions in analytic theory of ODEs (Poincare's Center-Focus problem) as well as in Approximation Theory and Signal Processing ("Algebraic Sampling"). While the solution space of any particular ODE admits such a bound, it will in the most general situation depend on the coefficients of this ODE. We believe that a good understanding of this dependence may provide a clue for attacking the problems mentioned above. In this paper we undertake an approach to the moment vanishing problem which utilizes the fact that the moment sequences under consideration satisfy a recurrence relation of fixed length, whose coefficients are polynomials in the index. For any given operator, we prove a general bound for its moment vanishing index. We also provide uniform bounds for several operator families.

Keywords Moment vanishing · Holonomic ODEs · Recurrence relations · Generalised exponential sums

1 Introduction

Let $f : [a, b] \to \mathbb{R}$ be a bounded piecewise-continuous function with points of discontinuity (of the first kind)

Batenkov is supported by the Adams Fellowship Program of the Israel Academy of Sciences and Humanities.

D. Batenkov (✉)
Department of Mathematics, Massachusetts Institute of Technology, Cambridge, MA, USA
e-mail: batenkov@mit.edu

G. Binyamini
Department of Mathematics, Weizmann Institute of Science, Rehovot, Israel
e-mail: galbin@gmail.com

L. Alsedà i Soler et al. (eds.), *Difference Equations, Discrete Dynamical Systems and Applications*, Springer Proceedings in Mathematics & Statistics 180, DOI 10.1007/978-3-662-52927-0_2

$$a = \xi_0 < \xi_1 < \cdots < \xi_p < \xi_{p+1} = b,$$

satisfying on each continuity interval $[\xi_j, \xi_{j+1}]$ a linear homogeneous ODE

$$\mathfrak{D} f = 0, \tag{1}$$

where $\mathfrak{D}$ is a linear differential operator of order n with polynomial coefficients:

$$\mathfrak{D} = p_n(x)\,\partial^n + \cdots + p_1(x)\,\partial + p_0(x)\,\mathrm{I}, \qquad \partial = \frac{\mathrm{d}}{\mathrm{d}x}, \ \deg p_j \le d_j. \tag{2}$$

We say that such f belongs to the class $\mathscr{PD}(\mathfrak{D}, p)$. The union of all such $\mathscr{PD}(\mathfrak{D}, p)$ is the class $\mathscr{PD}$ of *piecewise D-finite functions*, which was first studied in [2].

Any $f \in \mathscr{PD}$ has finite moments of all orders:

$$m_k(f) = \int_a^b x^k f(x)\,\mathrm{d}x, \qquad k = 0, 1, 2, \ldots \tag{3}$$

We consider the following questions.

Problem 1 Given $\mathfrak{D}$ and p, determine the *moment vanishing index* of $\mathscr{PD}(\mathfrak{D}, p)$, defined as

$$\sigma(\mathfrak{D}, p) \stackrel{\text{def}}{=} \sup_{f \in \mathscr{PD}(\mathfrak{D}, p),\ f \not\equiv 0} \{k : m_0(f) = \cdots = m_k(f) = 0\} + 1.$$

In Theorem 3 below we shall prove that the moment vanishing index is always finite. Consequently, the following problem becomes meaningful.

Problem 2 Find natural families $\mathscr{F} \subset \mathscr{PD}$ which admit a uniform bound on the moment vanishing index, i.e. for which

$$\sigma(\mathscr{F}) = \sup_{\mathscr{PD}(\mathfrak{D}, p) \subset \mathscr{F}} \sigma(\mathfrak{D}, p) < +\infty.$$

Our main results, presented in Sect. 4, provide a general bound for $\sigma(\mathfrak{D}, p)$ in terms of $\mathfrak{D}$. As a result, several examples of families $\mathscr{F}$ admitting uniform bound as above are given. The main technical tool is the recurrence relation satisfied by the moment sequence, established previously in [2].

Our main application is the problem of reconstructing functions $f \in \mathscr{PD}$ from a finite number of their moments. Inverse moment problems appear in some areas of mathematical physics, for instance heat conduction and inverse potential theory [1, 11], as well as in statistics. One particular reconstruction technique, introduced in [2] and further extended to two-dimensional setting in [7], can be regarded as a prototype for numerous "algebraic" reconstruction methods in signal processing, such as finite rate of innovation [17] and piecewise Fourier inversion [3, 5]. These methods, being essentially nonlinear, promise to achieve better reconstruction accuracy in some cases (as demonstrated recently in [3, 5]), and therefore we believe their study to be important. In Sect. 2 below we show that an answer to Problem 2 would in turn provide a bound on the minimal number of moments (measurements) required for unique reconstruction of any $f \in \mathscr{F}$. In essence, the results of this paper can be regarded as a step towards understanding the range of applicability of the piecewise D-finite reconstruction method to general signals in $\mathscr{PD}$. See Sect. 2 for further details.

Given a family $\mathscr{F} \subset \mathscr{PD}$, consider the corresponding family of moment generating functions $\{I_f(z)\}_{f \in \mathscr{F}}$, where $I_f(z) = \sum_{k=0}^{\infty} m_k(f) z^{-k-1}$. Obtaining information on the moment vanishing index is in fact an essential step towards studying the analytic properties of I_f, in particular a bound on its number of zeros near infinity (as provided by the notion of "Taylor Domination", see [4, 6]), as well as conditions for its identical vanishing. In turn, these questions play a central role in studies of the Center-Focus and Smale–Pugh problems for the Abel differential equation, see [8, 9, 16] and references therein.

The moment vanishing problem has been previously studied in the complex setting by V. Kisunko [14]. He showed that a uniform bound $\sigma(\mathscr{F})$ exists for families $\mathscr{F}$ consisting of non-singular operators, by using properties of Cauchy type integrals. In contrast, in this paper we consider the real setting only, while proving uniform bounds for some singular (as well as regular) operator families. Our method is based on the linear recurrence relation satisfied by the moment sequence. Using this method, in Sect. 5 we provide an alternative proof of Kisunko's result, stating that the moment generating function $I_f(z)$ of some $f \in \mathscr{PD}(\mathfrak{D}, p)$ satisfies a non-homogeneous ODE

$$\mathfrak{D} I_f(z) = R_f(z)$$

for a very special rational function $R_f(z)$, which depends on $\mathfrak{D}$ and on the values of f at the discontinuities.

In Sect. 6 we provide an interpretation of our main result in the language of Fuchsian theory of ODE.

2 Moment Reconstruction

We start by defining some preliminary notions.

Definition 1 The Pochhammer symbol $(i)_j$ denotes the falling factorial

$$(i)_j \stackrel{\text{def}}{=} i(i-1) \cdot \dots \cdot (i-j+1), \qquad i \in \mathbb{R},\ j \in \mathbb{N}$$

and the expression $(i)_j$ is defined to be zero for $i < j$.

Definition 2 Given $\mathfrak{D}$ of the form (2), the *bilinear concomitant* [13, p. 211] is the homogeneous bilinear form, defined for any pair of sufficiently smooth functions $u(x), v(x)$ as follows (all symbols depend on x):

$$\begin{aligned} P_{\mathfrak{D}}(u,v) \stackrel{\text{def}}{=} & \, u\left\{p_1 v - \partial(p_2 v) + \cdots + (-1)^{n-1}\partial^{n-1}(p_n v)\right\} \\ & + u'\left\{p_2 v - \partial(p_3 v) + \cdots + (-1)^{n-2}\partial^{n-2}(p_n v)\right\} \\ & + \ldots \\ & + u^{(n-1)} \cdot (p_n v). \end{aligned} \tag{4}$$

Proposition 1 (Green's formula, [13]) *Given $\mathfrak{D}$ of the form* (2), *let the formal adjoint operator be defined by*

$$\mathfrak{D}^*\{\cdot\} \stackrel{def}{=} \sum_{j=0}^{n} (-1)^j \partial^j \left\{p_j(x)\cdot\right\}.$$

Then for any pair of sufficiently smooth functions $u(x), v(x)$ the following identity holds:

$$\int_a^b v(x)(\mathfrak{D}u)(x)\,\mathrm{d}x - \int_a^b u(x)\left(\mathfrak{D}^* v\right)(x)\,\mathrm{d}x = P_{\mathfrak{D}}(u,v)(b) - P_{\mathfrak{D}}(u,v)(a). \tag{5}$$

Theorem 1 ([2]) *Let $f \in \mathscr{PD}(\mathfrak{D}, p)$ with $\mathfrak{D}$ of the form* (2). *Denote the discontinuities of f by $a = \xi_0 < \xi_1 < \cdots < \xi_p < \xi_{p+1} = b$. Then the moments $m_k = \int_a^b f(x)\,\mathrm{d}x$ satisfy*[1] *the recurrence relation*

$$\sum_{j=0}^{n}\sum_{i=0}^{d_j} a_{i,j}(-1)^j (i+k)_j\, m_{i-j+k} = \varepsilon_k, \quad k = 0, 1, \ldots, \tag{6}$$

where

$$\varepsilon_k = -\sum_{j=0}^{p}\left\{P_{\mathfrak{D}}\left(f, x^k\right)\left(\xi_{j+1}^-\right) - P_{\mathfrak{D}}\left(f, x^k\right)\left(\xi_j^+\right)\right\}. \tag{7}$$

Proof Apply Green's formula (5) to the identity

$$\int_{\xi_j}^{\xi_{j+1}} x^k (\mathfrak{D}f)(x)\,\mathrm{d}x \equiv 0$$

[1] For consistency of notation, the sequence $\{m_k\}$ is understood to be extended with zeros for negative k.

for each $j = 0, \ldots, p$ and sum up. The result is

$$\sum_{j=0}^{p} \int_{\xi_j}^{\xi_j+1} f(x) \mathfrak{D}^* \left\{x^k\right\} \mathrm{d}x = -\sum_{j=0}^{p} \left\{P_{\mathfrak{D}}\left(f, x^k\right)\left(\xi_{j+1}^-\right) - P_{\mathfrak{D}}\left(f, x^k\right)\left(\xi_j^+\right)\right\}$$

$$\int_a^b f(x) \mathfrak{D}^* \left\{x^k\right\} \mathrm{d}x = \varepsilon_k$$

The left-hand side of the last formula is precisely the linear combination of the moments given by the left-hand side of (6). This finishes the proof. □

Now consider the problem of recovering $f \in \mathscr{PD}(\mathfrak{D}, p) \subset \mathscr{PD}$ from the moments $\{m_0(f), \ldots, m_N(f)\}$ (the operator $\mathfrak{D}$ is assumed unknown in the most general setting). Based on the recurrence relation (6), we demonstrate in [2] that an exact recovery is possible, provided that the number $N \in \mathbb{N}$ is sufficiently large. However, the question of obtaining an upper bound for N turns out to be non-trivial, as we now demonstrate.

Definition 3 Given $\mathfrak{D}$ and p, the *moment uniqueness index* $\tau(\mathfrak{D}, p)$ is defined by

$$\tau(\mathfrak{D}, p) \stackrel{\text{def}}{=} \sup_{f, g \in \mathscr{PD}(\mathfrak{D}, p),\ f \neq g} \left\{k : m_j(f) = m_j(g),\ 0 \leqslant j \leqslant k\right\} + 1.$$

In other words, given $\mathfrak{D}$ and p, at least $\tau(\mathfrak{D}, p)$ first moments of $f \in \mathscr{PD}(\mathfrak{D}, p)$ are necessary for unique reconstruction of f.

Recalling boundedness of $\sigma(\mathfrak{D}, p)$ (see Theorem 3 below), we immediately obtain the following conclusion.

Lemma 1 *For any operator $\mathfrak{D}$ and any p*

$$\tau(\mathfrak{D}, p) \leq \sigma(\mathfrak{D}, 2p).$$

Proof Let $N = \sigma(\mathfrak{D}, 2p)$. Take f_1, f_2 having p jump points each, satisfying $\mathfrak{D} f_1 \equiv 0$, $\mathfrak{D} f_2 \equiv 0$ on each continuity interval such that

$$\begin{aligned} m_0(f_1) &= m_0(f_2) \\ &\ldots \\ m_N(f_1) &= m_N(f_2). \end{aligned}$$

The function $g = f_1 - f_2$ has at most $2p$ jumps, and it satisfies $\mathfrak{D} g \equiv 0$ on each continuity interval. The first N moments of g are zero, therefore $g \equiv 0$ and thus $f_1 \equiv f_2$. Therefore $\tau(\mathfrak{D}, p) \leq N$. □

Consequently, in order to uniquely reconstruct an unknown $f \in \mathscr{F} \subset \mathscr{PD}$, it is sufficient to get a uniform bound $\sigma(\mathscr{F})$ for the family $\mathscr{F}$. Perhaps the most natural choice for such families is when the parameters $p, n, \{d_j\}_{j=0}^{n}$ are fixed. Unfortunately, without making additional assumptions, the moment vanishing index of such families cannot be uniformly bounded. This can be seen from the following example.

Example 1 Let $\mathfrak{D}_m$ denote the Legendre differential operator

$$\mathfrak{D}_m = \left(1 - x^2\right) \frac{\mathrm{d}^2}{\mathrm{d}x^2} - 2x \frac{\mathrm{d}}{\mathrm{d}x} + m(m+1)\,\mathrm{I},$$

and consider $a = -1,\ b = 1$ and $p = 0$. It is well-known that for each $m \in \mathbb{N}$, the regular solution of $\mathfrak{D}_m f = 0$ is $\mathscr{L}_m$—the Legendre polynomial of degree m. Since the first $m - 1$ moments of $\mathscr{L}_m$ are zero, we conclude that

$$\sigma(\mathfrak{D}_m) = m$$

and therefore $\sigma(\mathfrak{D})$ cannot be uniformly bounded in terms of the combinatorial type of $\mathfrak{D}$ only.

Using the subsequent results, in Sect. 6 we shall in fact provide an explanation of this behaviour.

3 Generalized Power Sums

Proposition 2 *The sequence* $\{\varepsilon_k\}$, *given by Theorem 1, is of the form*

$$\varepsilon_k = \sum_{j=0}^{p+1} \sum_{\ell=0}^{n-1} \xi_j^{k-\ell} (k)_\ell\, c_{\ell,j}, \tag{8}$$

where each $c_{\ell,j}$ *is a homogeneous bilinear form in the two sets of variables*

$$\{p_m(\xi_j), p'_m(\xi_j), \ldots, p_m^{(n-1)}(\xi_j)\}_{m=0}^{n},$$
$$\{f(\xi_j^+) - f(\xi_j^-),\ f'(\xi_j^+) - f'(\xi_j^-), \ldots, f^{(n-1)}(\xi_j^+) - f^{(n-1)}(\xi_j^-)\}.$$

Proof Denote for convenience $f(a^-) = f(b^+) = 0$. Now consider the definition of $\{\varepsilon_k\}$ given by (7). Rearranging terms, we write

$$\varepsilon_k = \sum_{j=0}^{p+1} \left\{ P_{\mathfrak{D}}\left(f, x^k\right)\left(\xi_j^+\right) - P_{\mathfrak{D}}\left(f, x^k\right)\left(\xi_j^-\right) \right\}.$$

Furthermore, using the fact that the functions $\{p_m(x)\}_{m=0}^n$ and x^k are continuous at each ξ_j, we have

$$\begin{aligned} P_{\mathfrak{D}}\left(f, x^k\right)\left(\xi_j^+\right) - P_{\mathfrak{D}}\left(f, x^k\right)\left(\xi_j^-\right) = \left\{f\left(\xi_j^+\right) - f\left(\xi_j^-\right)\right\} \times \\ \times \left\{p_1(\xi_j)\xi_j^k - \left(p_2(\xi_j)k\xi_j^{k-1} + p_2'(\xi_j)\xi_j^k\right) + \dots\right\} \\ + \dots \\ + \left\{f^{(n-1)}\left(\xi_j^+\right) - f^{(n-1)}\left(\xi_j^-\right)\right\} p_n(\xi_j)\xi_j^k. \end{aligned} \tag{9}$$

Now using the definition (4), the claim is evident. □

The expression (8) for ε_k is nothing else but a generalized power sum. Let us recall several well-known facts about them (see e.g. [10, Sect. 2.3] or [15]).

Proposition 3 *Let the sequence s_k be of the form*

$$s_k = \sum_{j=0}^{p+1}\sum_{\ell=0}^{n-1} a_{\ell,j}(k)_\ell \xi_j^{k-\ell} \qquad a_{\ell,j}, \xi_j \in \mathbb{C}. \tag{10}$$

Then it satisfies the following linear recurrence relation with constant coefficients of length $n(p+2)+1$:

$$\left(\prod_{j=0}^{p+1}(E - \xi_j \mathrm{I})^n\right) s_k = 0 \tag{11}$$

where E is the forward shift operator in k and I is the identity operator.

Conversely, the fundamental set of solutions of the recurrence relation (11) is

$$\left\{\xi_0^k, k\xi_0^{k-1}, \dots, (k)_{n-1}\xi_0^{k-n+1}, \ \dots, \ \xi_{p+1}^k, k\xi_{p+1}^{k-1}, \dots, (k)_{n-1}\xi_{p+1}^{k-n+1}\right\}.$$

Corollary 1 *The sequence s_k as above, which is not identically zero, can have at most $n(p+2)-1$ first consecutive zero terms $s_0 = \dots = s_{n(p+2)-2} = 0$.*

Proof If $s_0 = \dots = s_{n(p+2)-1} = 0$, then by the recurrence relation (11) we would have automatically $s_{n(p+2)} = s_{n(p+2)+1} = \dots = 0$. □

Corollary 2 *Assume that the numbers $\{\xi_j\}_{j=0}^{p+1} \subset \mathbb{C}$ are pairwise distinct. Let the sequence s_k be given by (10), with a-priori unknown $\{a_{i,j}\}$. If $s_k = 0$ for all $k \in \mathbb{N}$, then necessarily all the coefficients $\{a_{i,j}\}$ are zero.*

4 Main Results

Let us now return to our main goal, namely, obtaining upper bounds on the moment vanishing index $\sigma(\mathfrak{D}, p)$.

Definition 4 Given $\mathfrak{D}$ of the form (2), denote for each $j = 0, \dots, n$

$$\alpha_j \stackrel{\text{def}}{=} d_j - j,$$

and also

$$\alpha = \alpha(\mathfrak{D}) \stackrel{\text{def}}{=} \max_{j=0,\dots,n} \alpha_j.$$

Proposition 4 *Let $f \in \mathscr{PD}(\mathfrak{D}, p)$. Then vanishing of the first $(p+2)n + \alpha(\mathfrak{D})$ moments of f (i.e. $m_0 = \dots = m_{(p+2)n+\alpha-1} = 0$) implies identical vanishing of the sequence $\{\varepsilon_k\}$ defined by Theorem 1.*

Proof Consider the recurrence relation (6). Denote its left-hand side by μ_k. Obviously, since each μ_k is a linear combination of the moments, we have

$$\mu_0 = \dots = \mu_{n(p+2)-1} = 0.$$

Consequently, the corresponding right-hand sides also vanish, i.e.

$$\varepsilon_0 = \dots = \varepsilon_{n(p+2)-1} = 0. \tag{12}$$

The conclusion follows immediately from Corollary 1. □

Now we establish our main result.

Theorem 2 *Let $f \in \mathscr{PD}(\mathfrak{D}, p)$, $f \not\equiv 0$ with discontinuity points*

$$a = \xi_0 < \xi_1 < \dots < \xi_p < \xi_{p+1} = b.$$

Assume that $p_n(\xi_j) \neq 0$ for at least one ξ_j as above. Then at most

$$(p+2)n + \alpha(\mathfrak{D}) - 1$$

first moments of f can vanish (i.e. $m_0 = \dots = m_{(p+2)n+\alpha-2} = 0$).

Proof Assume by contradiction that the first $(p+2)n + \alpha$ moments of f vanish, i.e.

$$m_0 = \dots = m_{n(p+2)+\alpha-1} = 0.$$

By Proposition 4 and Corollary 2 we immediately conclude that

$$c_{\ell,j} = 0, \quad j = 0, \dots, p+1, \ \ell = 0, \dots, n-1,$$

where $\{c_{\ell,j}\}$ are described by Proposition 2. Now we take the concrete j for which $p_n(\xi_j) \neq 0$. This means that the operator $\mathfrak{D}$ is regular at ξ_j, and consequently each solution to $\mathfrak{D} f = 0$ in the neighborhood of ξ_j is uniquely determined by the initial values $f(\xi_j), \dots, f^{(n-1)}(\xi_j)$. We claim that

$$f\left(\xi_j^+\right) - f\left(\xi_j^-\right) = f'\left(\xi_j^+\right) - f'\left(\xi_j^-\right) = \cdots = f^{(n-1)}\left(\xi_j^+\right) - f^{(n-1)}\left(\xi_j^-\right) = 0. \tag{13}$$

In this case, we would immediately conclude that the function f is analytic at ξ_j (being a solution of analytic ODE), contradicting the assumption that ξ_j is a point of discontinuity of f.

To prove (13), we proceed as follows. By Proposition 2 it is easy to see that the term $c_{n-1,j}\,(k)_{n-1}\,\xi_j^{k-n+1}$ is in fact equal to

$$\left\{f\left(\xi_j^+\right) - f\left(\xi_j^-\right)\right\}(k)_{n-1}\,p_n\left(\xi_j\right)$$

in the expression for ε_k. Since $p_n\left(\xi_j\right) \neq 0$, we conclude that $f\left(\xi_j^+\right) - f\left(\xi_j^-\right) = 0$. Substituting this into (9), we see that the next term $c_{n-2,j}\,(k)_{n-2}\,\xi_j^{k-n+2}$ equals

$$\left\{f'\left(\xi_j^+\right) - f'\left(\xi_j^-\right)\right\}(k)_{n-2}\,\xi_j^{k-n+2}\,p_n\left(\xi_j\right),$$

and thus $f'\left(\xi_j^+\right) - f'\left(\xi_j^-\right) = 0$. Proceeding in this manner, we arrive at (13). This finishes the proof of Theorem 2. □

As a first consequence, we have the real-valued version of the result by Kisunko [14].

Corollary 3 *For every $n, d > 0$ and $p \geqslant 0$ consider the family*

$$\mathscr{F}_{n,p,d}^{(1)} = \left\{ f \in \mathscr{PD}(\mathfrak{D}, p) : \quad \mathfrak{D} = \sum_{j=0}^{n} p_j(x)\,\partial^j, \ \alpha(\mathfrak{D}) = d, \ p_n(x) \neq 0 \text{ on } [a,b] \right\}.$$

Then

$$\sigma\left(\mathscr{F}_{n,p,d}^{(1)}\right) \leqslant (p+2)\,n + d - 1.$$

Since the leading coefficient $p_n(x)$ cannot vanish at more than $\deg p_n$ points, we also have the following result.

Corollary 4 *For every $n, d > 0$ and $p \geqslant 0$ consider the family*

$$\mathscr{F}_{n,p,d}^{(2)} = \left\{ f \in \mathscr{PD}(\mathfrak{D}, p) : \quad \mathfrak{D} = \sum_{j=0}^{n} p_j(x)\,\partial^j, \ \alpha(\mathfrak{D}) = d, \ \deg p_n < p+2 \right\}.$$

Then

$$\sigma\left(\mathscr{F}_{n,p,d}^{(2)}\right) \leqslant (p+2)\,n + d - 1.$$

Let us now try to establish what happens in the general case. Let $f \in \mathscr{PD}(\mathfrak{D}, p)$, $f \not\equiv 0$. Consider two possibilities.

1. The sequence $\{\varepsilon_k\}$ does not vanish identically. In this case, at least some of its initial terms $\left\{\varepsilon_0, \ldots, \varepsilon_{n(p+2)-1}\right\}$ must be nonzero (Corollary 1). But this immediately implies that some of the first $n\,(p+2)+\alpha-1$ moments must be nonzero as well (otherwise the equality (6) cannot hold).
2. The sequence $\{\varepsilon_k\}$ vanishes identically, but Theorem 2 is not applicable. In this case the recurrence relation (6) becomes homogeneous. We rewrite it in the form

$$\sum_{\ell=-n}^{\alpha} q_\ell\,(k)\, m_{k+\ell} = 0, \qquad k = 0, 1, \ldots, \tag{14}$$

where

$$q_\ell\,(k) \stackrel{\text{def}}{=} \sum_{j=0}^{n} (-1)^j\, a_{\ell+j,j}\, (k+\ell+j)_j\,. \tag{15}$$

The leading coefficient $q_\alpha\,(k)$ may have positive integer zeros. Let $\Lambda\,(\mathfrak{D})$ denote the largest such zero. Then we claim that no more than $\alpha + \Lambda\,(\mathfrak{D})$ moments can vanish. Indeed, starting with $k = \Lambda\,(\mathfrak{D}) + 1$ we can safely divide the recurrence (14) by $q_\alpha\,(k)$ and obtain

$$m_{k+\alpha} = \sum_{\ell=-n}^{\alpha-1} r_\ell\,(k)\, m_{k+\ell}, \quad k \geqslant \Lambda\,(\mathfrak{D}) + 1,$$

where $r_\ell\,(k)$ are some rational functions with non-vanishing denominators. Therefore if the first $\Lambda\,(\mathfrak{D}) + \alpha + 1$ moments are zero, then all the rest of the moments must vanish, implying vanishing of f itself.

Thus we have proved the following result.

Theorem 3 *For every* $\mathfrak{D}, p$ *we have*

$$\sigma\,(\mathfrak{D}, p) \leqslant \max\left\{n\,(p+2) - 1,\ \Lambda\,(\mathfrak{D})\right\} + \alpha\,(\mathfrak{D})\,.$$

In Sect. 6, we demonstrate that in the case of Fuchsian differential operators, the number $\Lambda\,(\mathfrak{D})$ has a well-known interpretation.

5 Moment Generating Function

In this section we provide an alternative proof for the result of Kisunko [14] concerning moment generating functions.

Proposition 5 Let $f \in \mathscr{P}\mathscr{D}$. *The formal power series*

$$I_f(z) \stackrel{def}{=} \sum_{k=0}^{\infty} \frac{m_k}{z^{k+1}}$$

is in fact the Laurent series of the Cauchy type integral

$$\int_a^b \frac{f(t)\,\mathrm{d}t}{z-t}.$$

Proof Write $\frac{1}{z-t} = \frac{1}{z}\left(\frac{1}{1-\frac{t}{z}}\right)$ and expand into geometric series. Convergence follows immediately for $z \to \infty$. □

The generalized power sums (Sect. 3) also have a well-known interpretation as the Taylor coefficients of rational functions. The following fact is well-known, and so we omit the proof.

Proposition 6 *Let the sequence $\{s_k\}$ be of the form (10). Then the formal generating function*

$$g(z) = \sum_{k=0}^{\infty} \frac{s_k}{z^{k+1}}$$

is a regular at infinity rational function, with poles $\{\xi_0, \ldots, \xi_{p+1}\}$, each with multiplicity at most n. In particular,

$$g(z) = \sum_{j=0}^{p+1} \sum_{\ell=0}^{n-1} \frac{(-1)^{\ell}\,\ell!\,a_{\ell,j}}{\left(z-\xi_j\right)^{\ell+1}}. \tag{16}$$

Theorem 4 *Let $f \in \mathscr{PD}(\mathfrak{D}, p)$. Then the Cauchy integral I_f satisfies in the neighborhood of ∞ the inhomogeneous ODE*

$$\mathfrak{D} I_f(z) = R_f(z), \tag{17}$$

where $R_f(z)$ is the rational function whose Taylor coefficients at infinity are given by the sequence ϵ_k as in (7). Consequently, $R_f(z)$ is given by the explicit expression (16), with $a_{\ell,j}$ replaced by $c_{\ell,j}$ from (8) (Proposition 2).

Proof Consider the asymptotic expansion of the function $\mathfrak{D} I_f$ at infinity

$$\mathfrak{D} I_f = \sum_{k=0}^{\infty} \frac{s_k}{z^{k+1}}.$$

By substituting $\mathfrak{D}$ as in (2) and $I_f = \sum_{k=0}^{\infty} \frac{m_k}{z^{k+1}}$ we get

$$
\begin{aligned}
\mathfrak{D} I_f &= \sum_{j=0}^{n} p_j(z) I_f^{(j)}(z) = \sum_{j=0}^{n} \sum_{i=0}^{d_j} \sum_{k=0}^{\infty} m_k \frac{(-1)^j (k+j)_j}{z^{k+1+j-i}} a_{i,j} \\
{}_{(k+j-i \to t)} &= \sum_{t=0}^{\infty} \frac{1}{z^{t+1}} \sum_{j=0}^{n} (-1)^j a_{i,j} (t+i)_j m_{t+i-j} \\
&= \sum_{s=0}^{\infty} \frac{s_k}{z^{k+1}}.
\end{aligned}
$$

Comparing powers of z we have that $s_k = \mu_k$ where μ_k denote the left-hand side of (6). From $\mu_k = \epsilon_k$ the conclusion follows. □

6 Fuchsian Operators

In this section we employ notions from the classical Fuchsian theory of linear ODEs in the complex domain (we used the Ref. [12]).

Assume that the sequence $\{\varepsilon_k\}$ vanishes identically. In this case, the Cauchy integral I_f satisfies in the neighborhood of ∞ the *homogeneous ODE*

$$\mathfrak{D} I_f = 0.$$

Definition 5 The operator $\mathfrak{D}$ is said to belong to the class $\mathfrak{R}$ if it has at most a regular singularity at ∞.

Lemma 2 *Let $\mathfrak{D} \in \mathfrak{R}$. Then*

1. *The numbers α_j (see Definition 4) satisfy*

$$\alpha_n \geqslant \alpha_j, \quad j = 0, \ldots, n-1. \tag{18}$$

2. *The characteristic exponents of $\mathfrak{D}$ at the point ∞ are the roots of the equation*

$$q_{\alpha_n}(s-1) = 0,$$

where $q_\ell(k)$ is defined by (15).

Proof Dividing the coefficients of $\mathfrak{D}$ by p_n, we get the operator

$$\partial^n + r_1(z) \partial^{n-1} + \cdots + r_n(z) \mathrm{I}, \qquad r_j(z) = \frac{p_{n-j}(z)}{p_n(z)}.$$

A necessary and sufficient condition for the point at infinity to be at most a regular singularity of this operator is that the function $r_j(z)$ is analytic at ∞ and has a zero there of order at least j [12, Theorem 9.8b]. That is,

$$\deg p_n - \deg p_{n-j} \geq j.$$

But this is equivalent to

$$\begin{aligned} \deg p_n - n &\geqslant \deg p_{n-j} - (n-j) \\ \alpha_n &\geqslant \alpha_{n-j}. \end{aligned}$$

To prove the second statement, substitute the formal Frobenius series at infinity

$$g(z) = \sum_{k=0}^{\infty} \frac{b_k}{z^{s+k}}$$

into $\mathfrak{D}g = 0$. By complete analogy with the calculation in Theorem 4 we get the recurrence relation

$$\sum_{j=0}^{n} \sum_{i=0}^{d_j} (-1)^j a_{i,j} (t+s+i-1)_j \, b_{t+i-j} = 0, \qquad t = 0, 1, \ldots.$$

For $t = 0$ we find the highest order coefficient in this recurrence to be equal to $(i - j = \alpha = \alpha_n)$

$$\sum_{j=0}^{n} (-1)^j (s+\alpha+j-1)_j \, a_{j+\alpha,j} = q_\alpha (s-1).$$

The proof is finished. □

Together with Theorem 3, this immediately implies the following bound.

Corollary 5 *Let $\mathfrak{D} \in \mathfrak{R}$, and let $\lambda(\mathfrak{D})$ denote its largest positive integer characteristic exponent at the point ∞. Then $\Lambda(\mathfrak{D}) = \lambda(\mathfrak{D}) - 1$, and consequently*

$$\sigma(\mathfrak{D}, p) \leqslant \max\{(p+2)n, \lambda(\mathfrak{D})\} + d_n - n - 1.$$

Now let us return to Example 1. The following fact is well-known (e.g. [12, Sect. 9.10].

Proposition 7 *The Legendre differential operator $\mathfrak{D}_m$ is of Fuchsian type with singularities $-1, 1, \infty$. The characteristic exponents at ∞ are $m+1$ and $-m$.*

Theorem 2 is clearly not applicable. Using the formula (7), it is easy to see that

$$P_{\mathfrak{D}}\left(f, x^{k}\right)(1) = P_{\mathfrak{D}}\left(f, x^{k}\right)(-1) = 0$$

for any $f \in \mathscr{P}\mathscr{D}$, and therefore the sequence $\{\varepsilon_k\}$ in this case is identically zero. Consequently, we conclude that

$$\sigma\left(\mathfrak{D}_m, 0\right) = m,$$

as expected.

Acknowledgments The authors would like to thank Y. Yomdin for useful discussions.

References

1. Ang, D.D.: Moment Theory and Some Inverse Problems in Potential Theory and Heat Conduction. Springer, Berlin (2002)
2. Batenkov, D.: Moment inversion problem for piecewise D-finite functions. Inverse Probl. **25**(10), 105001 (2009)
3. Batenkov, D.: Complete algebraic reconstruction of piecewise-smooth functions from fourier data. Math. Comput. **84**(295), 2329–2350 (2015)
4. Batenkov, D., Yomdin, Y.: Taylor Domination, Difference Equations, and Bautin Ideals. Submitted to this volume
5. Batenkov, D., Yomdin, Y.: Algebraic Fourier reconstruction of piecewise smooth functions. Math. Comput. **81**, 277–318 (2012)
6. Batenkov, D., Yomdin, Y.: Taylor domination, Turán lemma, and Poincaré-Perron sequences. Contemp. Math. **659**, 1–15 (2016)
7. Batenkov, D., Golubyatnikov, V., Yomdin, Y.: Reconstruction of planar domains from partial integral measurements. Contemp. Math. **591**, 51–56 (2013)
8. Batenkov, D., Binyamini, G.: Uniform upper bounds for the cyclicity of the zero solution of the Abel differential equation. J. Differ. Equ. **259**(11), 5769–5781. ISSN 0022-0396. http://dx.doi.org/10.1016/j.jde.2015.07.009. http://www.sciencedirect.com/science/article/pii/S002203961500368X (2015)
9. Briskin, M., Roytvarf, N., Yomdin, Y.: Center conditions at infinity for Abel differential equations. Ann. Math. **172**(1), 437–483 (2010)
10. Elaydi, S.: An Introduction to Difference Equations. Springer, Berlin (2005)
11. Gustafsson, B., He, C., Milanfar, P., Putinar, M.: Reconstructing planar domains from their moments. Inverse Probl. **16**(4), 1053–1070 (2000)
12. Henrici, P.: Applied and Computational Complex Analysis: Vol.: 2.: Special Functions: Integral Transforms: Asymptotics: Continued Fractions. Wiley, New York (1977)
13. Ince, E.L.: Ordinary Differential Equations. Courier Dover Publications, New York (1956)
14. Kisunko, V.: Cauchy type integrals and a D-moment problem. Math. Rep. Acad. Sci. R. Soc. Can. **29**(4), 115–122 (2008)
15. Myerson, G., van der Poorten, A.J.: Some problems concerning recurrence sequences. Am. Math. Mon. **102**(8), 698–705 (1995)
16. Pakovich, F., Roytvarf, N., Yomdin, Y.: Cauchy-type integrals of algebraic functions. Isr. J. Math. **144**(2), 221–291 (2004)
17. Vetterli, M., Marziliano, P., Blu, T.: Sampling signals with finite rate of innovation. IEEE Trans. Signal Process. **50**(6), 1417–1428 (2002)

Periodic Solutions of the Second Order Quadratic Rational Difference Equation $x_{n+1} = \frac{\alpha}{(1+x_n)x_{n-1}}$

Inese Bula

Abstract The aim of this article is to investigate the periodic nature of solutions of a rational difference equation

$$x_{n+1} = \frac{\alpha}{(1 + x_n)x_{n-1}}. \qquad (*)$$

We explore Open Problem 3.3 given in Amleh et al. (Int J Differ Equ 3(1):1–35, 2008, [2]) that requires to determine all periodic solutions of the equation (*). We conclude that for the equation (*) there are no periodic solution with prime period 3 and 4. Period 7 is first period for which exists nonnegative parameter α and nonnegative initial conditions.

Keywords Difference equation · Rational difference equation · Equilibrium point · Periodic solution

1 Introduction

Rational difference equations in last time have been studied very intensively. Many researchers have gained inspiration from the books [6, 7, 12, 13]. But rational difference equations with quadratic terms have been considered in [2, 3]. In this article we explore Open Problem 3.3 from [2] that requires to determine all periodic solutions of the equation

$$x_{n+1} = \frac{\alpha}{(1 + x_n)x_{n-1}}. \qquad (1)$$

I. Bula (✉)
Department of Mathematics, University of Latvia, Zeļļu iela 8, Rīga, Latvia
e-mail: ibula@lanet.lv

I. Bula
Institute of Mathematics and Computer Science of University of Latvia,
Raiņa bulv. 29, Rīga, Latvia

L. Alsedà i Soler et al. (eds.), *Difference Equations, Discrete Dynamical Systems and Applications*, Springer Proceedings in Mathematics & Statistics 180, DOI 10.1007/978-3-662-52927-0_3

Our goal is to clarify whether there exists periodic solutions with nonnegative parameter and nonnegative initial conditions. And if possible, then the study will be conducted more widely. We show that the rational difference equation (1) with negative initial conditions or/and negative parameter α has periodic solutions with prime period 2, 5, 6 but for nonnegative parameters and nonnegative initial conditions there are no periodic solutions with prime period 2, 5, 6. We conclude that for the equation (1) there are no periodic solution with prime period 3 and 4. Period 7 is first period for which exists nonnegative parameter α and nonnegative initial conditions.

About periodic solutions have been written numerous articles (for example, [1, 4, 8, 9, 11]) and a book [10] but they do not help to solve the mentioned problem.

The results about Open Problem 3.3 were presented in the 18th International Conference on Difference Equations and Applications (ICDEA), Casa de Convalescencia, Barcelona, 2012 [5].

2 Equilibrium Points

Let I be some interval of real numbers and let $f : I \times I \to I$ be a continuous function. Then for every set of initial conditions $x_{-1},\ x_0 \in I$ the difference equation

$$x_{n+1} = f(x_n, x_{n-1}), \quad n = 0, 1, \ldots, \tag{2}$$

has a unique solution $(x_n)_{n=-1}^{\infty}$.

Definition 1 A solution $(x_n)_{n=-1}^{\infty}$ of difference equation (2) is called periodic with period p if there exists an integer $p \geq 1$ such that

$$x_{n+p} = x_n \text{ for all } n \geq -1. \tag{3}$$

We say that the solution is periodic with prime period p if p is the smallest positive integer for which (3) holds.

A point $\bar{x} \in I$ is called an equilibrium point of the equation (2) if $\bar{x} = f(\bar{x}, \bar{x})$, that is, for all $n : x_n = \bar{x}$ is a solution of equation (2), or equivalently, $\bar{x}$ is a fixed point of f.

If we consider difference equation (1) with $\alpha > 0$, then it has a unique positive equilibrium $\bar{x}$, and $\bar{x}$ is the unique positive root of the cubic equation

$$\bar{x}^3 + \bar{x}^2 - \alpha = 0, \quad \bar{x} \neq -1, \quad \bar{x} \neq 0.$$

But from the graph of the function $\alpha = \bar{x}^3 + \bar{x}^2$ (Fig. 1) we see that equation (1) with $\alpha = \frac{4}{27}$ has two equilibrium points—one is positive and other is $-\frac{2}{3}$. If $0 < \alpha < \frac{4}{27}$, then equation (1) has three equilibrium points (one is positive and two are negative). If $\alpha < 0$, then equation (1) has one negative equilibrium point.

We conclude that for every positive α there exists positive equilibrium point.

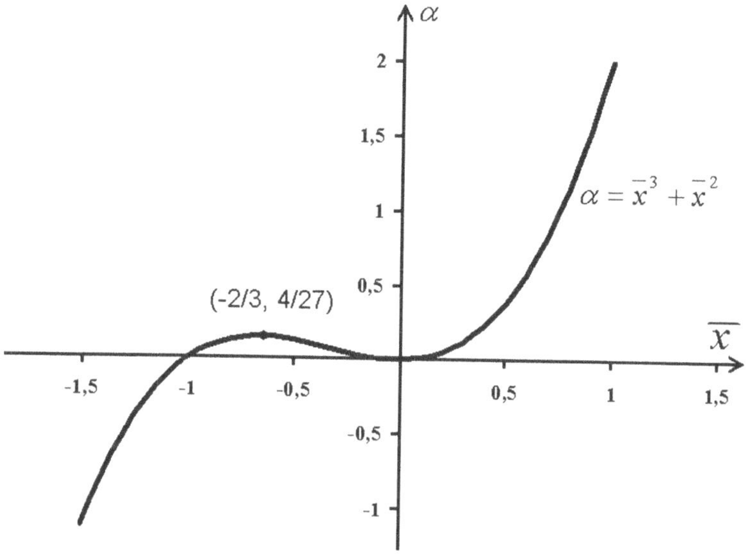

Fig. 1 The graph of the function $\alpha = \bar{x}^3 + \bar{x}^2$

3 Periodic Solutions with Period 2

Theorem 1 *If* $\alpha \in]-\infty, -4[$, *then the solution of the difference equation (1) with initial conditions* $x_{-1} = \frac{\alpha}{2} + \frac{\sqrt{\alpha^2+4\alpha}}{2}$ *and* $x_0 = \frac{\alpha}{2} - \frac{\sqrt{\alpha^2+4\alpha}}{2}$ *(or conversely* $x_{-1} = \frac{\alpha}{2} - \frac{\sqrt{\alpha^2+4\alpha}}{2}$ *and* $x_0 = \frac{\alpha}{2} + \frac{\sqrt{\alpha^2+4\alpha}}{2}$*) is periodic with prime period 2 and both initial conditions are negative.*

If $\alpha \in]0, +\infty[$, *then the solution of the difference equation (1) with initial conditions* $x_{-1} = \frac{\alpha}{2} + \frac{\sqrt{\alpha^2+4\alpha}}{2}$ *and* $x_0 = \frac{\alpha}{2} - \frac{\sqrt{\alpha^2+4\alpha}}{2}$ *(or contrariwise* $x_{-1} = \frac{\alpha}{2} - \frac{\sqrt{\alpha^2+4\alpha}}{2}$ *and* $x_0 = \frac{\alpha}{2} + \frac{\sqrt{\alpha^2+4\alpha}}{2}$*) is periodic with prime period 2 and one of the initial conditions is negative.*

If $\alpha \in [-4, 0]$, *then there are no solutions with prime period 2.*

Proof We fix an arbitrary α and we denote the initial conditions with x_{-1} and x_0. If a solution of difference equation (1) with given initial conditions forms a periodic solution with period 2, then

$$x_1 = \frac{\alpha}{(1+x_0)x_{-1}} = x_{-1},$$
$$x_2 = \frac{\alpha}{(1+x_1)x_0} = \frac{\alpha}{(1+x_{-1})x_0} = x_0,$$

therefore

$$\begin{cases} \frac{\alpha}{1+x_0} = x_{-1}^2 & \Rightarrow \quad \alpha = (1+x_0)x_{-1}^2, \\ \frac{\alpha}{1+x_{-1}} = x_0^2 & \Rightarrow \quad \alpha = (1+x_{-1})x_0^2. \end{cases}$$

We have an equality

$$(1+x_0)x_{-1}^2 = (1+x_{-1})x_0^2,$$

then we obtain a quadratic equation in respect to x_0

$$\frac{1+x_{-1}}{x_{-1}^2}x_0^2 - x_0 - 1 = 0.$$

This quadratic equation has two roots

$$\begin{aligned} x_0 &= \frac{1 \pm \sqrt{1 + \frac{4(1+x_{-1})}{x_{-1}^2}}}{\frac{2(1+x_{-1})}{x_{-1}^2}} = \frac{\left(1 \pm \sqrt{\frac{x_{-1}^2+4x_{-1}+4}{x_{-1}^2}}\right)x_{-1}^2}{2(1+x_{-1})} \\ &= \frac{\left(1 \pm \sqrt{\frac{(x_{-1}+2)^2}{x_{-1}^2}}\right)x_{-1}^2}{2(1+x_{-1})} = \frac{\left(1 \pm \frac{x_{-1}+2}{x_{-1}}\right)x_{-1}^2}{2(1+x_{-1})} \\ &= \frac{(x_{-1} \pm (x_{-1}+2))x_{-1}}{2(x_{-1}+1)}. \end{aligned}$$

There are two possibilities:

(1) $x_0 = \frac{(x_{-1}+x_{-1}+2)x_{-1}}{2(x_{-1}+1)} = x_{-1}$—in this case we obtain period 1 solution (equilibrium point).

(2) $x_0 = \frac{(x_{-1}-x_{-1}-2)x_{-1}}{2(x_{-1}+1)} = \frac{-x_{-1}}{x_{-1}+1}$—in this case we take an equality that determine a relevance between x_0 and x_{-1} for solution with period 2. Since $\alpha = (1+x_{-1})x_0^2$ then

$$\alpha = (1+x_{-1})\frac{x_{-1}^2}{(x_{-1}+1)^2} = \frac{x_{-1}^2}{x_{-1}+1} \quad \text{and}$$

$$x_{-1}^2 - \alpha x_{-1} - \alpha = 0. \tag{4}$$

The quadratic equation (4) has two roots

$$x_{-1} = \frac{\alpha \pm \sqrt{\alpha^2+4\alpha}}{2} = \frac{\alpha}{2} \pm \frac{\sqrt{\alpha^2+4\alpha}}{2}.$$

If $\alpha^2 + 4\alpha \geq 0$, then equation has two real roots. If $\alpha = 0$, then $x_0 = 0$ and difference equation with such initial condition is not defined. If $\alpha = -4$, then $x_{-1} = -2 = x_0$—we have a periodic solution with period 1 (equilibrium point). If $\alpha \in]-\infty, -4[\cup]0, +\infty[$, then exists solution with period 2. In this case

$$x_0 = \frac{-x_{-1}}{x_{-1} + 1} = \frac{-\left(\frac{\alpha}{2} \pm \frac{\sqrt{\alpha^2+4\alpha}}{2}\right)}{\frac{\alpha}{2} \pm \frac{\sqrt{\alpha^2+4\alpha}}{2} + 1} = \frac{-(\alpha \pm \sqrt{\alpha^2 + 4\alpha})}{\alpha \pm \sqrt{\alpha^2 + 4\alpha} + 2}.$$

If we consider a case $x_{-1} = \frac{\alpha}{2} + \frac{\sqrt{\alpha^2+4\alpha}}{2}$ (with "+" sign), then

$$\begin{aligned} x_0 &= \frac{-(\alpha + \sqrt{\alpha^2 + 4\alpha})}{\alpha + \sqrt{\alpha^2 + 4\alpha} + 2} = \frac{-(\alpha + \sqrt{\alpha^2 + 4\alpha})(\alpha + 2 - \sqrt{\alpha^2 + 4\alpha})}{(\alpha + 2 + \sqrt{\alpha^2 + 4\alpha})(\alpha + 2 - \sqrt{\alpha^2 + 4\alpha})} \\ &= \frac{-(\alpha^2 + 2\alpha - \alpha\sqrt{\alpha^2 + 4\alpha} + \alpha\sqrt{\alpha^2 + 4\alpha} + 2\sqrt{\alpha^2 + 4\alpha} - \alpha^2 - 4\alpha)}{\alpha^2 + 4\alpha + 4 - \alpha^2 - 4\alpha} \\ &= \frac{-(2\sqrt{\alpha^2 + 4\alpha} - 2\alpha)}{4} = \frac{\alpha}{2} - \frac{\sqrt{\alpha^2 + 4\alpha}}{2}. \end{aligned}$$

Similarly, if we consider a case $x_{-1} = \frac{\alpha}{2} - \frac{\sqrt{\alpha^2+4\alpha}}{2}$ (with "−" sign), then $x_0 = \frac{\alpha}{2} + \frac{\sqrt{\alpha^2+4\alpha}}{2}$. □

Corollary 1 *For the difference equation (1) with positive* α *there are no such positive initial conditions* x_{-1} *and* x_0 *that the solution of (1) is periodic with prime period 2.*

From Theorem 1 we can easily find periodic solutions with prime period 2 for difference equation (1). For example:

(1) if $\alpha = \frac{1}{2}$, then initial conditions $x_{-1} = 1$ and $x_0 = -\frac{1}{2}$ forms a periodic solution with period 2;
(2) if $\alpha = -\frac{64}{15}$, then a initial conditions $x_{-1} = -\frac{8}{5}$ and $x_0 = -\frac{8}{3}$ forms periodic solution with period 2;
(3) if $\alpha = 2$, then initial conditions $x_{-1} = 1 + \sqrt{3}$ and $x_0 = 1 - \sqrt{3}$ forms a periodic solution with period 2.

4 Periodic Solutions with Period 3 and 4

Theorem 2 *For the difference equation (1) there are no periodic solutions with prime period 3.*

Proof We fix an arbitrary α. If a solution of difference equation (1) with initial conditions x_{-1} and x_0 forms a periodic solution with period 3, then

$$x_1 = \tfrac{\alpha}{(1+x_0)x_{-1}}, \text{ and}$$

$$\begin{cases} x_2 = \tfrac{\alpha}{(1+x_1)x_0} = x_{-1}, \\ x_3 = \tfrac{\alpha}{(1+x_2)x_1} = \tfrac{\alpha}{(1+x_{-1})x_1} = x_0, \\ x_4 = \tfrac{\alpha}{(1+x_3)x_2} = \tfrac{\alpha}{(1+x_0)x_{-1}} = x_1. \end{cases}$$

Therefore

$$\begin{cases} \tfrac{\alpha}{(1+x_1)x_0} = x_{-1} \quad \Rightarrow \quad \alpha = (1+x_1)x_0x_{-1}, \\ \tfrac{\alpha}{(1+x_{-1})x_1} = x_0 \quad \Rightarrow \quad \alpha = (1+x_{-1})x_1x_0. \end{cases}$$

We have an equality

$$(1+x_1)x_0x_{-1} = (1+x_{-1})x_1x_0. \tag{5}$$

From (5) follows that (since for $x_0 = 0$ the term x_2 is not defined then $x_0 \neq 0$!)

$$(1+x_1)x_{-1} = (1+x_{-1})x_1 \quad \Rightarrow \quad x_{-1} + x_1x_{-1} = x_1 + x_{-1}x_1 \quad \Rightarrow \quad x_{-1} = x_1.$$

By definition of x_2 we have

$$x_2 = \frac{\alpha}{(1+x_1)x_0} = \frac{\alpha}{(1+\frac{\alpha}{(1+x_0)x_{-1}})x_0} = \frac{\alpha(1+x_0)x_{-1}}{((1+x_0)x_{-1}+\alpha)x_0} = x_{-1},$$

therefore

$$\alpha(1+x_0) = ((1+x_0)x_{-1}+\alpha)x_0 \quad \Rightarrow \quad \alpha + \alpha x_0 = (1+x_0)x_{-1}x_0 + \alpha x_0 \quad \Rightarrow$$

$$\Rightarrow \quad \alpha = (1+x_0)x_{-1}x_0 \quad \Rightarrow \quad x_0 = \tfrac{\alpha}{(1+x_0)x_{-1}}.$$

But $\frac{\alpha}{(1+x_0)x_{-1}} = x_1$. Finally we have obtained that $x_{-1} = x_0 = x_1$—this condition forms only the equilibrium solution. □

Theorem 3 *For the difference equation (1) there are no periodic solutions with prime period 4.*

Proof We fix an arbitrary α. If a solution of difference equation (1) with initial conditions x_{-1} and x_0 forms a periodic solution with period 4, then

$$x_1 = \tfrac{\alpha}{(1+x_0)x_{-1}}, \ x_2 = \tfrac{\alpha}{(1+x_1)x_0} \text{ and}$$

$$\begin{cases} x_3 = \tfrac{\alpha}{(1+x_2)x_1} = x_{-1}, \\ x_4 = \tfrac{\alpha}{(1+x_3)x_2} = \tfrac{\alpha}{(1+x_{-1})x_2} = x_0, \\ x_5 = \tfrac{\alpha}{(1+x_4)x_3} = \tfrac{\alpha}{(1+x_0)x_{-1}} = x_1\text{—follows by the assumption}, \\ x_6 = \tfrac{\alpha}{(1+x_5)x_4} = \tfrac{\alpha}{(1+x_1)x_0} = x_2\text{—follows by the assumption}. \end{cases}$$

From the definition of x_1 we get $\frac{\alpha}{1+x_0} = x_1 x_{-1}$and from the definition of x_3: $\frac{\alpha}{1+x_2} = x_1 x_{-1}$, therefore

$$\frac{\alpha}{1+x_0} = \frac{\alpha}{1+x_2} \quad \Rightarrow \quad x_0 = x_2.$$

From the definition of x_2 we obtain $\frac{\alpha}{1+x_1} = x_0 x_2$and from the definition of x_4: $\frac{\alpha}{1+x_{-1}} = x_0 x_2$, therefore

$$\frac{\alpha}{1+x_1} = \frac{\alpha}{1+x_{-1}} \quad \Rightarrow \quad x_1 = x_{-1}.$$

Consequently, $x_{-1}, x_0, x_{-1}, x_0, x_{-1}, x_0, \ldots$—this sequence is not periodic with prime period 4. □

5 Periodic Solutions with Period 5

Theorem 4 *For the difference equation (1) with positive α there are no such positive initial conditions x_{-1} and x_0 that the solution of (1) is periodic with prime period 5.*

For every $x_{-1} = x_0 = x$ we can find α:

$$\alpha = \frac{-x(1+x)(1-x-x^2) \pm \sqrt{x^2(1+x)^2(1-x-x^2)^2 + 4(2+x)(1+x)^3x^3}}{2(2+x)} \tag{6}$$

such that the solution of (1) is periodic with period 5: if we consider the equality (6) with "+", then we obtain equilibrium points, if we consider the equality (6) with "−", then we obtain the periodic solution of (1) with prime period 5 and $\alpha < 0$.

For every α and every $x_{-1} > 0$ we can find x_0:

$$x_0 = \frac{-(x_{-1}^2 + x_{-1}(1+\alpha) + \alpha) \pm \sqrt{(x_{-1}^2 + x_{-1}(1+\alpha) + \alpha)^2 - 4\alpha(x_{-1}^2 + x_{-1})x_{-1}}}{2(x_{-1}^2 + x_{-1})}$$

such that the solution of (1) is periodic with prime period 5 (if $\alpha > 0$, then $x_0 < 0$).

Proof We fix an arbitrary α. If a solution of difference equation (1) with initial conditions x_{-1} and x_0 forms a periodic solution with period 5, then

$$x_1 = \frac{\alpha}{(1+x_0)x_{-1}}, \quad x_2 = \frac{\alpha}{(1+x_1)x_0}, \quad x_3 = \frac{\alpha}{(1+x_2)x_1} \text{ and}$$

$$\begin{cases} x_4 = \frac{\alpha}{(1+x_3)x_2} = x_{-1}, \\ x_5 = \frac{\alpha}{(1+x_4)x_3} = \frac{\alpha}{(1+x_{-1})x_3} = x_0, \\ x_6 = \frac{\alpha}{(1+x_5)x_4} = \frac{\alpha}{(1+x_0)x_{-1}} = x_1\text{—follows by the assumption,} \\ x_7 = \frac{\alpha}{(1+x_6)x_5} = \frac{\alpha}{(1+x_1)x_0} = x_2\text{—follows by the assumption,} \\ x_8 = \frac{\alpha}{(1+x_7)x_6} = \frac{\alpha}{(1+x_2)x_1} = x_3 \text{—follows by the assumption.} \end{cases}$$

From x_4 we express x_3:

$$x_3 = \frac{\alpha - x_2x_{-1}}{x_2x_{-1}} = \frac{\alpha}{x_2x_{-1}} - 1.$$

From x_5 we express x_3:

$$x_3 = \frac{\alpha}{(1+x_{-1})x_0}.$$

Therefore

$$\frac{\alpha}{x_2x_{-1}} = \frac{\alpha}{(1+x_{-1})x_0} + 1$$

and we can express x_2:

$$\frac{1}{x_2} = \frac{x_{-1}}{(1+x_{-1})x_0} + \frac{x_{-1}}{\alpha} = \frac{\alpha x_{-1} + (1+x_{-1})x_0x_{-1}}{(1+x_{-1})x_0\alpha} \quad \Rightarrow$$

$$x_2 = \frac{(1+x_{-1})x_0\alpha}{x_{-1}(\alpha + (1+x_{-1})x_0)}. \tag{7}$$

On the other hand we can express x_2 using initial conditions and the definition of the given sequence

$$x_2 = \frac{\alpha}{(1+x_1)x_0} = \frac{\alpha}{\left(1 + \frac{\alpha}{(1+x_0)x_{-1}}\right)x_0} = \frac{\alpha(1+x_0)x_{-1}}{((1+x_0)x_{-1}+\alpha)x_0}. \tag{8}$$

Then from (7) and (8) follow the equality:

$$\frac{\alpha(1+x_0)x_{-1}}{((1+x_0)x_{-1}+\alpha)x_0} = \frac{(1+x_{-1})x_0\alpha}{x_{-1}(\alpha + (1+x_{-1})x_0)}.$$

And we have obtained an expression that contains only α and initial conditions

$$x_{-1}^2(1+x_0)(\alpha + (1+x_{-1})x_0) = x_0^2(1+x_{-1})((1+x_0)x_{-1}+\alpha) \quad \Rightarrow$$

$$(x_{-1}-x_0)(\alpha(x_{-1}+x_0) + \alpha x_0x_{-1} + x_0x_{-1} + x_0x_{-1}(x_{-1}+x_0) + x_0^2x_{-1}^2) = 0.$$

We have two possibilities.

Case 1. $x_{-1} = x_0$. We denote the initial conditions with x: $x_{-1} = x_0 = x$. We remark that

$$x_1 = \frac{\alpha}{(1+x)x}, \; x_4 = x \text{ and } x_5 = \frac{\alpha}{(1+x)x_3} = x.$$

From the last equality we obtain

$$x_3 = \frac{\alpha}{(1+x)x} = x_1.$$

On the other hand

$$\begin{aligned} x_3 &= \frac{\alpha}{(1+x_2)x_1} = \frac{\alpha}{(1+\frac{\alpha}{(1+x_1)x})x_1} = \frac{\alpha(1+x_1)x}{((1+x_1)x+\alpha)x_1} \\ &= \frac{\alpha\left(1+\frac{\alpha}{(1+x)x}\right)x}{\left(\left(1+\frac{\alpha}{(1+x)x}\right)x+\alpha\right)\frac{\alpha}{(1+x)x}} = \frac{(1+x)x((1+x)x+\alpha)}{(1+x)x+\alpha+\alpha(1+x)}. \end{aligned}$$

Then we obtain an equality

$$\frac{(1+x)x((1+x)x+\alpha)}{(1+x)x+\alpha+\alpha(1+x)} = \frac{\alpha}{(1+x)x}.$$

Therefore

$$(1+x)^2x^2((1+x)x+\alpha) = \alpha((1+x)x+\alpha+\alpha(1+x)) \quad \Rightarrow$$

$$\alpha^2(2+x) + \alpha(1+x)x(1-x-x^2) - (1+x)^3x^3 = 0. \tag{9}$$

The Eq. (9) is a polynomial of order 6 in respect to x but it is only a second order equation in respect to α. We solve this equation in respect to α

$$\alpha = \frac{-x(1+x)(1-x-x^2) \pm \sqrt{x^2(1+x)^2(1-x-x^2)^2 + 4(2+x)(1+x)^3x^3}}{2(2+x)}. \tag{10}$$

At first we note that the roots (10) are defined for every $x \in \mathbf{R}$. This follows from the following equalities

$$\begin{aligned} &x^2(1+x)^2(1-x-x^2)^2 + 4(2+x)(1+x)^3x^3 \\ &= x^2(1+x)^2((1-x(1+x))^2 + 4(2+x)(1+x)x) \\ &= x^2(1+x)^2(1-2x(1+x)+x^2(1+x)^2 + 4(2+x)(1+x)x) \\ &= x^2(1+x)^2(1+x(1+x)(-2+x+x^2+8+4x) \\ &= x^2(1+x)^2(1+x(1+x)(2+x)(3+x)). \end{aligned}$$

Since the derivative of the function $g(x) = x(1+x)(2+x)(3+x)$ is

$$\begin{aligned}
g'(x) &= (1+x)(2+x)(3+x) + x(2+x)(3+x) + x(1+x)(3+x) + x(1+x)(2+x)\\
&= (1+x)(2+x)(3+2x) + x(3+x)(3+2x)\\
&= (3+2x)(2+3x+x^2+3x+x^2) = 2(3+2x)(x^2+3x+1),
\end{aligned}$$

then the stationary points of $g(x)$ are $-\frac{3}{2}, \frac{-3-\sqrt{5}}{2}, \frac{-3+\sqrt{5}}{2}$. Signs of derivatives of the function $g(x)$ shows that points $(\frac{-3-\sqrt{5}}{2}, -1)$ and $(\frac{-3+\sqrt{5}}{2}, -1)$ are global minima points, therefore

$$\forall x \in \mathbf{R} \quad g(x) \geq -1$$

and consequently the root of (10) is greater or equal to 0.

If we consider the equality (10) with "+", then we obtain equilibrium points. Indeed, since the equilibrium equation is in the form $x^3 + x^2 - \alpha = 0$, then $\alpha = x^3 + x^2$. We verify that following equality is true

$$\begin{aligned}
&2(2+x)\alpha = 2(x^3+x^2)(2+x) =\\
&\quad = -x(1+x)(1-x-x^2) + \sqrt{x^2(1+x)^2(1-x-x^2)^2 + 4(2+x)(1+x)^3x^3},\\
&2x^2(1+x)(2+x) + x(1+x)(1-x-x^2) =\\
&\quad = \sqrt{x^2(1+x)^2(1-x-x^2)^2 + 4(2+x)(1+x)^3x^3},\\
&4x^4(1+x)^2(2+x)^2 + 4x^3(1+x)^2(2+x)(1-x-x^2) + x^2(1+x)^2(1-x-x^2)^2 =\\
&\quad = x^2(1+x)^2(1-x-x^2)^2 + 4(2+x)(1+x)^3x^3,\\
&4x^3(1+x)^2(2+x)(x(2+x)+1-x-x^2) = 4(2+x)(1+x)^3x^3,\\
&2x + x^2 + 1 - x - x^2 = 1 + x,\\
&1 + x = 1 + x.
\end{aligned}$$

Since

$$\forall a\, \forall c > 0 \quad -a - \sqrt{a^2 + c} < 0$$

then from the equality (10) with "−" and $x > 0$ we obtain that $\alpha < 0$. But the root of (10) is defined for every x therefore for $x < 0$ exists α such that the solution of the difference equation (1) is periodic with prime period 5.

Case 2. We consider second case:

$$\alpha(x_{-1} + x_0) + \alpha x_0 x_{-1} + x_0 x_{-1} + x_0 x_{-1}(x_{-1} + x_0) + x_0^2 x_{-1}^2 = 0.$$

In this case

$$\alpha = -x_{-1}x_0\left(\frac{1}{x_{-1} + x_0 + x_{-1}x_0} + 1\right). \tag{11}$$

From the equality (11) directly follows that if $x_{-1} > 0$ and $x_0 > 0$ then $\alpha < 0$. On the other hand for fixed α and x_{-1} we can find x_0:

$$x_0 = \frac{-(x_{-1}^2 + x_{-1}(1+\alpha) + \alpha) \pm \sqrt{(x_{-1}^2 + x_{-1}(1+\alpha) + \alpha)^2 - 4\alpha(x_{-1}^2 + x_{-1})x_{-1}}}{2(x_{-1}^2 + x_{-1})}.$$

Since

$$\begin{aligned}&(x_{-1}^2 + x_{-1}(1+\alpha) + \alpha)^2 - 4\alpha(x_{-1}^2 + x_{-1})x_{-1}\\ &= (x_{-1}^2 + \alpha x_{-1})^2 + (x_{-1} - \alpha)^2 + 2x_{-1}(x_{-1}^2 + \alpha^2)\end{aligned}$$

then the root is defined for every α and $x_{-1} > 0$ (if $\alpha > 0$, then $x_0 < 0$). □

6 Periodic Solutions with Period 6

Theorem 5 *For every pair of initial conditions $x_{-1} \neq x_0$ we can find*

$$\alpha = \frac{x_{-1}x_0(1+x_0)(1+x_{-1})(x_{-1}^2(1+x_0) - x_0^2(1+x_{-1}))}{(1+x_0)x_{-1}(x_0 - x_{-1}^2(1+2x_0)) - (1+x_{-1})x_0(x_{-1} - x_0^2(1+2x_{-1}))}$$

such that the solution of the difference equation (1) is periodic with prime period 6 (if $x_{-1} > 0$ and $x_0 > 0$, then α is negative).

If $x_{-1} = x_0 \equiv x$, $x \notin \{-1, -\frac{1}{2}, 0\}$ and $\alpha = -\frac{(1+x)x}{2}$, then the solution of (1) is periodic with prime period 6; the solution is in the form

$$x, x, -\frac{1}{2}, -(1+x), -(1+x), -\frac{1}{2}, x, x, \ldots$$

Proof We fix an arbitrary α. If the solution of the difference equation (1) with initial conditions x_{-1} and x_0 forms a periodic solution with period 6, then

$$x_1 = \frac{\alpha}{(1+x_0)x_{-1}},\ x_2 = \frac{\alpha}{(1+x_1)x_0},\ x_3 = \frac{\alpha}{(1+x_2)x_1},\ x_4 = \frac{\alpha}{(1+x_3)x_2} \text{ and}$$

$$\begin{cases} x_5 = \frac{\alpha}{(1+x_4)x_3} = x_{-1},\\ x_6 = \frac{\alpha}{(1+x_5)x_4} = \frac{\alpha}{(1+x_{-1})x_4} = x_0,\\ x_7 = \frac{\alpha}{(1+x_6)x_5} = x_1\text{—follows by the assumption},\\ x_8 = \frac{\alpha}{(1+x_7)x_6} = x_2\text{—follows by the assumption},\\ x_9 = \frac{\alpha}{(1+x_8)x_7} = x_3\text{—follows by the assumption},\\ x_{10} = \frac{\alpha}{(1+x_9)x_8} = x_4\text{—follows by the assumption}.\end{cases} \tag{12}$$

From x_5 we express x_4

$$\frac{\alpha}{1+x_4} = x_3 x_{-1} \Rightarrow x_4 = \frac{\alpha}{x_3 x_{-1}} - 1. \tag{13}$$

From x_6 we express x_4

$$x_4 = \frac{\alpha}{(1+x_{-1})x_0}. \tag{14}$$

Therefore from (13) and (20) follow

$$\frac{\alpha}{x_3 x_{-1}} - 1 = \frac{\alpha}{(1+x_{-1})x_0}$$

and we can express x_3:

$$x_3 = \frac{(1+x_{-1})x_0\alpha}{(\alpha + (1+x_{-1})x_0)x_{-1}}. \tag{15}$$

On the other hand we can express x_3 in terms of the initial conditions

$$\begin{aligned} x_3 &= \frac{\alpha}{(1+x_2)x_1} = \frac{\alpha}{\left(1+\frac{\alpha}{(1+x_1)x_0}\right)x_1} = \frac{\alpha(1+x_1)x_0}{((1+x_1)x_0+\alpha)x_1} = \\ &= \frac{\alpha(1+\frac{\alpha}{(1+x_0)x_{-1}})x_0}{\left((1+\frac{\alpha}{(1+x_0)x_{-1}})x_0+\alpha\right)\frac{\alpha}{(1+x_0)x_{-1}}} = \frac{((1+x_0)x_{-1}+\alpha)x_0(1+x_0)x_{-1}}{x_0(1+x_0)x_{-1}+\alpha x_0+\alpha(1+x_0)x_{-1}}. \end{aligned}$$

Then

$$\frac{((1+x_0)x_{-1}+\alpha)x_0(1+x_0)x_{-1}}{x_0(1+x_0)x_{-1}+\alpha x_0+\alpha(1+x_0)x_{-1}} = \frac{(1+x_{-1})x_0\alpha}{(\alpha+(1+x_{-1})x_0)x_{-1}}.$$

And we have an expression that contains only α and initial conditions

$$\begin{aligned} &(\alpha+(1+x_0)x_{-1})(1+x_0)x_{-1}^2(\alpha+(1+x_{-1})x_0) \\ &= \alpha(1+x_{-1})(x_0(1+x_0)x_{-1}+\alpha(x_0+(1+x_0)x_{-1}). \end{aligned}$$

This equation in respect to α is a quadratic polynomial but in respect to x_0—a polynomial of degree 3 and in respect to x_{-1}—polynomial of degree 4. As a quadratic polynomial in respect to α it is in the form

$$\alpha^2(x_{-1}+x_0+2x_{-1}x_0)+\alpha(1+x_0)x_{-1}(x_0-x_{-1}^2-2x_0x_{-1}^2)-(1+x_0)^2x_{-1}^3(1+x_{-1})x_0 = 0. \tag{16}$$

If we solve the equation and replace x_{-1} and x_0 with concrete values, then it is possible to find periodic solution with period 6 but not always. Why? This equation does not include all the conditions that are required in the system (12).

We can express x_4 from x_6 in the system (12) and then compare it to definition of x_4:

$$x_4 = \frac{\alpha}{(1+x_{-1})x_0} = \frac{\alpha}{(1+x_3)x_2}.$$

From this last equality we express x_2 and replace x_3 with (15)

$$x_2 = \frac{(1+x_{-1})x_0}{1+x_3} = \frac{(1+x_{-1})x_0}{1+\frac{(1+x_{-1})x_0\alpha}{(\alpha+(1+x_{-1})x_0)x_{-1}}} = \frac{((1+x_{-1})x_0+\alpha)x_{-1}(1+x_{-1})x_0}{((1+x_{-1})x_0+\alpha)x_{-1}+\alpha(1+x_{-1})x_0}.$$

We can express x_2 in terms of initial conditions by the definition of the given sequence

$$x_2 = \frac{\alpha}{(1+x_1)x_0} = \frac{\alpha}{\left(1+\frac{\alpha}{(1+x_0)x_{-1}}\right)x_0} = \frac{\alpha(1+x_0)x_{-1}}{((1+x_0)x_{-1}+\alpha)x_0}.$$

Both expressions of x_2 must be the same

$$\frac{\alpha(1+x_0)x_{-1}}{((1+x_0)x_{-1}+\alpha)x_0} = \frac{((1+x_{-1})x_0+\alpha)x_{-1}(1+x_{-1})x_0}{((1+x_{-1})x_0+\alpha)x_{-1}+\alpha(1+x_{-1})x_0}.$$

We again obtain a quadratic polynomial in respect to α

$$\alpha^2(x_{-1}+x_0+2x_{-1}x_0)+\alpha(1+x_{-1})x_0(x_{-1}-x_0^2-2x_0^2x_{-1})-(1+x_{-1})^2x_0^3(1+x_0)x_{-1}=0. \tag{17}$$

Since (16) and (17) have to be equal then we get a new equality

$$\begin{aligned}&\alpha(1+x_0)x_{-1}(x_0-x_{-1}^2-2x_0x_{-1}^2)-(1+x_0)^2x_{-1}^3(1+x_{-1})x_0\\&=\alpha(1+x_{-1})x_0(x_{-1}-x_0^2-2x_0^2x_{-1})-(1+x_{-1})^2x_0^3(1+x_0)x_{-1}.\end{aligned}$$

Therefore

$$\alpha = \frac{x_{-1}x_0(1+x_0)(1+x_{-1})(x_{-1}^2(1+x_0)-x_0^2(1+x_{-1}))}{(1+x_0)x_{-1}(x_0-x_{-1}^2(1+2x_0))-(1+x_{-1})x_0(x_{-1}-x_0^2(1+2x_{-1}))}. \tag{18}$$

This formula is not valid if $x_{-1}=x_0$ and if $x_{-1}>0$ and $x_0>0$, then $\alpha<0$. If $x_{-1}=x_0\equiv x$, then from (16) and (17) we obtain

$$\begin{aligned}\alpha &= \frac{-(1+x)x(x-x^2-2x^3) \pm \sqrt{(1+x)^2x^2(x-x^2-2x^3)^2+4(2x+2x^2)(1+x)^3x^4}}{2(2x+2x^2)}\\ &= \frac{-x(1+x)(1-2x) \pm \sqrt{(1+x)^4x^4(1-2x)^2+8x^5(1+x)^4}}{4x(1+x)}\\ &= \frac{-x^2(1+x)^2(1-2x) \pm (1+x)^2x^2\sqrt{(1-2x)^2+8x}}{4x(1+x)}\\ &= \frac{1}{4}(-x(1+x)(1-2x) \pm (1+x)x(1+2x)).\end{aligned}$$

If we consider this equality with "+" sign, then we have equilibrium points:

$$\alpha_+ = \frac{x(1+x)}{4}(-1+2x+1+2x) = x^2(1+x).$$

If we consider this equality with "−" sign, then

$$\alpha_- = \frac{x(1+x)}{4}(-1+2x-1-2x) = \frac{-x(1+x)}{2}$$

and in this case if $x \neq -1$ or $x \neq 0$, or $x \neq -\frac{1}{2}$ we obtain a periodic solution with prime period 6, the orbit is

$$x,\ x,\ -\frac{1}{2},\ -(1+x),\ -(1+x),\ -\frac{1}{2}.$$

□

Corollary 2 *For the difference equation (1) with positive α there are no positive initial conditions x_{-1} and x_0 such that the solution of (1) is periodic with prime period 6.*

7 Periodic Solutions with Period 7

Theorem 6 *For every pair of initial conditions x_{-1}, x_0 for which is well defined*

$$\alpha = \frac{-K \pm \sqrt{K^2+4(x_0+x_{-1}+x_0x_{-1})x_{-1}x_0(1+x_{-1})(1+x_0)}}{2(x_0+x_{-1}+x_0x_{-1})},$$

where

$$K = x_{-1}^2x_0 + x_{-1}x_0^2 + x_{-1}^2x_0^2 - x_0 - x_{-1} - x_{-1}x_0,$$

we obtain the solution of the difference equation (1) such is periodic with prime period 7.

Proof We fix an arbitrary α. If the solution of the difference equation (1) with initial conditions x_{-1} and x_0 forms a periodic solution with period 7, then

$$x_1 = \frac{\alpha}{(1+x_0)x_{-1}},\ x_2 = \frac{\alpha}{(1+x_1)x_0},\ x_3 = \frac{\alpha}{(1+x_2)x_1},\ x_4 = \frac{\alpha}{(1+x_3)x_2},\ x_5 = \frac{\alpha}{(1+x_4)x_3} \quad \text{and}$$

$$\begin{cases} x_6 = \frac{\alpha}{(1+x_5)x_4} = x_{-1}, \\ x_7 = \frac{\alpha}{(1+x_6)x_5} = \frac{\alpha}{(1+x_{-1})x_4} = x_0, \\ x_8 = \frac{\alpha}{(1+x_7)x_6} = x_1\text{—follows by the assumption,} \\ x_9 = \frac{\alpha}{(1+x_8)x_7} = x_2\text{—follows by the assumption,} \\ x_{10} = \frac{\alpha}{(1+x_9)x_8} = x_3\text{—follows by the assumption,} \\ x_{11} = \frac{\alpha}{(1+x_{10})x_9} = x_4\text{—follows by the assumption,} \\ x_{12} = \frac{\alpha}{(1+x_{11})x_{10}} = x_5\text{—follows by the assumption.} \end{cases}$$

From x_7 we express x_5, and from x_5 we express x_3

$$x_5 = \frac{\alpha}{(1+x_{-1})x_0} = \frac{\alpha}{(1+x_4)x_3} \Rightarrow x_3 = \frac{(1+x_{-1})x_0}{1+x_4}. \tag{19}$$

From x_6 we express x_4 and replace expressed x_5

$$x_4 = \frac{\alpha}{(1+x_5)x_{-1}} = \frac{\alpha}{\left(1+\frac{\alpha}{(1+x_{-1})x_0}\right)x_{-1}} = \frac{\alpha(1+x_{-1})x_0}{((1+x_{-1})x_0+\alpha)x_{-1}}. \tag{20}$$

The expression (20) we place in equality (19) and find x_3 expressed only with initial conditions who conside with x_3 by definition

$$\begin{aligned} x_3 &= \frac{(1+x_{-1})x_0}{1+\frac{\alpha(1+x_{-1})x_0}{((1+x_{-1})x_0+\alpha)x_{-1}}} = \frac{(1+x_{-1})x_0x_{-1}((1+x_{-1})x_0+\alpha)}{((1+x_{-1})x_0+\alpha)x_{-1}+\alpha(1+x_{-1})x_0} \\ &= \frac{(1+x_0)x_0x_{-1}((1+x_0)x_{-1}+\alpha)}{((1+x_0)x_{-1}+\alpha)x_0+\alpha(1+x_0)x_{-1}}. \end{aligned}$$

And we have an expression only with α and initial conditions

$$\begin{aligned} &(1+x_{-1})((1+x_{-1})x_0+\alpha)(((1+x_0)x_{-1}+\alpha)x_0+\alpha(1+x_0)x_{-1}) \\ &= (1+x_0)((1+x_0)x_{-1}+\alpha)(((1+x_{-1})x_0+\alpha)x_{-1}+\alpha(1+x_{-1})x_0). \end{aligned}$$

This equation in respect to α is a quadratic polynomial but in respect to x_0 or x_{-1}—a polynomial of degree 3. As a quadratic polynomial in respect to α it is in the form

$$\begin{aligned} &(x_{-1}-x_0)\big(\alpha^2(x_0+x_{-1}+x_0x_{-1})+ \\ &+\alpha(x_{-1}^2x_0+x_{-1}x_0^2+x_{-1}^2x_0^2-x_0-x_{-1}-x_0x_{-1})-x_{-1}x_0(1+x_{-1})(1+x_0)\big) = 0. \end{aligned}$$

We need to analyze two cases.

Case 1. $x_{-1} = x_0$. We denote $x_{-1} = x_0 = x$. We remark that

$$x_1 = \frac{\alpha}{(1+x)x}, \ x_2 = \frac{\alpha}{(1+x_1)x} = \frac{\alpha(1+x)}{(1+x)x+\alpha}, \ x_3 = \frac{\alpha}{(1+x_2)x_1} = \frac{x(1+x)(\alpha+x+x^2)}{x(1+x)+\alpha(2+x)}.$$

Therefore

$$x_6 = \frac{\alpha}{(1+x_5)x_4} = x \Rightarrow x_4 = \frac{\alpha}{(1+x_5)x}, \tag{21}$$

$$x_7 = \frac{\alpha}{(1+x_6)x_5} = \frac{\alpha}{(1+x)x_5} = x \Rightarrow x_5 = \frac{\alpha}{(1+x)x} = x_1. \tag{22}$$

From the equalities (21) and (22) we obtain that

$$x_4 = \frac{\alpha}{(1+x_1)x} = x_2.$$

On the other hand

$$x_4 = \frac{\alpha}{(1+x_3)x_2} = x_2 \Rightarrow \frac{\alpha}{1+x_3} = x_2^2.$$

Therefore

$$\frac{\alpha}{1+\frac{x(1+x)(\alpha+x+x^2)}{x(1+x)+\alpha(2+x)}} = \frac{\alpha^2(1+x)^2}{((1+x)x+\alpha)^2}.$$

We shorten α and obtain the equality

$$\begin{aligned}(x(1+x)+\alpha(2+x))(x+x^2+\alpha)^2 \\ = \alpha(1+x)^2(x(1+x)+\alpha(2+x)+x(1+x)(\alpha+x+x^2))\end{aligned}$$

which is a polynomial of order 3 in respect to α and of order 6 in respect to x. As a polynomial of α it is in the form

$$\begin{aligned}\alpha^3(2+x)+\alpha^2(1+x)(2+x)(x-1-x^2) \\ +\alpha x(1+x)^2(2x-x^3-x^2-1)+x^3(1+x)^3 = 0.\end{aligned} \tag{23}$$

Since $\alpha = x^2+x^3$ is an equilibrium point then the Eq. (23) is divisible by $\alpha - (x^2 + x^3)$ and we obtain a polynomial of second order

$$\alpha^2(2+x)+\alpha(1+x)(2+x)(x-1)-(1+x)^2 x = 0. \tag{24}$$

The roots of the Eq. (24) are

$$\begin{aligned}\alpha &= \frac{-(1+x)(2+x)(x-1) \pm \sqrt{(1+x)^2(2+x)^2(x-1)^2 + 4(2+x)(1+x)^2 x}}{2(2+x)} \\ &= \frac{-(1+x)(2+x)(x-1) \pm \sqrt{(1+x)^3(2+x)(x^2-x+2)}}{2(2+x)} \\ &= \frac{-(1+x)(2+x)(x-1) \pm (1+x)\sqrt{(1+x)(2+x)(x^2-x+2)}}{2(2+x)}\end{aligned}$$

The minimum point of parable $x^2 - x + 2$ is $(\frac{1}{2}, 1\frac{3}{4})$ therefore α is defined for all positive x and $\alpha_+ > 0$.

Case 2. We consider second case:

$$\begin{aligned}&\alpha^2(x_0 + x_{-1} + x_0 x_{-1}) + \\ &+\alpha(x_{-1}^2 x_0 + x_{-1}x_0^2 + x_{-1}^2 x_0^2 - x_0 - x_{-1} - x_0 x_{-1}) - x_{-1}x_0(1+x_{-1})(1+x_0) = 0.\end{aligned} \tag{25}$$

Then

$$\alpha = \frac{-K \pm \sqrt{K^2 + 4(x_0 + x_{-1} + x_0 x_{-1})x_{-1}x_0(1+x_{-1})(1+x_0)}}{2(x_0 + x_{-1} + x_0 x_{-1})}, \tag{26}$$

where

$$K = x_{-1}^2 x_0 + x_{-1}x_0^2 + x_{-1}^2 x_0^2 - x_0 - x_{-1} - x_{-1}x_0.$$

Calculations with a computer show that for all fixed initial conditions for which α is well defined with the formula (26) we obtain a periodic solution of the difference equation (1) with prime period 7 (it is clear that $x_{-1} \notin \{-1, 0\}$ and $x_0 \notin \{\frac{-x_{-1}}{1+x_{-1}}, -1, 0\}$ but α is not defined, for example, $x_{-1} = -0.7$ and $x_0 = 4$). If $x_{-1} > 0$ and $x_0 > 0$, then $\alpha > 0$—we conclude that the prime period 7 is the first for which exists positive initial conditions and positive parameter α such that the solution of the difference equation (1) is periodic with prime period 7.

For example, if $x_{-1} = 8$, $x_0 = 0.5$, then $\alpha_+ = 1.063201$ and the solution is

$$8,\ 0.5,\ 0.0886,\ 1.953336,\ 4.063201,\ 0.107501,\ 0.236267,\ 8,\ 0.5, \ldots.$$

if $x_{-1} = 8$, $x_0 = 0.5$, then $\alpha_- = -4,0632$ and the solution is

$$8,\ 0.5,\ -0.3386,\ -12.2867,\ -1.0632,\ -5.23246,\ -0.90294,\ 8,\ 0.5, \ldots.$$

We remark that if $x_{-1} = x_0 = x$, then equality (25) is equal to (24). □

Corollary 3 *For the difference equation (1) with some positive α there exist positive initial conditions x_{-1} and x_0 such that the solution of (1) is periodic with prime period 7.*

8 Conclusions

The Open Problem 3.3 given in [2] is not answered fully. Even for periodic solutions with period 7 we can not say that for every α exists initial conditions such that the solution of the difference equation (1) is periodic with prime period 7. Finding the periodic solution with a period of 8 is more complicated than the considered cases. We conclude that for solving the problem completely we need to seek other ways of analysis. Looking at the arbitrary solution of difference equation (1) we have the feeling that solution is periodic. But it is possible that period is so large that we can not say exactly the size of period. The solution is sensitive to the calculations—it was observed on testing a periodic solutions even with the period 2.

The Eq. (1) is gifted with the invariant [2]

$$x_{n-1} + x_n + x_{n-1}x_n + \alpha\left(\frac{1}{x_{n-1}} + \frac{1}{x_n}\right) = \text{constant}, \quad \forall n \geq 0.$$

It is possible that we can use this property to solve the Open Problem 3.3.

Acknowledgments This work was partially supported by the Latvian Council of Science research project 09.1220.

References

1. Agarwal, R.P., Popenda, J.: Periodic solutions of first order linear difference equations. Math. Comput. Model. **22**(1), 11–19 (1995)
2. Amleh, A.M., Camouzis, E., Ladas, G.: On the dynamics of a rational difference equations, part 1. Int. J. Differ. Equ. **3**(1), 1–35 (2008)
3. Amleh, A.M., Camouzis, E., Ladas, G.: On the dynamics of a rational difference equations, part 2. Int. J. Differ. Equ. **3**(2), 195–225 (2008)
4. Bajo, I., Litz, E.: Periodicity on discrete dynamical systems generated by a class of rational mappings. J. Differ. Equ. Appl. **12**(12), 1201–1212 (2006)
5. Bula, I.: On the second order quadratic rational difference equation x(n+1)=alpha/((1+x(n))x(n-1)). In: Abstracts book of ICDEA2012, 18th International Conference on Difference Equations and Applications, Casa de Convalescencia, Barcelona. http://www.gsd.uab.cat/icdea2012/scientific-program_abstracts.php (2012). Accessed 23–27 July 2012
6. Camouzis, E., Ladas, G.: Dynamics of Third Order Rational Difference Equations with Open Problems and Conjectures. Advances in Discrete mathematics and Applications, vol. 5. Chapman & Hall/CRC Press, Boca Raton (2008)
7. Cull, P., Flahive, M., Robson, R.: Difference Equations. From Rabbits to Chaos. Undergraduate Texts in Mathematics. Springer Science+Business Media, Inc., New York (2005)

8. El-Metwally, H., Grove, E.A., Ladas, G.: A global convergence result with applications to periodic solutions. J. Math. Anal. Appl. **245**, 161–170 (2000)
9. El-Owaidy, H., Mohamed, H.Y.: On the periodic solutions for an nth-order difference equations. Appl. Math. Comput. **131**, 461–467 (2002)
10. Grove, E.A., Ladas, G.: Periodicities in Nonlinear Difference Equations. Chapman & Hall/CRC, Boca Raton (2005)
11. Janowski, E.J., Kulenovič, M.R.S., Silič, E.: Periodic solutions of linearizable difference equations. Int. J. Differ. Equ. **6**(2), 113–125 (2011)
12. Kocic, V.L., Ladas, G.: Global Behavior of Nonlinear Difference Equations of Higher Order with Applications. Mathematics and Its Applications, vol. 256, Kluwer Academic Publishers, Dordrecht (1993)
13. Kulenovič, M.R.S., Ladas, G.: Dynamics of Second Order Rational Difference Equations, with Open Problems and Conjectures. Chapman & Hall/CRC, Boca Raton (2002)

Li–Yorke Chaos in Perturbed Rational Difference Equations

Francisco Balibrea and Antonio Cascales

Abstract In this paper we continue and complete with numerical examples, the study of chaos, in Li–Yorke and Marotto sense of difference equations of order two polynomials and rational, initiated in Balibrea and Cascales (Li-Yorke chaos in difference equations of order two polynomials and rationals [3]). Such equations arise, for example, in the application of the Newton's method to polynomials equations, models of population dynamics and control problems. We use and review Marotto's ideas from Marotto (J. Math. Anal. Appl. 63:199, 1978 [14]) based on a subtle study of the dynamics near a special kind of equilibrium (snap-back repeller). We claim that in the second order setting, there are cases which can be seen as two dimensional perturbations of equations of first order which are easier to handle. In this sense we continue applying results from Marotto (J. Math. Anal. Appl. 72:716–729, 1979 [15]). As an example we compute numerically the snap-back repellers of the inverse logistic first order equation whose perturbation allow to obtain some rational two order equations with chaotic dynamics. We apply the stated results and conjectures to the rational competition model of Hassell and Comins (Theor. Popul. Biol. 9:202–221, 1976 [6]).

Keywords Li–Yorke chaos · Difference equation · Perturbation · Competition model

1 Introduction, First Definitions and Terminology

Let (X, d) be a *metric space* and $f : X \to X$ a *continuous map* from X into itself. In this paper we will consider discrete dynamical systems (X, f) in the form of *difference equations*, DE, defined by the recurrence

F. Balibrea · A. Cascales (✉)
Departamento de Matemáticas, Universidad de Murcia,
Campus de Espinardo, 30100 Murcia, Spain
e-mail: antoniocascales@yahoo.es

F. Balibrea
e-mail: balibrea@um.es

L. Alsedà i Soler et al. (eds.), *Difference Equations, Discrete Dynamical Systems and Applications*, Springer Proceedings in Mathematics & Statistics 180, DOI 10.1007/978-3-662-52927-0_4

$$x_{k+1} = f(x_k) \tag{1}$$

where $k = 0, 1, 2, \ldots$. A *solution* of (1) starting at $x_0 \in X$ or having x_0 as initial point is the sequence $(x_k)_{k=0}^{+\infty} = (f^k(x_0))_{k=0}^{+\infty}$ obtained by using the recurrence and where $f^k = f^{k-1} \circ f$ for $k > 0$ and $f^0(x) = x$ for all $x \in X$. Such sequence is also called the *trajectory of the point* x_0 or the *forward orbit of* x_0. A solution of (1) is *periodic of prime period p* or x_0 is a *periodic of period p*, if p is the first positive integer for which $x_{k+p} = x_k$ for all $k > 0$. When $p = 1$ then the solution is *constant* or x_0 is a fixed point of f, that is $f(x_0) = x_0$. If $f^m(x_0)$ is periodic for some $m > 1$ then we say that x_0 is *eventually periodic* or that the corresponding solution starting in such point is *eventually periodic*.

When $X = A$, with $A = \mathbb{R}$ or a subinterval of $\mathbb{R}$, in [12] was introduced for the first time in the mathematical literature the notion of *chaos* for the corresponding scalar difference equations (1). It was said that (1) is *chaotic in the sense of Li–Yorke* or *Li–Yorke chaotic* if the following conditions are fulfilled

LY1 There exists $N \in \mathbb{N}$ such that $\forall p \geq N$ there are periodic solutions with period p.

LY2 There exists an invariant uncountable set $S \subset A$ ($f(S) \subset S$) non containing periodic points, such that for every $x \neq y \in S$,

$$\limsup_{k\to\infty} |f^k(x) - f^k(y)| > 0 \qquad (a)$$

and for any periodic point y_p of f

$$\limsup_{k\to\infty} |f^k(x) - f^k(y_p)| > 0 \qquad (b)$$

LY3 There is an uncountable subset $S_0 \subset S$, such that for every $x_0 \neq y_0 \in S_0$ is

$$\liminf_{k\to\infty} |f^k(x_0) - f^k(y_0)| = 0$$

S_0 is a *scrambled set* of the DE and every pair of distinct points in S_0 is a *Li–Yorke pair*.

In [12] it is proved that when the DE has a periodic point of period *three*, then condition LY1 is satisfied with $N = 1$ and also LY2 and LY3. Besides, it was proved (see for example [1]) that condition LY2(b) is redundant. Some time later, in [9] was proved that if there is one Li–Yorke pair, then there is an uncountable number of them. Finally it was said that when $X = A$, a DE is chaotic in the sense of Li–Yorke if there is a Li–Yorke pair.

In current literature it is said also that when X is a metric space, DE (1) is chaotic in the sense of Li–Yorke if there exists an uncountable set $S \subseteq X$ such that for every pair of distinct points $x, y \in S$ the following conditions are fulfilled

ly1 $\limsup_{k\to\infty} d(f^k(x), f^k(y)) > 0$

ly2 $\liminf_{k\to\infty} d(f^k(x), f^k(y)) = 0$

This definition is applicable in particular to the case $X = \mathbb{R}^n$ for $n > 1$.

But in [14] and for $f \in C^1$ on $\mathbb{R}^n$, a sufficient condition was considered to obtain properties LY1, LY2 and LY3 (replacing the absolute value by any norm in $\mathbb{R}^n$). In such situation, we say that there exists *chaos in the sense of Marotto* or *Marotto chaos* of a DE with $n > 1$. By other hand, Marotto chaos has been introduced in [13] for DE with $n > 1$ and $f \in C(\mathbb{R}^n)$, that is, for general continuous maps holding the three former properties.

The sufficient condition given by Marotto was the existence of one particular type of homoclinic solution (see [2] for a general definition). In such type of solutions there is a relevant fixed point called by Marotto, *snap-back repeller*. Then Marotto's result from [14] says that when DE has a snap-back repeller, then DE is Marotto chaotic.

In the literature has been a controversy concerning the definition of snap-back repeller, particularly see [10, 11, 16]. In this paper we have adopted the following definition which is a restriction of the original Marotto's definition. First we start recalling the definition of expanding neighborhood.

Definition 1 Let $f \in C^1(\mathbb{R}^n)$. A set $A \subset \mathbb{R}^n$ is expanding with respect to a norm $\| \cdot \|$ if

$$\| f(x) - f(y) \| > s \| x - y \|$$

for some $s > 1$ and all $x \neq y \in A$.

Definition 2 A fixed point z of f is a snap-back repeller if:

SBR1 There exists a neighborhood A of z expanding with respect to a norm $\| \cdot \|$

SBR2 There exists a finite sequence of points $x_0, x_1, \ldots, x_M = z$ such that $x_0 \neq z$ belongs to A, $x_k = f^k(x_0)$ and $|Df(x_k)| \neq 0$ for $1 \leq k \leq M$.

(See Fig. 1.)

In the original Marotto's definition of snap-back repeller, instead of condition SBR1 was stated the following:

SBR1a There exists a neighborhood of z in which all of the eigenvalues of $Df(x)$ have absolute values larger than one.

Relationship between SBR1 and SBR1a is studied, in the general case, in [25]. The main idea is that when SBR1a is true, then there exists a norm and a local open set where condition SBR1 is satisfied. Also, for a linear operator, the expansivity condition implies that all the eigenvalues must be outside the disk of center zero and radius one.

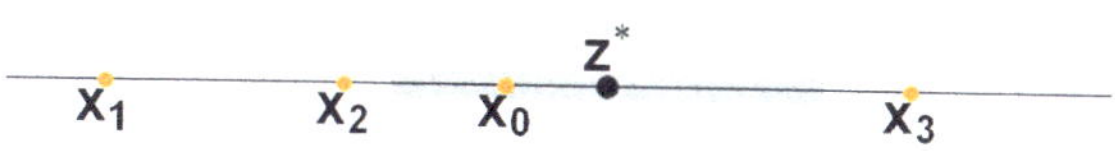

Fig. 1 A snap-back repeller: x_0 belonging to a expanding neighborhood of the fixed point $z^* = x_4$ hits the equilibrium after a finite number of iterations

However, the question about the original formulation of Marotto's criterion for chaos is not yet clarified. In the cited controversy, there was an emphasis in the error of the original proof of Marotto, but, as far as we know, there is no example of a *non chaotic* DE with a snap-back repeller (as Marotto originally defined it). Moreover, the popularity of Marotto's condition is due to that a standard technique to search snap-back repellers consists on the location of a repelling fixed point for which we take points in its backward trajectory. If those points belong to a critical neighborhood of the equilibrium, then a snap-back repeller (and therefore chaos) is founded. But the word *critical* means, originally, SBR1a, a condition easy to verify in numerical examples. In the current state of the question, the new meaning is related to the expansive property, something awkward to compute.

When $n = 1$ a simple application of Rolle's theorem shows that SBR1a implies SBR1, and the previous question can be skipped. But the problem about the eigenvalues condition for greater dimensions is still a central question in the theory of chaotic difference equations.

In this paper we will deal with the case $X = \mathbb{R}^n$ for $n > 1$ and f a rational map. This case will be denoted by RDE. Rational difference equations have been widely considered in the literature since many model from population dynamics have been formulated in such way, mainly by the quotient of two polynomials. In particular, one interesting example is given by

$$x_{k+1} = \frac{P_1(x_k, y_k)}{Q_1(x_k, y_k)}$$

$$y_{k+1} = \frac{P_2(x_k, y_k)}{Q_2(x_k, y_k)}$$

where P_1, Q_1, P_2, Q_2 are quadratic polynomials in x_k, y_k containing real parameters. In numerical simulations (see [13, 17]) it is observed some subsets of the phase space where we have some complicated structure of solutions. It motivates to look for conditions for the existence of chaotic behavior for different values of initial conditions and/or of parameters of the RDE. Moreover, for some particular values of the parameters, in families considered in [14–17] it is possible to prove analytically the existence of snap-back repellers. In general it is not easy to get.

Using rational models has been also common in applications to population dynamics (see for example the book [4]) and also in the application of the Newton's method to polynomial equations like $p(x) = 0$ (see also the book [7]).

Sometimes the construction of solutions of the equations can not be done because x_k for some particular k can not be obtained, because we reach points where the map f is not defined. In such cases, if we denote by x one of such points, we say that x is an element of the *forbidden set of the RDE*. It is evident that the forbidden set of the RDE contains at least the poles of the rational components of the map.

Forbidden sets can have a complicated geometric structure, not well understood in most cases. The main problem is that their points can be spread out over the phase space.

In [3] it is proved the existence of Marotto's chaos for RDEs. The proof is made analogously to the broader Marotto's proof, but it is more difficult since it is necessary in some steps to overcome the problem which appears when some points belong to the forbidden set of the RDE.

In the last section we recall the results of [3] for the family of one dimensional RDEs $x_{k+1} = \frac{1}{x_k^2 - r}$, giving sufficient conditions over the parameter r which guarantee the presence of chaos. Also we apply the previous ideas to the family $x_{k+1} = \frac{\lambda x_k}{(1+x_k)^2}$. As an application, the main conjecture of this paper allow to state a range of parameters for the bidimensional competition model of Hassell and Comins (see [6]) to be chaotic.

In fact there is a great interest to consider one dimensional RDE, say $x_{k+1} = f(x_k)$ and prove the existence of snap-back repellers, because in most cases, this one dimensional system can be seen as a two dimensional RDE given by $F(x_k, x_{k+1}) = (f(x_k), x_k)$. In such case, in [15] it is proved that if the one dimensional DE represented by f has a snap-back repeller, then the system $F(x, y) = (f(x), x)$ has a transversal homoclinic orbit and as a consequence, the main theorem from [20] applies, obtaining the existence of chaotic behavior in a Cantor set in a stronger sense. Also in [15], more general two dimensional RDEs are proved to be chaotic in a range of parameters using the former remark and a perturbation of the system given by F.

For example, using the ideas from [15] we claim that in a range of values of the real parameter b, the following two dimensional RDE is chaotic in a stronger sense:

$$x_{k+1} = \frac{1 + b \cdot g(x_{k-1})}{x_k^2 - r} \tag{2}$$

where $r > \frac{1}{\sqrt[3]{2}}$ and g is any sufficiently smooth real function.

2 Transversal Homoclinic Orbits and Snap-Back Repellers

Suppose that $F \in C^1(\mathbb{R}^n)$ be a diffeomorphism with a hyperbolic fixed point x_0. Then it is well known that there exist stable and unstable manifolds of F at x_0 denoted by $W^s(x_0)$ and $W^u(x_0)$. A point in the intersection of such manifolds is said to be *homoclinic to* x_0. When both manifolds intersect transversely in a point x_1, then it is called a *transverse homoclinic point*. Then the solution $(x_k)_{k=-\infty}^{+\infty}$ verifies $x_k \to x_1$ and $x_{-k} \to x_1$ when $k \to \infty$. In a neighborhood of a transverse homoclinic point, there exists an invariant Cantor set where the dynamics of F^n for some n is topologically conjugate to a full shift of N symbols. These are the contents of the following result, proved by Smale in [19].

Theorem 1 *Let F be an $C^1(\mathbb{R}^n)$ diffeomorphism. If F has a transversal homoclinic orbit, then there exists a Cantor set in $\mathbb{R}^n$ where for some positive integer N, F^N is topologically equivalent to a shift automorphism.*

The proof, definitions and more details can be seen in [21]. What is important here is to remark that in such Cantor set, F has a dense set of periodic orbits and a dense orbit. It implies that there is a relationship of analogy between the existence of a transversal homoclinic orbit and that of snap-back repeller. In fact a snap-back repeller can be seen as a particular example of fixed point with a transversal homoclinic orbit when we extend the notion of homoclinic orbit to maps which are not necessarily one to one. To see it, consider the unstable manifold in $\mathbb{R}^n$ which transversely intersects the zero-dimensional stable manifold composed of the fixed point Z which produces the orbit $(X_k)_{k=-\infty}^{+\infty}$ with $X_M = Z$ and $X_k \to Z$ when $k \to -\infty$.

Although either type of homoclinic behavior implies chaos, there are practical differences between them, particularly in what concerns computation. Generally speaking, snap-back repellers are easier to compute than transversal homoclinic orbits.

Given $F(x, y) = (f(x), x)$ a two dimensional problem where $f : \mathbb{R} \to \mathbb{R}$ is differentiable, in [15] it is proved that there is a closed connection between such problem and the one-dimensional $x_{k+1} = f(x_k)$ given by

Theorem 2 *If the former one-dimensional problem has a snap-back repeller, then the two dimensional has a transversal homoclinic orbit.*

In fact, in [15] the two dimensional problem is seen as a two dimensional perturbation of the one dimensional. In the same paper, such idea is extended to the family of two dimensional problems of second order difference equations

$$x_{k+1} = f(x_k, bx_{k-1}) \tag{3}$$

with f differentiable and b a real parameter. It is immediate that such equation can be obtained as the system

$$x_{k+1} = f(x_k, by_k)$$

$$y_{k+1} = x_k$$

from which when $b = 0$, then we have the associate one dimensional problem

$$x_{k+1} = f(x_k, 0) \tag{4}$$

Then we have the result (see [15]).

Theorem 3 *If (4) has a snap-back repeller, then (3) has a transversal homoclinic orbit for all $|b| < \varepsilon$ for some $\varepsilon > 0$.*

Let f, g be real differentiable maps and $G(x, y) = (f(x), g(y))$. For them we have (see [15]):

Theorem 4 *1. If one of the maps f or g has a snap-back repeller and the other has an unstable fixed point, then $G(x, y)$ has a snap-back repeller*

2. *If one of the maps f or g has a snap-back repeller and the other a stable fixed point, then $G(x, y)$ has a transversal homoclinic orbit.*

The former ideas can be used to be applied to the more general case

$$\begin{aligned} x_{k+1} &= f(x_k, by_k) \\ y_{k+1} &= g(cx_k, y_k) \end{aligned} \tag{5}$$

with f, g differentiable maps and b, c real parameters. When $b = c = 0$, (5) reduces to the uncoupled system

$$x_{k+1} = f(x_k, 0) \tag{6}$$

$$y_{k+1} = g(0, y_k) \tag{7}$$

In this case the map $G(x, y, b, c) = (f(x, by), g(cx, y))$ can be seen as a perturbation of the map $G(x, y, 0, 0)$ former treated.

We have the following result

Theorem 5 *1. If one of the problems (6) or (7) has a snap-back repeller and the other an unstable fixed point, then (5) has a snap-back repeller for all $|b|, |c| < \varepsilon$ for some $\varepsilon > 0$*

2. If one of (6) or (7) has a snap-back repeller and the other a stable fixed point, then (5) has a transversal homoclinic orbit for all $|b|, |c| < \varepsilon$ for some $\varepsilon > 0$.

In a series of papers by Zgliczynski and Misiurewicz–Zgliczynski (see [18, 22–24]) has been proved in the first three papers that if an unperturbed one dimensional map has a point of period n, then sufficiently close multidimensional perturbations of this map have periodic points of all periods allowed by Sharkovskii's theorem and applied it to prove that the Rössler equations have an infinite number of periodic points. While in [18] it is proved that if an interval map of positive entropy is perturbed to a compact multidimensional map, then the topological entropy can not drop down considerably if the perturbation is small.

We claim that similar results could be obtained for rational difference equations. In fact we have proved using numerical approach and simulation that in such setting it is such situation. Currently it is necessary to prove rigorously them.

3 Numerical Examples

Let the following family of one dimensional rational difference equations:

$$x_{k+1} = \frac{1}{x_k^2 - r} \tag{8}$$

where r is a real number. In [3] the chaotic character of Eq. (8) was proved for $r = 1$ and it was conjectured that for $r \geq \frac{1}{\sqrt[3]{2}}$ the corresponding equation has a snap-back repeller and therefore is Li–Yorke chaotic (Fig. 2).

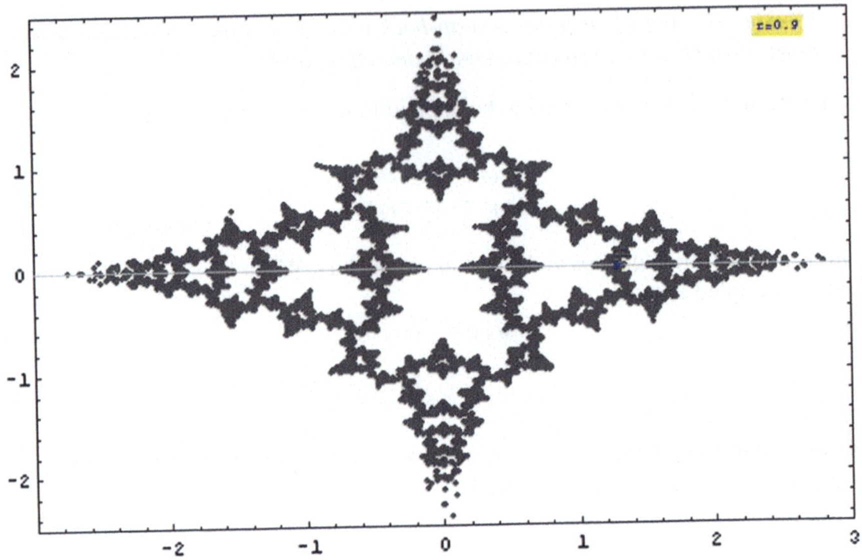

Fig. 2 Inverse solution of the snap back repeller of Eq. (8) for $r = 0.9$. The real component of this fractal is also a numerical estimate of the forbidden set as it was discussed in [3]

Now we study the following two dimensional rational system:

$$
\begin{aligned}
x_{k+1} &= \frac{\lambda x_k}{[1 + a(x_k + \alpha y_k)]^b} \\
y_{k+1} &= \frac{\mu y_k}{[1 + c(\beta x_k + y_k)]^b}
\end{aligned}
\tag{9}
$$

This is a competition model introduced by Hassell and Comins which also appear in [6] where parameters and initial values are nonnegative. If $b = 2$ and $\alpha = \beta = 0$ the system reduces to:

$$
\begin{aligned}
x_{k+1} &= \frac{\lambda x_k}{(1 + a x_k)^2} \\
y_{k+1} &= \frac{\mu y_k}{(1 + c y_k)^2}
\end{aligned}
\tag{10}
$$

Moreover, the change of variables $u_k = a x_k$, $v_k = c y_k$ eliminates parameters a and c. Therefore the system is simplified to the following one parameter family:

$$
x_{k+1} = \frac{\lambda x_k}{(1 + x_k)^2}
\tag{11}
$$

We are going to locate snap back repellers in (11) for a range of values of λ. We will consider arbitrary parameters and initial conditions. Therefore using perturbation theory, we will conjecture that the system (9) has chaotic behavior for sufficiently small values of α and β and adequately chosen λ and μ.

We remark that negative values of the variables x_k and y_k can not be interpreted in the original biological model, but this is a way to study the difference system (9) in a wider mathematical setting.

Let $f : \mathbb{R} \to \mathbb{R}$ be a real function. Recall that a sufficient condition for an equilibrium point $\bar{x}$ of the difference equation $x_{k+1} = f(x_k)$ to be a repeller is that $|f'(\bar{x})| > 1$. Also, in that case $\bar{x}$ is an unstable equilibrium point. If $|f'(\bar{x})| < 1$ the equilibrium is stable. A simple computation gives the next result.

Lemma 1 *If $\lambda > 0$, Eq. (11) has three equilibrium points:*

1. $\bar{x} = 0$, *a repeller if* $\lambda > 1$
2. $\bar{x} = \sqrt{\lambda} - 1$, *a repeller if* $\lambda \in (0, 1)$
3. $\bar{x} = -\sqrt{\lambda} - 1$, *a repeller if* $\lambda > 0$

Remark The equilibrium $\bar{x} = \sqrt{\lambda} - 1$ is stable if $\lambda > 1$ and is unstable if $\lambda \in (0, 1)$.

The iteration function associated to difference equation (11) is $f(x) = \frac{\lambda x}{(1+x)^2}$. The inverse iteration is a multi-valued function splitted in two branches:

$$
\begin{aligned}
g_+(x) &= \tfrac{\lambda}{2x} - 1 + \tfrac{\sqrt{\lambda^2 - 4\lambda x}}{2x} \\
g_-(x) &= \tfrac{\lambda}{2x} - 1 - \tfrac{\sqrt{\lambda^2 - 4\lambda x}}{2x}
\end{aligned}
\qquad (12)
$$

To locate a snap back repeller in Eq. (11) we look at point $\bar{x} = \sqrt{\lambda} - 1$ when $\lambda \in (0, 1)$. In that case a critical interval is $(-1, \rho)$ where ρ is the real root of $-x^3 - 3x^2 - (3 + \lambda)x - 1 + \lambda$. A critical interval is a neighborhood of the equilibrium point in which the repelling condition $|f'(x)| > 1$ is satisfied. The presence of a non trivial pre-image of $\bar{x}$ in $(-1, \rho)$ means that $\bar{x}$ is a snap back repeller. This is precisely the situation for every $\lambda \in (0, 1)$ as it is shown in Fig. 3. We have obtained:

Proposition 1 *If $\lambda \in (0, 1)$, the difference equation (11) has a snap back repeller and therefore is Li–Yorke chaotic.*

Let $F : \mathbb{R}^2 \to \mathbb{R}$ and $G : \mathbb{R}^2 \to \mathbb{R}$ be C^1 functions. Given $\alpha, \beta \in \mathbb{R}$, consider the system of difference equations

$$
\begin{aligned}
x_{k+1} &= F(x_k, \alpha y_k) \\
y_{k+1} &= G(\beta x_k, y_k)
\end{aligned}
\qquad (13)
$$

when $\alpha = \beta = 0$ the system reduces to the pair of one dimensional difference equations:

$$
\begin{aligned}
x_{k+1} &= F(x_k, 0) \\
y_{k+1} &= G(0, y_k)
\end{aligned}
\qquad (14)
$$

We claim that Theorem 5 is also true if F and G are rational functions. In that case the model of Hassell and Comins (see [6]) is Li–Yorke chaotic when $\lambda, \mu \in (0, 1)$ when α and β are small enough.

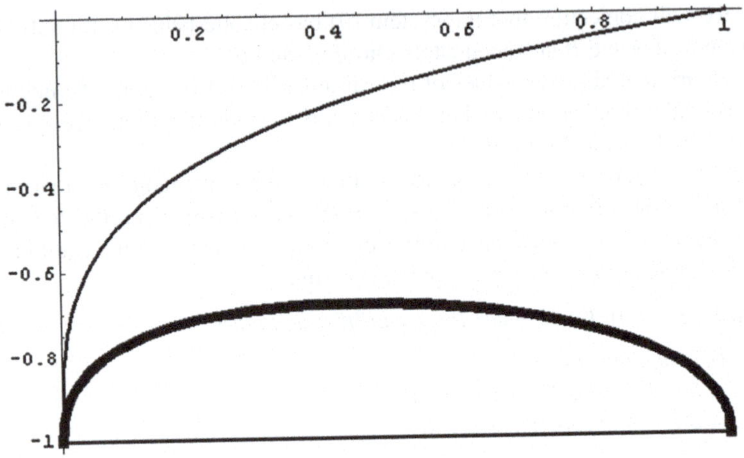

Fig. 3 Function $g_+ \circ g_-(\bar{x}(\lambda))$. For every $\lambda \in (0, 1)$ the function is bounded by -1 and $\rho(\lambda)$, the real root of $-x^3 - 3x^2 - (3 + \lambda)x - 1 + \lambda$ such that $(-1, \rho(\lambda))$ is a critical interval

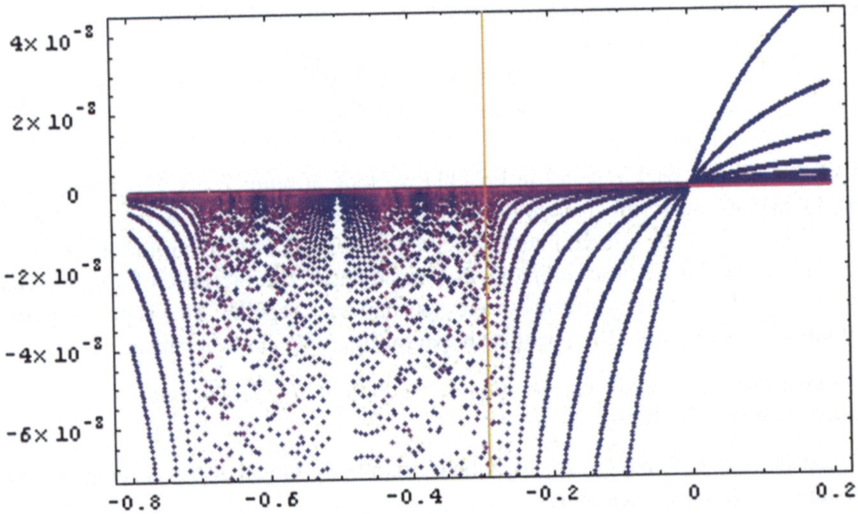

Fig. 4 The orbits of (11) when $\lambda = 0.5$ are drawn over the vertical of each initial condition x_0 lying in the horizontal axis. A vertical line indicates the equilibrium. 40 iterations were realized, casting aside the first 20 elements of each orbit

Now we try to visualize our conjecture with some numerical experiments. In Fig. 4 a different behavior is observed in the region neighboring the equilibrium. To construct this graphic we have to avoid the problem of the forbidden set using a truncated function that approximates the iteration function $f(x) = \frac{\lambda x}{(1+x)^2}$.

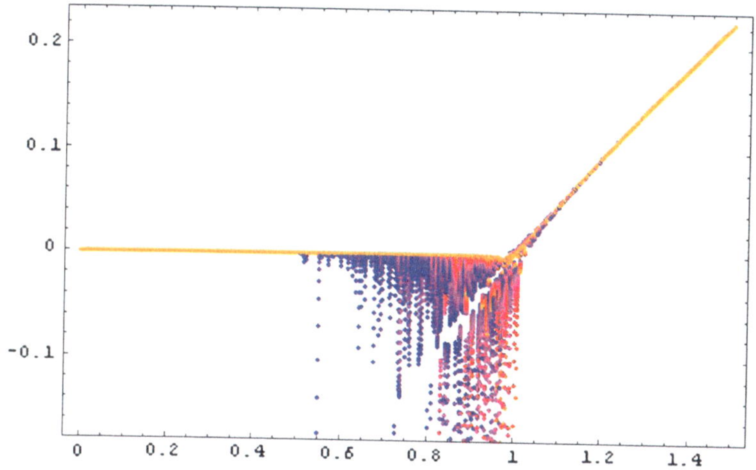

Fig. 5 Bifurcation diagram of $x_{k+1} = \frac{\lambda x_k}{(1+x_k)^2}$. For each λ in the horizontal axis a suitable orbit is drawed over the vertical

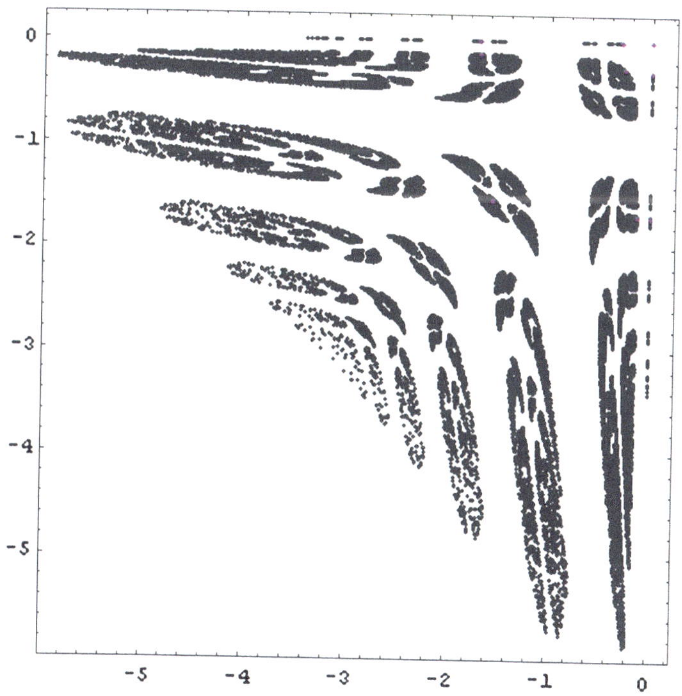

Fig. 6 Six inverse multi-valued iterates of the equilibria in Eq. (9) taking $\lambda = \mu = 0.5, \alpha = \beta = 0.1$

A bifurcation diagram is showed in Fig. 5. Here we use again the truncated approximation and take as initial condition a random perturbation of the equilibrium $\bar{x} = \sqrt{\lambda} - 1$. The interesting dynamical behavior is still in the region $\lambda \in (0, 1)$.

Finally we draw in Fig. 6 the multi-valued inverse orbit of the set of fixed points of system (9) when $\lambda = \mu = 0.5$ and $\alpha = \beta = 0.1$.

It should be noted that the previous diagrams could indicate also non chaotic dynamics. The scale in Fig. 4 could be regarded as global convergence to zero and the bifurcation of Fig. 5 is not observed if a greater number of iterations is taken. We think that this may be due to the use of the truncated approximation or to a more subtle fact: the scrambled set in which the dynamic behaves chaotically can have Lebesgue measure equal to zero. In the late case the former figures may be showing statistical traces of the theoretical conjecture.

Acknowledgments This paper has been partially supported by Grants MTM2011-232211 and CGL2008-05688-C02-02 from Ministerio de Ciencia e Innovación (Spain), Project 08667/PI-08 Fundación Séneca de la Comunidad Autónoma de Murcia (Spain) and Grant PEII09-0220-0222 from Junta de Comunidades de Castilla La Mancha (Spain).

References

1. Aulbach, B., Kieninger, B.: On three definitions of chaos. Nonlinear Dyn. Syst. Theory **1**(1), 23–37 (2001)
2. Block, L.: Homoclinic points of mappings of the interval. Proc. Am. Math. Soc. **72**, 3 (1978)
3. Cascales, A.: Ecuaciones en diferencias racionales. Ph.D dissertation. Universidad de Murcia (2014). http://hdl.handle.net/10803/132965
4. Camouzis, E., Ladas., G.: Dynamics of Third-Order Rational Difference Equations with Open Problems and Conjectures, Chapman & Hall, United Kingdom (2008)
5. Diamond, P.: Chaotic behaviour of systems of difference equations. Int. J. Syst. Sci. **7**(8), 953–956 (1976)
6. Hassell, M.P., Comins, H.N.: Discrete time models for two-species competition. Theor. Popul. Biol. **9**, 202–221 (1976)
7. Holmgrem, R.: A First Course in Discrete Dynamical Systems. Universitext, 2nd edn. Springer, Heidelberg (1996)
8. Hirsch, M., Smale, S.: Differential Equations, Dynamical Systems and Linear Algebra. Academic Press, NY (1974)
9. Kuchta, M., Smital, J.: Two point scrambled set implies chaos. Proceedings of the European Conference on Iteration Theory, pp. 427–430. World Scientific, Singapore (1989)
10. Li, C., Chen, G.: An improved version of Marotto theorem. Chaos, Solitons and Fractals **18**, 69–77 (2003). Erratum: **20**, 655 (2004)
11. Lin, W., Ruan, J., Zhao, W.: On the mathematical clarification of the snap-back repeller in high-dimensional systems and chaos in a discrete neural network model. Int. J. Bifurc. Chaos **12**(5), 1129–1139 (2002)
12. Li, T.-Y., Yorke, J.: Period three implies chaos. Am. Math. Mon. **82**, 985–992 (1975)
13. Kloeden, P., Li, Z.: Li-Yorke in higher dimensions: a review. J. Differ. Equ. Appl. **12**, 247–269 (2006)
14. Marotto, F.R.: Snap-Back repellers imply chaos in $\mathbb{R}^n$. J. Math. Anal. Appl. **63**, 199 (1978)
15. Marotto, F.R.: Perturbations of stable and chaotic difference equations. J. Math. Anal. Appl. **72**, 716–729 (1979)

16. Marotto, F.R.: On redefining a snap-back repeller. Chaos, Solitons and Fractals **25**, 25 (2005)
17. Mazrooei-Sebdani R.: Chaos in rational systems in the plane containing quadratic. Communications in Nonlinear Science and Numerical Simulation (to appear)
18. Misiurewicz, M., Zgliczynski, P.: Topological entropy for multidimensional perturbations of one-dimensional maps. Int. J. Bifurc. Chaos **11**(5), 1443–1446 (2001)
19. Smale, S.: Diffeomorphisms with many periodic points. In: Cairns, S.S (ed.) Differential and Combinatorial Topology, pp. 63–80. Princeton University Press, Princeton (1963)
20. Smale, S.: Differentiable dynamical systems. Bull. Am. Math. Soc. **73**, 747–817 (1967)
21. Wiggins, S.: Introduction to Applied Nonlinear Dynamical Systems and Chaos. Texts in Applied Mathematics, vol. 2. Springer, Heidelberg (1990)
22. Zgliczynski, P.: Sharkovskii's theorem for multidimensional perturbations of one-dimensional maps. Ergod. Theory Dyn. Syst. **19**, 1655–1684 (1999)
23. Zgliczynski, P.: Sharkovskii's theorem for multidimensional perturbations of one-dimensional maps II. Topol. Methods Nonlinear Anal. (Journal of the Juliusz Schauder Center) **14**, 169–182 (1999)
24. Zgliczynski, P.: Multidimensional perturbations of one-dimensional maps and stability of Sharkovskii ordering. Int. J. Bifurc. Chaos **9**(9), 1867–1876 (1999)
25. Zhao, Y., Xie, L., Yiu, C.: An improvement on Marotto's theorem and its applications to chaotification of switching systems. Chaos, Solitons and Fractas **39**, 2225–2232 (2009)

Simple Mandelpinski Necklaces for $z^2 + \lambda/z^2$

Daniel Cuzzocreo and Robert L. Devaney

Abstract For the family of maps $F_\lambda(z) = z^n + \lambda/z^n$ where $n \geq 3$, it is known that there is a McMullen domain surrounding the origin in the parameter plane. This domain is then surrounded by infinitely many "Mandelpinski" necklaces $\mathscr{S}_k$ for $k = 0, 1, 2, \ldots$. These are simple closed curves surrounding the McMullen domain and passing through exactly $(n-2)n^k + 1$ centers of baby Mandelbrot sets and the same number of centers of Sierpinski holes. When $n = 2$ there is no such McMullen domain in the parameter plane. However, we show in this paper that there do exist Mandelpinski necklaces $\mathscr{S}_k$ in this case. Now these necklaces converge down to the origin. And, consistent with the formula for higher values of n, each $\mathscr{S}_k$ passes through the centers of only one Mandelbrot set and one Sierpinski hole.

Keywords Julia set · Mandelpinski necklace · Mandelbrot set · Rational map · Sierpinski hole · Complex dynamics

1 Introduction

In recent years a number of papers have appeared that deal with the dynamics of singularly perturbed maps of the form $F_\lambda(z) = z^n + \lambda/z^n$ where $n \geq 2$. It turns out that the case $n = 2$ is very different from the case $n > 2$. One reason for this is that, when $n = 2$, as $\lambda \to 0$, the Julia sets of F_λ converge to the closed unit disk (the filled Julia set of z^2), but this does not occur when $n > 2$ [6]. A second difference is that there is a *McMullen domain* in the parameter plane when $n > 2$. This is a punctured open disk surrounding the origin that consists of parameters for which the Julia sets of the corresponding maps are Cantor sets of simple closed curves, all of which are

D. Cuzzocreo
Department of Mathematics, Northwestern University, Evanston, IL, USA

R.L. Devaney (✉)
Department of Mathematics, Boston University,
111 Cummington Street, Boston, MA 02215, USA
e-mail: bob@bu.edu

L. Alsedà i Soler et al. (eds.), *Difference Equations, Discrete Dynamical Systems and Applications*, Springer Proceedings in Mathematics & Statistics 180, DOI 10.1007/978-3-662-52927-0_5

dynamically and topologically the same [10]. There is no such region when $n = 2$. Rather, in any neighborhood of 0 in the parameter plane, there are infinitely many different topological types of Julia sets [4]. And a third difference is that, when $n > 2$, the second images of the free critical points all tend to ∞ as $\lambda \to 0$ (this is what generates the McMullen domain), whereas when $n = 2$, the second images of the free critical points tend to 1/4 as $\lambda \to 0$ (and 1/4 is not in the basin of ∞ when λ is small).

In the case $n > 2$, there is an interesting structure that surrounds the McMullen domain. In [2, 3, 8] it is shown that this domain in the parameter plane is surrounded by infinitely many disjoint simple closed curves $\mathscr{S}_k$ for $k = 0, 1, 2, \ldots$ called Mandelpinski necklaces. Each $\mathscr{S}_k$ passes alternately through the centers of $(n-2)n^k + 1$ Mandelbrot sets with base period $k + 1$ (with one slight exception when $k = 1$) and the same number of centers of Sierpinski holes with escape time $k + 3$. See Fig. 1. A center of a Mandelbrot set of base period ℓ is a parameter in the main cardioid for which a critical point is periodic with prime period ℓ. A Sierpinski hole with escape time ℓ is a collection of parameters for which the critical orbits all land in the immediate basin of ∞ at iteration ℓ. A center of a Sierpinski hole is a parameter for which the critical orbits actually land at ∞.

These Mandelpinski necklaces provide a great deal of structure around the McMullen domain when $n > 2$. For example, when $n = 3$, the necklace $\mathscr{S}_{16}$ passes through exactly 43,046,722 Mandelbrot sets and Sierpinski holes. When the parameter lies in one of these Mandelbrot sets there are infinitely many small copies of quadratic Julia sets embedded in the much larger Julia set of F_λ. And when the parameter lies in a Sierpinski hole, the Julia set of F_λ is a Sierpinski curve, i.e., a set that is homeomorphic to the Sierpinski carpet fractal.

Because of the different behaviors of the critical orbits as $\lambda \to 0$ and the lack of a McMullen domain when $n = 2$, it was always assumed that there were no such Mandelpinski necklaces when $n = 2$. However, note that the above formula says that each necklace $\mathscr{S}_k$ should pass through exactly $(2-2)2^k + 1 = 1$ Mandelbrot set and 1 Sierpinski hole when $n = 2$ for each k. In fact, as we show in this paper, this does indeed happen. So we do have some simplified Mandelpinski necklaces in this case. These necklaces no longer surround a McMullen domain; rather, they converge to the origin as $k \to \infty$. So the structure of the parameter plane around the origin when $n = 2$ is very different from the case $n > 2$. See Fig. 2. We conjecture that the existence of these simple necklaces will allow us to begin to understand the very complicated structure of the parameter plane for $n = 2$ around the origin, just as the Mandelpinski necklaces did in the case $n > 2$.

2 Preliminaries

In this paper we shall concentrate on the family of complex rational maps given by

$$F_\lambda(z) = z^2 + \frac{\lambda}{z^2}$$

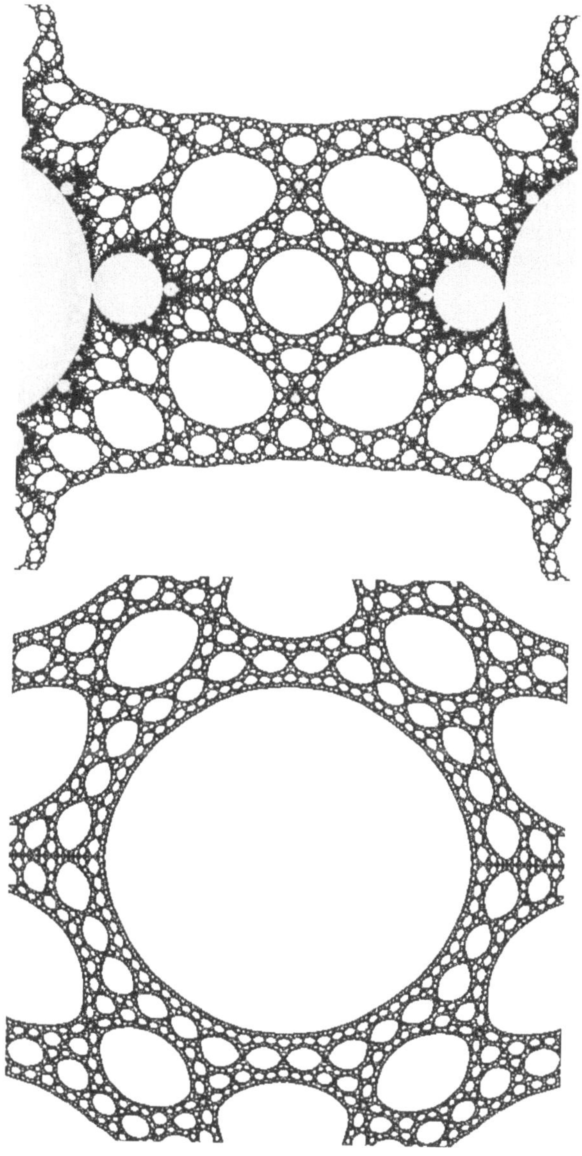

Fig. 1 Magnifications of the parameter plane for the family $z^3 + \lambda/z^3$ around the McMullen domain (the central *white* disk)

where $\lambda \in \mathbb{C}$. One checks easily that the point at ∞ is fixed in the Riemann sphere and $F'_\lambda(\infty) = 0$ so ∞ is a superattracting fixed point. We denote the immediate basin of attraction at ∞ by B_λ. Since 0 is a pole there is an open set about 0 that is mapped to B_λ. This set may or may not be disjoint from B_λ, but in the cases we consider in this paper, these two sets will be disjoint. We then call the preimage of B_λ surrounding 0 the *trap door* and denote this set by T_λ.

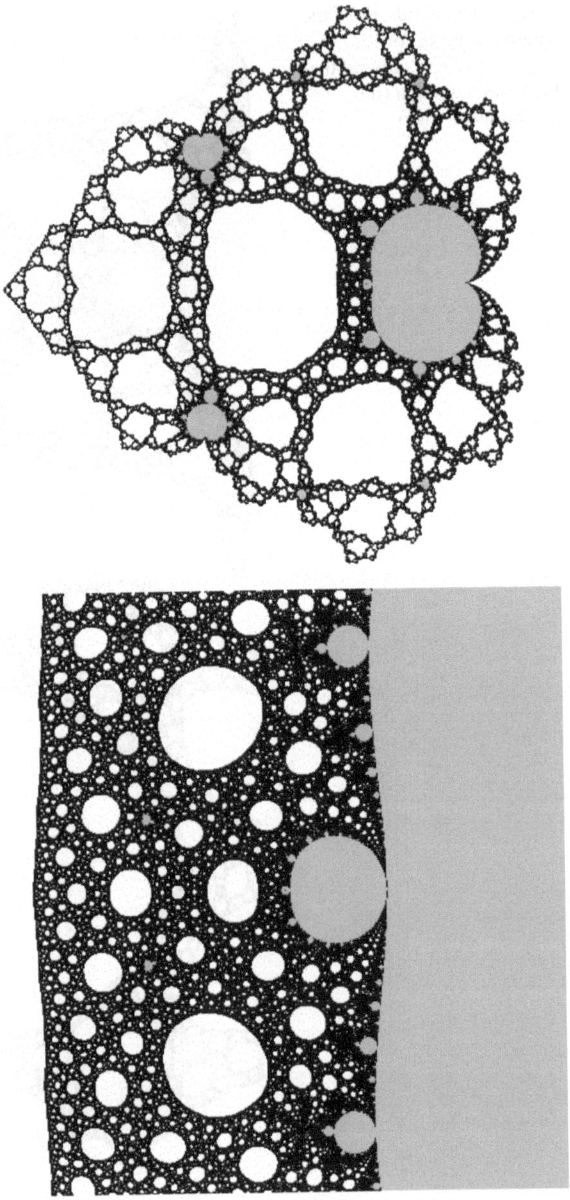

Fig. 2 The parameter plane and a magnification around the origin for the family $z^2 + \lambda/z^2$. The large central disk is a Sierpinski hole, not the McMullen domain. The origin is located at the "tip of the tail" of the Mandelbrot set that appears to straddle the positive real axis

It is well known that the *Julia set* of F_λ, denoted by $J(F_\lambda)$, has several equivalent definitions [11]. One definition is that $J(F_\lambda)$ consists of all points at which the family of iterates of F_λ fails to be a normal family in the sense of Montel. A second definition is that the Julia set is the closure of the set of repelling periodic points of F_λ. And a third definition is that the Julia set is the boundary of the set of all points whose orbits tend to ∞ (not just those in B_λ). These definitions then imply that $J(F_\lambda)$ is the chaotic regime since arbitrarily close to any point in the Julia set there are points whose orbits tend to ∞ and other points whose orbits are periodic. More importantly, Montel's Theorem implies that any neighborhood of a point in $J(F_\lambda)$ is mapped over the entire Riemann sphere by the set of all iterates of maps in this family. So, on the Julia set, F_λ exhibits extreme sensitive dependence on initial conditions.

There are several symmetries in the dynamical and parameter planes for these maps. We have $F_\lambda(-z) = F_\lambda(z)$ and $F_\lambda(iz) = -F_\lambda(z)$. Therefore the orbits of z and iz are the same after two iterations. As a consequence, the Julia set is symmetric under the map $z \to iz$, i.e., $J(F_\lambda)$ has fourfold symmetry. Also, let $H_\lambda(z) = \lambda^{1/2}/z$. Then $F_\lambda(H_\lambda(z)) = F_\lambda(z)$, so the Julia set is also symmetric under the involution H_λ. We also have that F_λ is conjugate to $F_{\overline{\lambda}}$ via the map $z \to \overline{z}$, so the parameter plane is symmetric under complex conjugation.

A straightforward computation shows that there are four free critical points for F_λ that are given by $\lambda^{1/4}$. We call these critical points "free" since there are two other critical points at ∞ and 0, but ∞ is fixed and 0 maps directly to ∞ for each λ. However, there are only two critical values given by $\pm 2\sqrt{\lambda}$ since two of the free critical points are mapped to $+2\sqrt{\lambda}$ and the other two are mapped to $-2\sqrt{\lambda}$. In fact, just like the quadratic polynomial family $z^2 + c$, there really is only one free critical orbit as both critical values are then mapped to $4\lambda + 1/4$, so all of the critical points end up on the same orbit after two iterations.

There are also four prepoles for F_λ given by $(-\lambda)^{1/4}$. So the prepoles and critical points all lie on the circle of radius $|\lambda|^{1/4}$ centered at the origin. We call this circle the *critical circle* and denote it by C_0^λ. Another easy computation shows that F_λ maps the critical circle 4-to-1 onto the line segment connecting the two critical values $\pm 2\sqrt{\lambda}$ and passing through the origin. We call this line the *critical segment*. Any other circle centered at the origin is then mapped as a 2-to-1 covering onto an ellipse whose foci are the critical values. In particular, the region in the exterior of the critical circle is then mapped as a 2-to-1 covering onto the complement of the critical segment in the Riemann sphere and so too is the interior of the critical circle.

We shall assume for the remainder of this paper that the critical values both lie on or inside the critical circle, so the critical segment will always lie in the the disk bounded by the critical circle. It is known [1, 9] that, in this case, $J(F_\lambda)$ is connected and that ∂B_λ is a simple closed curve lying in the exterior of C_0^λ. Since $H_\lambda(B_\lambda) = T_\lambda$, we have that ∂T_λ is also a simple closed curve that lies inside C_0^λ.

Let $\mathcal{O}$ be the punctured disk in the parameter plane that consists of all nonzero parameters for which the critical segment lies strictly inside the critical circle. When λ lies on the boundary of $\mathcal{O}$, we must have $2|\sqrt{\lambda}| = |\lambda|^{1/4}$, so it follows that $|\lambda| = 1/16$. Therefore the boundary of $\mathcal{O}$ is the circle of radius $1/16$ centered at the origin in the parameter plane. For $\lambda \in \mathcal{O}$, F_λ maps the exterior of the critical circle as a 2-to-1 covering onto the exterior of the critical segment. Thus there is a simple closed curve in the exterior of C_0^λ that is mapped 2-to-1 onto C_0^λ. Call this curve C_1^λ. Since C_0^λ contains four critical points and four prepoles, C_1^λ contains eight pre-critical points and eight pre-prepoles. Since the exterior of C_1^λ is then mapped onto the exterior of

C_0^λ as a 2-to-1 covering, there is another simple closed curve C_2^λ that lies outside C_1^λ and is mapped 2-to-1 onto C_1^λ. Continuing in this fashion, we find an infinite collection of simple closed curves C_k^λ for $k > 0$ satisfying $F_\lambda(C_k^\lambda) = C_{k-1}^\lambda$ and hence $F_\lambda^k(C_k^\lambda) = C_0^\lambda$. Note that the C_k^λ are all disjoint and these curves converge outward toward ∂B_λ as $k \to \infty$. This follows since, if this were not the case, the limiting set of the C_k^λ would be a closed, invariant set, say Λ_λ. If $\Lambda_\lambda \neq \partial B_\lambda$, then the region bounded by ∂B_λ and Λ_λ would also be invariant. But this cannot happen since there would then be points in ∂B_λ that have neighborhoods on which the family of functions $\{F_\lambda^k\}$ would be normal, which cannot happen since $\partial B_\lambda \subset J(F_\lambda)$. In addition, C_k^λ contains 2^{k+2} points that are mapped by F_λ^k to critical points and the same number of points that are mapped to the prepoles on C_0^λ. The points that map to critical points and to prepoles are arranged alternately around C_λ^k.

Since the interior of the critical circle is also mapped as a 2-to-1 covering of the exterior of the critical segment, there are other simple closed curves C_{-k}^λ for $k = 1, 2, \ldots$ such that F_λ maps C_{-k}^λ as a 2-to-1 covering of C_{k-1}^λ just as above. We have $H_\lambda(C_{-k}^\lambda) = C_k^\lambda$. The C_{-k}^λ now converge down to ∂T_λ as $k \to \infty$. And, just as above, C_{-k}^λ contains exactly 2^{k+2} points that are mapped to critical points and the same number to prepoles by F_λ^k.

3 Rings in the Parameter Plane

In this section, we prove that the origin in the parameter plane is surrounded by infinitely many disjoint simple closed curves $\mathscr{S}_k$ with the $\mathscr{S}_k$ converging to 0 as $k \to \infty$. The curve $\mathscr{S}_k$ will consist of parameters for which the critical orbit lands on the critical circle after exactly $k+1$ iterations in a manner specified below. We shall show that $\mathscr{S}_k$ contains exactly one parameter for which one of the critical points is periodic with period $k+1$. Results in [5] shows that this parameter is a center of the main cardioid of a Mandelbrot set in the parameter plane (with two exceptions noted at the end of this section). And we shall show that there is one other parameter in $\mathscr{S}_k$ for which the critical orbits all land on ∞ at iteration $k+3$. It is known [12] that this parameter is then the center of a Sierpinski hole with escape time $k+3$. The parameters that are centers of a main cardioid of a Mandelbrot set will lie in $\mathbb{R}^+$ while the parameters that are centers of a Sierpinski hole will lie in $\mathbb{R}^-$.

We first describe the ring $\mathscr{S}_0$ in the parameter plane. This curve consists of λ-values for which the critical values lie on the critical circle C_0^λ in the dynamical plane. So, on this set, we must have $|\lambda|^{1/4} = 2|\sqrt{\lambda}|$. Solving this equation shows that $\mathscr{S}_0$ is the circle of radius $1/16$ centered at the origin in the parameter plane, i.e., the boundary of $\mathscr{O}$. When $\lambda \in \mathscr{S}_0$, the critical circle C_0^λ is the circle of radius $1/2$ centered at the origin. Note that, as λ rotates around $\mathscr{S}_0$, the critical points and prepoles each rotate around C_0^λ by a quarter of a turn while the critical values rotate by half a turn. It then follows that there is exactly one parameter in $\mathscr{S}_0$ for which the critical values land on a critical point, namely $\lambda = 1/16$, and one other parameter for which they land on a prepole, namely $\lambda = -1/16$. So, for $\lambda = 1/16$, F_λ has a superattracting fixed point while, for $\lambda = -16$, the critical orbit escapes at iteration 3. This gives the result for $\mathscr{S}_0$.

For $\lambda \in \mathscr{O}$ with $0 \leq \text{Arg}\,\lambda < 2\pi$, let $c_0^\lambda = \lambda^{1/4}$ denote the critical point satisfying $0 \leq \text{Arg}\,c_0^\lambda < \pi/2$ and let c_j^λ, $j = 1, 2, 3$ denote the other three critical points where the c_j^λ are arranged in the counterclockwise direction around the origin. Let I_0^λ denote the closed sector in $\mathbb{C}$ bounded by the two critical point rays that are given by tc_0^λ and tc_3^λ with $t \geq 0$. Let I_j^λ denote the similar sector bounded by tc_{j-1}^λ and tc_j^λ. Note that the interior of each I_j^λ is mapped one-to-one onto $\mathbb{C}$ minus the two critical value rays given by $\pm tv_\lambda$, $t \geq 1$. One of the critical point rays that bounds each I_j^λ is mapped onto one of these critical value rays while the other critical point ray is mapped to the other critical value ray. Note also that, when $\lambda \in \mathbb{R}^+$, the critical value rays lie in $I_0^\lambda \cap I_1^\lambda = \mathbb{R}^+$ and $I_2^\lambda \cap I_3^\lambda = \mathbb{R}^-$. For all other λ-values, one of the critical value rays lies in the interior of I_1^λ while the other lies in the interior of I_3^λ. See Fig. 3.

Since C_0^λ is an actual circle, we may define a natural parametrization $C_0^\lambda(\theta)$ of this curve by setting $C_0^\lambda(0) = c_0^\lambda$. We choose this parameterization so that $C_0^\lambda(\theta)$ rotates in the clockwise direction as θ increases. Here we again assume that $0 \leq \text{Arg}\,\lambda < 2\pi$. Let γ_0^λ be the portion of C_0^λ that lies inside I_0^λ, i.e., $\gamma_0^\lambda(\theta) = C_0^\lambda(\theta)$ where $0 \leq \theta \leq$

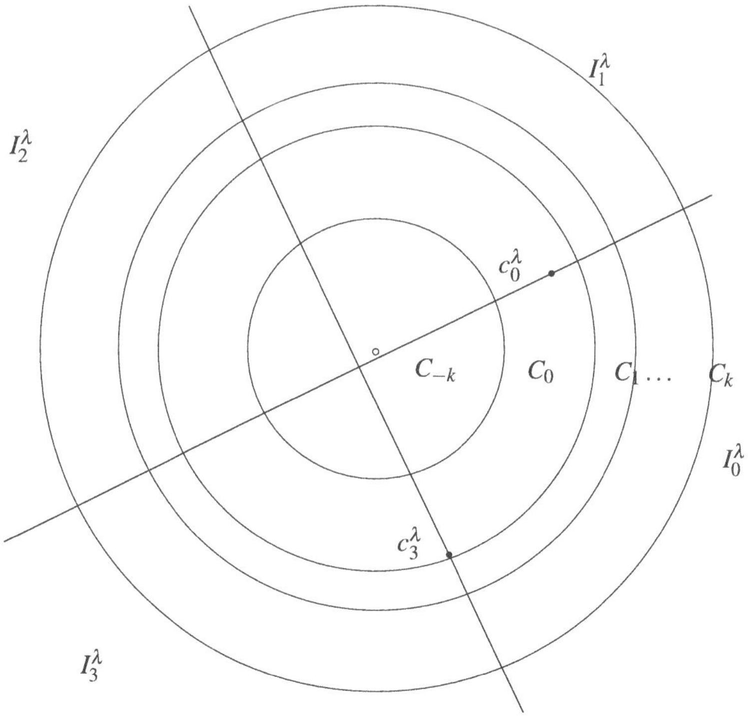

Fig. 3 The critical circle and its preimages together with the sectors I_j

$\pi/2$. Now the sector I_0^λ is mapped over itself univalently (except when $\lambda \in \mathbb{R}^+$ in which case one boundary curve is mapped 2-to-1 to a portion of $\mathbb{R}^+$). In all cases there is then a smooth curve γ_1^λ lying in $C_1^\lambda \cap I_0^\lambda$ that is mapped univalently onto γ_0^λ. We define $\gamma_1^\lambda(\theta)$ to be the point on this portion of C_1^λ that is mapped to $\gamma_0^\lambda(\theta)$. Inductively, we then define $\gamma_k^\lambda(\theta)$ to be the point in $C_k^\lambda \cap I_0^\lambda$ for which $F_\lambda(\gamma_k^\lambda(\theta)) = \gamma_{k-1}^\lambda(\theta)$ for each $k \geq 1$. Then we let $\gamma_{-k}^\lambda(\theta) = H_\lambda(\gamma_k^\lambda(\theta))$ where H_λ is the involution $z \to \lambda^{1/2}/z$ that fixes the critical points $\pm c_0^\lambda$. So $F_\lambda(\gamma_{-k}^\lambda(\theta)) = \gamma_{k-1}^\lambda(\theta)$. One checks easily that H_λ interchanges I_0^λ and I_1^λ, so $\gamma_{-k}^\lambda(\theta)$ lies in $I_1^\lambda \cap C_{-k}^\lambda$ for each $k > 0$.

Lemma *Given $k > 0$, there exists $\lambda^* > 0$ such that, if $|\lambda| \leq \lambda^*$, then both critical values of F_λ lie strictly inside the curve C_{-k}^λ.*

Proof For $|\lambda|$ sufficiently small, the critical circle C_0^λ has magnitude that is very small. Since $F_\lambda \approx z^2$ away from the origin when $|\lambda|$ is small, we may choose λ_1 so that, if $|\lambda| < \lambda_1$, then the closed curve C_{k-1}^λ lies strictly inside the circle of radius $1/8$ surrounding the origin. Also, since $F_\lambda(v_\lambda) = 1/4 + 4\lambda$, we may choose λ_2 so that, if $|\lambda| < \lambda_2$, then $|F_\lambda(v_\lambda)| > 1/8$. Let $\lambda^* = \min(\lambda_1, \lambda_2)$. Then we have that, for each λ inside the circle of radius λ^*, the image of the critical value lies outside the circle of radius $1/8$ and hence outside C_{k-1}^λ. Therefore $\pm v_\lambda$ lies strictly inside the closed curve C_{-k}^λ. □

We now define the rings $\mathcal{S}_k$ for $k \geq 1$ in the parameter plane. Recall that $\mathcal{O}$ is the set of nonzero parameters for which v_λ lies inside the critical circle.

Proposition *Suppose $\lambda \in \mathcal{O}$. Fix $k \geq 1$ and θ in the interval* $[0, \pi/2]$. *Then there is a unique parameter $\lambda = \lambda_\theta^k$ in $\mathcal{O}$ for which a critical value lies at the point $\gamma_{-k}^\lambda(\theta)$. Moreover, λ_θ^k varies continuously with θ and $\lambda_0^k = \lambda_{\pi/2}^k$ is a parameter in $\mathbb{R}^+$.*

Proof Since $\mathcal{O}$ is the open disk of radius $1/16$ with the origin removed, we have the universal covering half-plane $\tilde{\mathcal{O}}$ given by $\operatorname{Re} z < \log(1/16)$. We then have two maps defined on $\tilde{\mathcal{O}}$.

The first is a map that we shall denote by $\tilde{V}(\lambda)$. To define this map, let v_λ be the critical value that lies in the upper half plane when $0 < \operatorname{Arg}\lambda < 2\pi$. Clearly, the map $\lambda \mapsto v_\lambda$ is not well-defined on $\mathcal{O}$ since v_λ moves to $-v_\lambda$ as $\operatorname{Arg}\lambda$ rotates from 0 to 2π. However, we can lift this map to a new map $V : \tilde{\mathcal{O}} \to \mathcal{X}$ where $\mathcal{X}$ is the annulus $0 < |z| < 1/2$ so that V agrees with the map $\lambda \mapsto v_\lambda$ when $0 < \operatorname{Arg}\lambda < 2\pi$. Let $\tilde{\mathcal{X}}$ be the universal covering of $\mathcal{X}$. Then we can lift V to a map $\tilde{V} : \tilde{\mathcal{O}} \to \tilde{\mathcal{X}}$. Note that $\tilde{V}$ is an analytic, invertible map on $\tilde{\mathcal{O}}$, and, as the argument of λ increases by 2π in $\mathcal{O}$, the imaginary part of $\tilde{V}$ in $\tilde{\mathcal{X}}$ increases by exactly π.

For fixed values of $k \geq 1$ and $\theta \in [0, \pi/2]$, we also have the map $\lambda \mapsto \gamma_{-k}^\lambda(\theta)$ defined when $0 \leq \operatorname{Arg}\lambda < 2\pi$. Again this map is not well-defined on $\mathcal{O}$, but we can lift it to a new map $L : \tilde{\mathcal{O}} \to \mathbb{C}$ as above. By construction, $L(\lambda)$ is strictly contained inside the annulus $\mathcal{X}$. So we may lift L to a map $\tilde{L} : \tilde{\mathcal{O}} \to \tilde{\mathcal{X}}$.

As we have shown, the point $\gamma_{-k}^\lambda(\theta)$ is contained in the sector I_1^λ as long as $\theta \in [0, \pi/2]$. And this sector rotates by exactly $\pi/2$ radians as $\operatorname{Arg}\lambda$ increases from

0 to 2π. Moreover, the argument of $\gamma_{-k}^{\lambda}(\theta)$ never increases by π as λ rotates around the origin, since this would imply that this point visited both the positive and negative real axis enroute. Hence the argument of $\tilde{L}(\lambda)$ increases by an amount strictly less than π as the argument of λ increases by 2π.

Now, since $\tilde{V}$ is invertible, we may consider the composition $\Phi = \tilde{V}^{-1} \circ \tilde{L} : \tilde{\mathcal{O}} \to \tilde{\mathcal{O}}$. We claim that there is a unique fixed point for Φ in $\tilde{\mathcal{O}}$. To see this, first note that we may extend both $\tilde{V}$ and $\tilde{L}$ to the boundary $\text{Re}\, z = 1/16$ of $\tilde{\mathcal{O}}$ and Φ maps this boundary strictly inside $\tilde{\mathcal{O}}$. This follows since γ_{-k}^{λ} always lies strictly inside the critical circle for F_λ, which then lies inside the critical circle for F_μ when μ lies on the boundary of $\mathcal{O}$. But this is the circle $r = 1/2$. Hence there are no fixed points on the boundary of $\tilde{\mathcal{O}}$. Next note that there are no fixed points in the far left half-plane in $\tilde{\mathcal{O}}$. This follows immediately from the previous Lemma. Finally, since the argument of $\tilde{V}$ increases by more than the argument of $\tilde{L}$ as λ rotates around the origin, it follows that Φ must have a fixed point in $\tilde{\mathcal{O}}$ and, by the Schwarz Lemma, this fixed point must be unique. Then the projection of this point into $\mathcal{O}$ is a parameter λ for which a critical value lands on the point $\gamma_{-k}^{\lambda}(\theta)$. This is the parameter λ_θ^k. Since all of the above varies continuously with θ, it follows that $\theta \mapsto \lambda_\theta^k$ traces out a continuous curve in $\mathcal{O}$.

Now when $\lambda \in \mathbb{R}^+$, elementary arguments using real dynamics shows that there is a superstable parameter value for which

$$v_\lambda < c_0^\lambda = F_\lambda^{k+1}(c_0^\lambda) < F_\lambda^k(c_0^\lambda) < \cdots < F_\lambda^2(c_0^\lambda).$$

This is then the parameter λ_0^k. Similarly, when $\text{Arg}\,\lambda = 2\pi$, the above result shows that we have a similar unique parameter for which v_λ lies in $\mathbb{R}^-$ and then maps onto the same superattracting cycle. This is now the parameter $\lambda_{\pi/2}^k$. But this then implies that $\lambda_0^k = \lambda_{\pi/2}^k$. Note that these are the only two θ-values for which two "different" λ_θ^k's coincide. Therefore the parameters λ_θ^k lie along a simple closed curve surrounding the origin in the parameter plane. □

Thus we may define the ring $\mathcal{S}_k$ to be the simple closed curve parametrized by $\theta \mapsto \lambda_\theta^k$.

Corollary *There is a unique parameter in $\mathcal{S}_k$ for which a critical point lies on a superattracting cycle of period $k+1$ and another unique parameter for which the critical orbits escape at iteration $k+3$.*

Proof As shown above, the parameter $\lambda_0^k = \lambda_{\pi/2}^k$ is the unique parameter in $\mathcal{S}_k$ for which $F_\lambda^k(v_\lambda)$ lands on a critical point in the curve γ_0^λ. There is also a unique parameter for which $F_\lambda^k(v_\lambda)$ lands on the prepole in γ_0^λ and hence the critical orbit escapes at iteration $k+3$. The graph of the real function F_λ shows that this parameter lies in $\mathbb{R}^-$. □

Remarks:

1. Note that the parameter λ_0^0 is the parameter for which c_0^λ is a superattracting fixed point and this parameter appears to lie at the center of the main cardioid of a Mandelbrot set that straddles the positive real axis. However, this is not quite a "full" Mandelbrot set, as the tip of the tail (i.e., the parameter corresponding to $c = -2$ for the quadratic Mandelbrot set) lies at the origin, so the dynamics associated to this parameter do not correspond to those for the parameter $c = -2$. We conjecture that this is the only portion of the Mandelbrot set that is missing.
2. Also, the parameter value λ_0^1 does not lie at the center of a main cardioid of a baby Mandelbrot set; rather, this parameter lies at the center of the period 2 bulb of the above Mandelbrot set.
3. All other parameters λ_0^k do lie at the center of a baby Mandelbrot set that lies inside the main Mandelbrot set on the real axis. This follows from a polynomial-like map construction. See [5] for details.

References

1. Blanchard, P., Devaney, R.L., Look, D.M., Seal, P., Shapiro, Y.: Sierpinski curve julia sets and singular perturbations of complex polynomials. Ergod. Theory Dyn. Syst. **25**, 1047–1055 (2005)
2. Devaney, R.L.: The McMullen domain: satellite mandelbrot sets and Sierpiński holes. Conform. Geom. Dyn. **11**, 164–190 (2007)
3. Devaney, R.L.: Structure of the McMullen domain in the parameter space of rational maps. Fund. Math. **185**, 267–285 (2005)
4. Devaney, R.L.: Dynamics of $z^n + \lambda/z^n$; why the case $n = 2$ is crazy. In: Conformal Dynamics and Hyperbolic Geometry. AMS Contemporary Mathematics, vol. 573, pp. 49–65 (2012)
5. Devaney, R.L.: The McMullen domain: satellite mandelbrot sets and Sierpinski holes. Conform. Geom. Dyn. **11**, 164–190 (2007)
6. Devaney, R.L., Garijo, A.: Julia sets converging to the unit disk. Proc. AMS **136**, 981–988 (2008)
7. Devaney, R.L., Look, D.M., Uminsky, D.: The escape trichotomy for singularly perturbed rational maps. Indiana Univ. Math. J. **54**, 1621–1634 (2005)
8. Devaney, R.L., Marotta, S.: The McMullen domain: rings around the boundary. Trans. AMS **359**, 3251–3273 (2007)
9. Devaney, R.L., Russell, E.M.: Connectivity of Julia Sets for Singularly Perturbed Rational Maps. To appear
10. McMullen, C.: Automorphisms of rational maps. In: Holomorphic Functions and Moduli, vols. 1, 10. Mathematical Sciences Research Institute Publications Springer, New York (1988)
11. Milnor, J.: Dynamics in One Complex Variable. Princeton University Press, Princeton (2006)
12. Roesch, P.: On capture zones for the family $f_\lambda(z) = z^2 + \lambda/z^2$. In: Dynamics on the Riemann Sphere, pp. 121–130. European Mathematical Society (2006)

R&D Dynamics with Asymmetric Efficiency

M. Ferreira, J.P. Almeida, B.M.P.M. Oliveira and Alberto A. Pinto

Abstract We consider an R&D investment function in a Cournot duopoly competition model inspired in the logistic equation. We study the economical effects resulting from the firms having different R&D efficiencies. We present three cases: (1) both firms are efficient and have the same degree of efficiency; (2) both firms are less efficient and have the same degree of efficiency; (3) firms are asymmetric in terms of the efficiency of their R&D investment programs. We study the myopic dynamics on the production costs obtained from investing the Nash investment equilibria.

Keywords Strategic R&D · Cournot duopoly model · Patents

1 Introduction

We consider a Cournot duopoly competition model where two firms invest in R&D projects to reduce their production costs. This competition is modeled, as usual, by a two stages game (see [3, 6]). In the first subgame, two firms choose, simultaneously, R&D investment strategies to reduce their initial production costs. In the second

M. Ferreira
LIAAD-INESC TEC and ESEIG-IPP, Rua D. Sancho I, 981,
4480-876 Vila Do Conde, Portugal
e-mail: migferreira2@gmail.com

J.P. Almeida (✉)
LIAAD-INESC TEC and Polytechnic Institute of Bragança, Campus de Santa Apolónia,
5300-253 Braganca, Portugal
e-mail: jpa@ipb.pt

B.M.P.M. Oliveira
LIAAD-INESC TEC and FCNAUP, Rua Dr. Roberto Frias, 4200-465 Porto, Portugal
e-mail: bmpmo@fcna.up.pt

A.A. Pinto
LIAAD-INESC TEC and Department of Mathematics, FCUP, Rua Do Campo Alegre 687,
4169-007 Porto, Portugal
e-mail: aapinto@fc.up.pt

L. Alsedà i Soler et al. (eds.), *Difference Equations, Discrete Dynamical Systems and Applications*, Springer Proceedings in Mathematics & Statistics 180, DOI 10.1007/978-3-662-52927-0_6

subgame, the two firms are involved in a Cournot competition with production costs equal to the reduced cost determined by the R&D investment strategies chosen in the first stage. We use an R&D cost reduction function inspired in the logistic equation that was first introduced in [5].

We consider two firms that are identical except, at most, in their R&D investment programs efficiency. Concerning their R&D efficiency, we study the economical effects of three different scenarios: (1) both firms R&D investment programs are efficient and both firms hold the same degree of efficiency. We will refer to this case as the symmetric efficient (SE) case; (2) both firms R&D investment programs are less efficient and both have the same degree of efficiency. We will refer to this case as the symmetric inefficient (SI); (3) firms are asymmetric in terms of their R&D investment programs efficiency, i.e. one of the firms has a more efficient R&D investment program allowing bigger cost reduction under equal investment whereas the other firm has a less efficient R&D investment program forcing it to invest more in order to achieve the same cost reduction as the other firm. We will refer to this case as the Asymmetric (A) case.

We present the Perfect Nash equilibria of this two stages game and we study the economical effects of these equilibria. The second subgame, consisting of a Cournot competition, has a unique perfect Nash equilibrium. For the first subgame, consisting of an R&D cost reduction investment program, we exhibit four different regions of Nash investment equilibria that we characterize as follows: a competitive Nash investment region C where both firms invest, a single Nash investment region S_1 for firm F_1, where just firm F_1 invests, a single Nash investment region S_2 for firm F_2, where just firm F_2 invests, and a nil Nash investment region N, where neither of the firms invest (see [5, 6]).

The Nash investment equilibria are not necessarily unique. The non uniqueness leads to an economical complexity in the choice of the best R&D investment strategies by the firms. For high production costs, that can correspond to the production of new technologies, there are subregions of production costs where there are multiple Nash investment equilibria: a region $R_{S_i \cap C}$ where the intersection between the single Nash investment region S_i and the competitive Nash investment region C is non-empty; a region $R_{S_1 \cap S_2}$ where the intersection between the single Nash investment regions S_1 and S_2 is non-empty; a region $R_{S_1 \cap C \cap S_2}$ where the intersection between the single Nash investment regions S_1 and S_2 and the competitive Nash investment region C is non-empty. When we compare the cases symmetric efficient (SE) and symmetric inefficient (SI), we observe that, in the SI-scenario, the single Nash investment regions S_1 and S_2 increase in size and so the competitive Nash investment region C becomes smaller. In the asymmetric case (A), we observe that the single Nash investment region S_2 of firm F_2 is considerably bigger due to its advantage in the R&D cost reduction program efficiency.

We present the R&D deterministic dynamics on the production costs of the Cournot competition, based on the R&D investment strategies of the firms, as follows: at every period of time, the firms choose the investment corresponding to one of the Nash investment equilibria that determines the new production costs of the firms. Hence, the implicit equations determining the R&D deterministic dynamics are distinct in the competitive Nash investment region C and in the single Nash investment regions S_1 and S_2 (see Theorems 1 and 2 in [5]). The nil Nash investment region N determines the set of all production costs that are fixed by the dynamics. Depending upon the initial production costs of both firms and upon their R&D investment strategies, the nil Nash investment region N is the set of equilibria for the R&D deterministic dynamics. It is unusual in dynamical systems to have a non-isolated set of equilibrium points. This is due to the complex investment structure that we have to deal in these problems and to the lower bound in investment (economically, it must be non negative). The competitive Nash investment region determines the region where the production costs of both firms evolve, for both firms, along the time. The single Nash investment region S_1 (resp. S_2) determines the set of production costs where the production cost of firm F_2 (resp. F_1) is constant, along the time, and just the production costs of firm F_1 (resp. F_2) evolve that may lead Firm F_2 (resp. F_1) to bankruptcy or that may allow it to recover from an initial disadvantage.

2 R&D Investments on Costs

The Cournot duopoly competition with R&D investments on the reduction of the initial production costs consists of two subgames in one period of time. The first subgame is an R&D investment program, where both firms have initial production costs and choose, simultaneously, their R&D investment strategies to obtain new production costs. The second subgame is a Cournot competition with production costs equal to the reduced cost determined by the R&D investment program. As it is well known, the second subgame has a unique perfect Nash equilibrium.

2.1 The R&D Program

We consider an economy with a monopolistic sector with two firms, F_1 and F_2, each one producing a differentiated good. The inverse demands p_i are linear:

$$p_i = \alpha - \beta q_i - \gamma q_j, \tag{1}$$

with parameters $\alpha > 0$, $\beta > 0$ and γ. We assume that $\gamma > 0$ and thus the goods are substitutes. The firm F_i invests an amount v_i in an R&D program $a_i : \mathbb{R}_0^+ \to [b_i, c_i]$ that reduces its production cost to

$$a_i(v_i) = c_i - \frac{\varepsilon(c_i - c_L)v_i}{\lambda_i + v_i}. \tag{2}$$

Next, we explain the parameters of the R&D program: (i) the parameter c_i is the unitary production cost of firm F_i at the beginning of the period satisfying $c_L \leq c_i \leq$

α; (ii) the parameter c_L is the minimum attainable production cost; (iii) the parameter $0 < \varepsilon < 1$ has the following meaning: since $b_i = a_i(+\infty) = c_i - \varepsilon(c_i - c_L)$, the maximum reduction $\eta_i = \varepsilon(c_i - c_L)$ of the production cost is a percentage $0 < \varepsilon < 1$ of the difference between the current cost c_i and the lowest possible production cost c_L; (iv) the parameter $\lambda_i > 0$ can be seen as a measure of the inverse of the quality of the R&D investment program for firm F_i and is directly related to what we call efficiency of the R&D investment program that we define next (a smaller λ_i will result in a bigger reduction of the production costs for the same investment). The R&D investment program of firm F_1 is more efficient than the R&D investment program of firm F_2 if and only if with the same investment $v_1 = v_2 = v$, the new cost obtained by firm F_1, a_1, is smaller or equal to the new cost obtained by firm F_2, a_2, i.e. $a_1(v) \le a_2(v)$. This R&D program was first introduced in [5].

2.2 Optimal Output Levels

The profit $\pi_i(q_i, q_j)$ of firm F_i is given by

$$\pi_i(q_i, q_j) = q_i(\alpha - \beta q_i - \gamma q_j - a_i) - v_i, \tag{3}$$

for $i, j \in \{1, 2\}$ and $i \neq j$. The Nash equilibrium output (q_1^*, q_2^*) is given by

$$q_i^* = \begin{cases} 0, \ if \ R_i \le 0 \\ R_i, \ if \ 0 < R_i < \frac{\alpha - a_j}{\gamma} \\ \frac{\alpha - a_i}{2\beta}, \ if \ R_i \ge \frac{\alpha - a_j}{\gamma} \end{cases}, \tag{4}$$

where

$$R_i = \frac{2\beta(\alpha - a_i) - \gamma(\alpha - a_j)}{4\beta^2 - \gamma^2},$$

with $i, j \in \{1, 2\}$ and $i \neq j$. Hence, if $R_i \le 0$ the firm F_j is at monopoly output level and, conversely, if $R_i \ge (\alpha - a_j)/\gamma$ the firm F_i is at monopoly output level and for intermediate values $0 \le R_i < (\alpha - a_j)/\gamma$, both firms have positive optimal output levels and so we are in the presence of duopoly competition. From now on, we will always consider that both firms choose their Nash equilibrium output (q_1^*, q_2^*).

2.3 New Production Costs

The sets of possible new production costs for firms F_1 and F_2, given initial production costs c_1 and c_2 are, respectively,

$$A_1 = A_1(c_1, c_2) = [b_1, c_1] \quad \text{and} \quad A_2 = A_2(c_1, c_2) = [b_2, c_2],$$

where $b_i = c_i - \varepsilon(c_i - c_L)$, for $i \in \{1, 2\}$. The R&D programs a_1 and a_2 of the firms determine a bijection between the investment region $\mathbb{R}_0^+ \times \mathbb{R}_0^+$ of both firms and the new production costs region $A_1 \times A_2$, given by the map

$$\begin{aligned} \mathbf{a} = (a_1, a_2) : \mathbb{R}_0^+ \times \mathbb{R}_0^+ &\to \quad A_1 \times A_2 \\ (v_1, v_2) &\mapsto (a_1(v_1), a_2(v_2)) \end{aligned}$$

where

$$a_i(v_i) = c_i - \frac{\eta_i v_i}{\lambda_i + v_i}.$$

We denote by $W = (W_1, W_2) : \mathbf{a}(\mathbb{R}_0^+ \times \mathbb{R}_0^+) \to \mathbb{R}_0^+ \times \mathbb{R}_0^+$

$$W_i(a_i) = \frac{\lambda_i(c_i - a_i)}{\eta_i - (c_i - a_i)}$$

the inverse map of $\mathbf{a}$.

The new production costs region can be decomposed, at most, in three disconnected economical regions characterized by the optimal output level of the firms:

M_1 The *monopoly region* M_1 of firm F_1 that is characterized by the optimal output level of firm F_1 being the monopoly output and, so, the optimal output level of firm F_2 is zero;

D The *duopoly region* D that is characterized by the optimal output levels of both firms being non-zero and, so, below their monopoly output levels;

M_2 The *monopoly region* M_2 of firm F_2 that is characterized by the optimal output level of firm F_2 being the monopoly output and, so, the optimal output level of firm F_1 is zero.

The boundaries between the duopoly region D and the monopoly region M_i are l_{M_i} with $i \in \{1, 2\}$ and are presented, explicitly in [5].

In equilibrium, i.e. when both firms choose their optimal output levels, the profit function $\pi_i : A_i \times A_j \to \mathbb{R}$ of firm F_i, in terms of its new production costs (a_1, a_2), is a piecewise smooth continuous function given by

$$\pi_i(a_1, a_2) = \begin{cases} \pi_{i,M_i}, & \text{if } (a_1, a_2) \in M_i \\ \pi_{i,D}, & \text{if } (a_1, a_2) \in D \\ -W_i(a_1, a_2), & \text{if } (a_1, a_2) \in M_j \end{cases},$$

where

$$\pi_{i,M_i} = \pi_{i,M_i}(a_1, a_2; c_1, c_2) = \frac{(\alpha - a_i)^2}{4\beta} - W_i(a_1, a_2),$$

$$\pi_{i,D} = \pi_{i,D}(a_1, a_2; c_1, c_2) = \beta \left(\frac{2\beta(\alpha - a_i) - \gamma(\alpha - a_j)}{4\beta^2 - \gamma^2} \right)^2 - W_i(a_1, a_2).$$

2.4 Nash Investment Regions

Let $V_i(v_j)$ be the best investment response function of firm F_i to a given investment v_j of firm F_j. The best investment response function $V_i : \mathbb{R}_0^+ \to \mathbb{R}_0^+$ of firm F_i is explicitly computed in [5]. Note that the best investment response function $V_i : \mathbb{R}_0^+ \to \mathbb{R}_0^+$ can be a multi-valued function.

Let c_L be the minimum attainable production cost and α the value to buyers. Given production costs $(c_1, c_2) \in [c_L, \alpha] \times [c_L, \alpha]$, the *Nash investment equilibria* $(v_1, v_2) \in \mathbb{R}_0^+ \times \mathbb{R}_0^+$ are the solutions of the system

$$\begin{cases} v_1 = V_1(v_2) \\ v_2 = V_2(v_1) \end{cases}$$

where V_1 and V_2 are the best investment response functions computed in the previous sections.

All the results presented, hold in an open region of parameters $(c_L, \varepsilon, \alpha, \beta, \gamma)$ containing the point $(4, 0.2, 10, 0.013, 0.013)$. The parameter λ_i that measures the efficiency of the R&D investment program, i.e. the smaller the λ_i, the more efficient the R&D investment program, is the parameter we are interested in studying. In the case we referred to as symmetric efficient, λ_i is equal to 10; in symmetric inefficient case, λ_i is equal to 20; in the asymmetric case, λ_1 is equal to 30 and λ_2 is equal to 10.

We observe that the Nash investment equilibria consists of a unique, or two, or three points depending upon the pair of initial production costs, as we will explain throughout the chapter. The set of all Nash investment equilibria form the *Nash investment equilibrium set*. We discuss the Nash investment equilibria by considering the following three regions of production costs:

- C the *competitive Nash investment region* C that is characterized by both firms investing;
- S_i the *single Nash investment region* S_i that is characterized by only one of the firms investing;
- N the *nil Nash investment region* N that is characterized by neither of the firms investing.

3 Nash Investment Equilibria Under Asymmetric Efficiency

In this chapter we compare the Nash investment equilibria choices of the two firms under three different scenarios: (i) symmetric efficient (SE) scenario where both firms R&D investment programs' are efficient and both firms hold the same degree of efficiency; (ii) symmetric inefficient (SI) scenario where both firms R&D investment programs' are less efficient and both firms hold the same degree of efficiency;

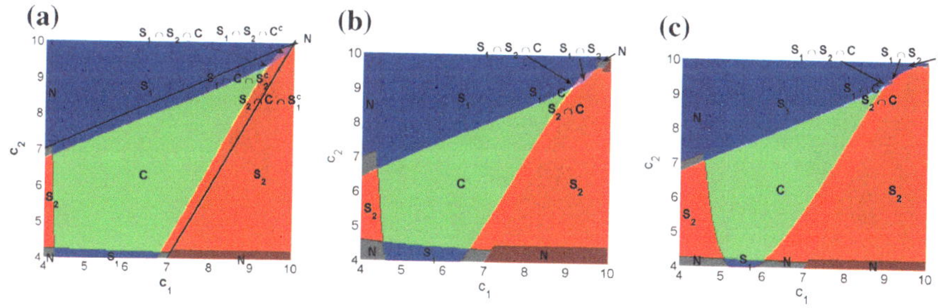

Fig. 1 Full characterization of the Nash investment regions in terms of the firms' initial production costs (c_1, c_2). The monopoly lines l_{M_i} are colored *black*. The nil Nash investment region N is colored *grey*. The single Nash investment regions S_1 and S_2 are colored *blue* and *red*, respectively. The competitive Nash investment region C is colored *green*. The region where S_1 and S_2 intersect are colored *pink*, the region where S_1 and C intersect are colored light *blue* and the region where S_2 and C intersect are colored *yellow*. The region where the regions S_1, S_2 and C intersect are colored light *grey*. (**a**) Symmetric efficient case. (**b**) Symmetric inefficient case. (**c**) Asymmetric case

asymmetric (A) scenario where one of the firms possesses a more efficient R&D program and the other firm possesses a less efficient R&D investment program.

We observe the existence, in the three distinct cases, of four different regions of Nash investment equilibria: a competitive Nash investment region C where both firms invest, a single Nash investment region S_1 for firm F_1, where just firm F_1 invests, a single Nash investment region S_2 for firm F_2, where just firm F_2 invests, and a nil Nash investment region N, where neither of the firms invest.

Let $R = [c_L, \alpha] \times [c_L, \alpha]$ be the region of all possible pairs of productions costs (c_1, c_2). Let $A^c = R - A$ be the complementary of A in R. The intersection between different Nash investment regions can be non-empty: (i) the intersection $R_{S_1 \cap S_2} = S_1 \cap S_2 \cap C^c$ between the *single Nash investments regions* S_1 and S_2 can be non empty; (ii) the intersection $R_{C \cap S_i} = C \cap S_i \cap S_j^c$ with $i \neq j$ between the *competitive Nash investment region* C and the *single Nash investment region* S_i can be non-empty; (iii) the intersection $R_{S_1 \cap C \cap S_2} = S_1 \cap C \cap S_2$ between the *competitive Nash investment region* C and the *single Nash investment regions* S_1 and S_2 can be non-empty (Fig. 1).

Let us consider the region of high production costs, that can correspond to the production of new technologies, where there are multiple Nash investment equilibria. In this section, we exhibit the production costs that correspond to the existence of multiple Nash investment equilibria.

We observe that the intersection $R_{S_1 \cap S_2} = S_1 \cap S_2 \cap C^c$ between the *single Nash investments regions* S_1 and S_2 is non empty. Thus, in this region we have two equilibria: a single Nash investment equilibrium to firm F_1 and a single Nash investment equilibrium to firm F_2. We also observe that the intersection $R_{C \cap S_i} = C \cap S_i \cap S_j^c$ with $i \neq j$ between the *competitive Nash investment region* C and the *single Nash investment region* S_i is non-empty. Therefore, in this region we have two Nash investment equilibria, one single Nash investment equilibrium for firm F_1 and a competitive Nash investment equilibrium. Finally, we see that the intersection

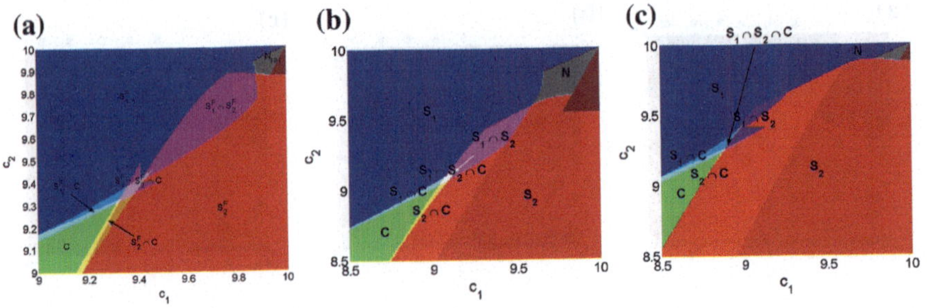

Fig. 2 Nash investment regions in the high production costs region, $c_i \in [9, 10]$, with $i \neq j$; (**a**) Symmetric efficient case. (**b**) Symmetric inefficient case. (**c**) Asymmetric case

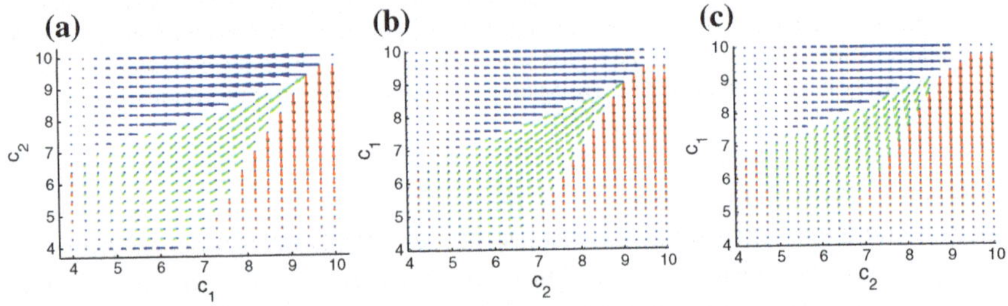

Fig. 3 Dynamics on the production costs in terms of the initial production costs (c_1, c_2): in *blue*, the dynamics in the single Nash investment region for firm F_1, S_1 where just firm F_1 invests; in *red* the dynamics in the single Nash investment region for firm F_2, S_2 where just firm F_2 invests; and in *green* the dynamics in the competitive Nash investment region C where both firms invest. (**a**) Symmetric efficient case. (**b**) Symmetric inefficient case. (**c**) Asymmetric case

$R_{S_1 \cap C \cap S_2} = S_1 \cap C \cap S_2$ between the *competitive Nash investment region* C and the *single Nash investment regions* S_1 and S_2 is non-empty. Thus, we have, simultaneously, a competitive equilibrium, a single favorable Nash investment equilibrium for firm F_1 and a single Nash investment equilibrium for firm F_2. This aspect enhances the high complexity of the R&D strategies of the firms, for high values of initial production costs (Figs. 2 and 3).

4 R&D Deterministic Dynamics

The R&D deterministic dynamics on the production costs of the duopoly competition appear from the firms deciding to play a perfect Nash equilibrium in the Cournot competition with R&D investment programs, period after period. The nil Nash investment region is the set of equilibria for these dynamics (Figs. 4 and 5).

The R&D deterministic dynamics in the single Nash investment region are implicitly determined by Theorems 1 and 2 in [5]. Let $S_1 = S_1^F \cup S_1^R$ be the single Nash

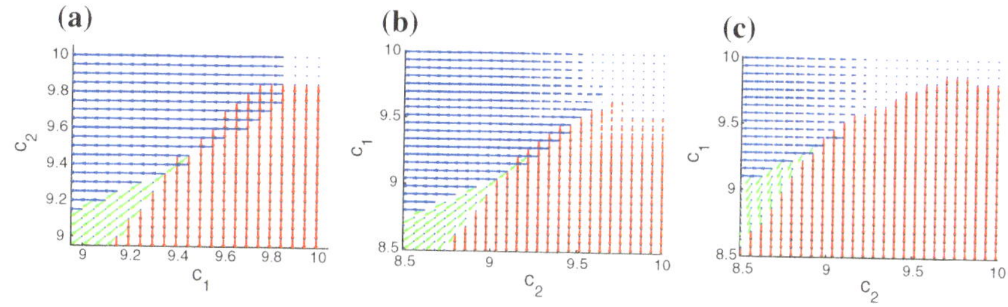

Fig. 4 Dynamics on the production costs in the high production costs region, $c_i \in [9, 10]$, with $i \neq j$: in *blue*, the dynamics in the single Nash investment region for firm F_1, S_1 where just firm F_1 invests; in *red* the dynamics in the single Nash investment region for firm F_2, S_2 where just firm F_2 invests; and in *green* the dynamics in the competitive Nash investment region C where both firms invest. (**a**) Symmetric efficient case. (**b**) Symmetric inefficient case. (**c**) Asymmetric case

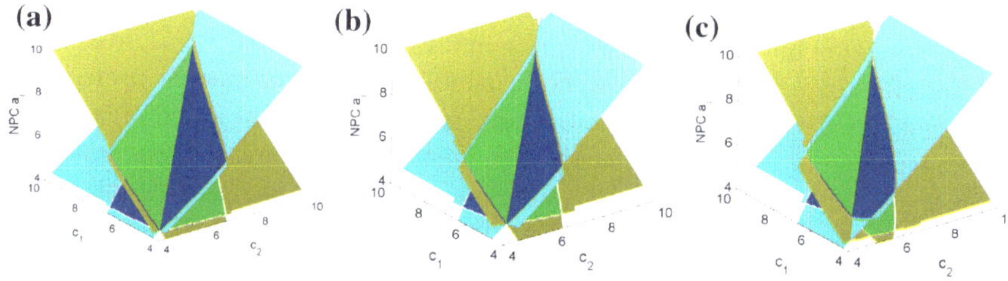

Fig. 5 New Production costs (NPC) for both firms F_1 and F_2 (a_1, a_2) in terms of the initial production costs (c_1, c_2): in *green* the NPC for firm F_2 and *blue* the NPC for firm F_1; in strong *green* the NPC in the competitive Nash investment region C and lighter *green*, the NPC in the single Nash investment region S_2; in strong blue the NPC in the competitive Nash investment region C and lighter *blue*, the NPC in the single Nash investment region S_1; (**a**) Symmetric efficient case. (**b**) Symmetric inefficient case. (**c**) Asymmetric case

investment region of firm F_1. If $(c_1, c_2) \in S_1^F$, then just firm F_1 invests along the time. Furthermore, at some period of time, the pair of new production costs falls in the monopoly region and so firm F_2 is driven out of the market by firm F_1. The production costs approach, along the time, the region N_{LH}. Hence, the production costs of firm F_1 approach low costs of production but the production costs of firm F_2 are always fixed at high production costs. If $(c_1, c_2) \in S_1^R$ then just firm F_1 invests along the time. So, firm F_1 will recover, along the time, from its disadvantageous position. The production costs approach, along the time, the region N_{LL}. Hence firm F_1 is able to recover, along the time, to the region where both firms have low production costs. The R&D deterministic dynamics in the competitive Nash investment region are implicitly determined by Theorems 1 and 2 in [5]. In the competitive Nash investment region both firms invest along the time and the production costs converge to the nil equilibrium region N_{LL}. Hence, the production costs of both firms are driven by the R&D deterministic dynamics to low production costs.

5 Conclusions

We presented R&D deterministic dynamics on the production costs of Cournot competitions based on perfect Nash equilibria of R&D investment strategies of the firms at every period. The following conclusions are valid in some parameter region of our model. We used an R&D investment function inspired in the logistic equation introduced in [5] and found all Perfect Nash investment equilibria of the Cournot competition model with R&D programs.

We described four main economic regions for the R&D deterministic dynamics corresponding to distinct perfect Nash equilibria: a competitive Nash investment region C where both firms invest, a single Nash investment region for firm F_1, S_1, where just firm F_1 invests, a single Nash investment region for firm F_2, S_2, where just firm F_2 invests, and a nil Nash investment region N where neither of the firms invest.

We considered three different scenarios in terms of the firms' R&D investment program efficiency: a first scenario, corresponding to the one studied in detail in [5], denominated symmetric efficient, where both firms possess an efficient R&D investment program; a second scenario, denominated symmetric inefficient, where both firms have an inefficient R&D program; and a third scenario, denominated asymmetric, where one of the firms has an efficient R&D program and the other one possesses and R&D program that is less efficient.

We showed, following [5], the existence of regions where the Nash investment equilibrium are not unique: the intersection $R_{S_1 \cap S_2}$ between the single Nash investment region S_1 and the single Nash investment region S_2 is non empty; the intersection $R_{S_i \cap C}$, with between the single Nash investment region S_i and the competitive Nash investment region C is non empty; the intersection $R_{S_1 \cap C \cap S_2}$ between the single Nash investment region S_1, the single Nash investment region S_2 and the competitive Nash investment region C is non empty. In this chapter we observed the persistence of these regions and described how these regions change as we change the efficiency of the R&D programs of both firms.

We presented the R&D deterministic dynamics on the production costs of Cournot competitions based on R&D investment strategies of the firms and we illustrated the transients and the asymptotic limits of the R&D deterministic dynamics on the production costs.

Acknowledgments We would like to thank LIAAD – INESC TEC and to acknowledge the financial support received by the ERDF – European Regional Development Fund through the Operational Programme for Competitiveness and Internationalisation – COMPETE 2020 Programme within project POCI-01-0145-FEDER-006961, and by National Funds through the FCT – Fundação para a Ciência e a Tecnologia (Portuguese Foundation for Science and Technology) as part of project UID/EEA/50014/2013 and by National Funds through the FCT within Project "Dynamics, optimization and modelling" with reference PTDC/MAT-NAN/6890/2014. Alberto Adrego Pinto also acknowledges the financial support received through the Special Visiting Researcher Program (Bolsa Pesquisador Visitante Especial – PVE) "Dynamics, Games and Applications" with reference 401068/2014-5 (call: MEC/MCTI/ CAPES/CNPQ/FAPS), at IMPA, Brazil. Bruno Oliveira gratefully acknowledges financial support from PRODEP III by FSE and EU and Miguel Ferreira gratefully acknowledges financial support from Fundação para a Ciência e a Tecnologia (FCT) given in the form of a PhD scholarship.

References

1. Brander, J.A., Spencer, B.J.: Strategic commitment with R&D: the symmetric case. Bell J. Econ. **14**, 225–235 (1983)
2. Cournot, A.: Recherches sur les Principes Mathématiques de la Théorie des Richesses. Paris, 1838. English edition: Researches into the Mathematical Principles of the Theory of Wealth. Edited by N. Bacon. New York: Macmillan (1897)
3. d'Aspremont, C., Jacquemin, A.: Cooperative and noncooperative R and D in duopoly with spillovers. In: American Economic Review, vol. 78, pp. 1133–1137 (1988). (Erratum. 80, 641–642)
4. DeBondt, R.: Spillovers and innovative activities. Int. J. Industr. Organ. **15**, 1–28 (1997)
5. Ferreira, M., Oliveira, B.M.P.M., Pinto, A.A.: Patents in new technologies. J. Diff. Equ. Appl. **15**, 1135–1149 (2009)
6. Ferreira, M., Oliveira, B.M.P.M., Pinto, A.A.: Piecewise R and D dynamics on costs. Fasciculi Mathematici **44**, 1–13 (2010)
7. Kamien, M., Muller, E., Zang, I.: Research joint ventures and R and D cartels. Am. Econ. Rev. **82**, 1293–1306 (1992)
8. Kamien, M., Zang, I.: Competing research joint ventures. J. Econ. Manag. Strateg. **2**, 23–40 (1993)
9. Katz, M.: An analysis of cooperative research and development. Rand J. Econ. **17**, 527–543 (1986)
10. Tirole, J.: The Theory of Industrial Organization. MIT Press, Massachusetts (1988)
11. Singh, N., Vives, X.: Price and quantity competition in a differentiated duopoly. RAND J. Econ. **15**, 546–554 (1984)

Different Approaches to the Global Periodicity Problem

Anna Cima, Armengol Gasull, Víctor Mañosa and Francesc Mañosas

Abstract Let F be a real or complex n-dimensional map. It is said that F is *globally periodic* if there exists some $p \in \mathbb{N}^+$ such that $F^p(x) = x$ for all x, where $F^k = F \circ F^{k-1}$, $k \geq 2$. The minimal p satisfying this property is called the *period* of F. Given a m-dimensional parametric family of maps, say F_λ, a problem of current interest is to determine all the values of λ such that F_λ is globally periodic, together with their corresponding periods. The aim of this paper is to show some techniques that we use to face this question, as well as some recent results that we have obtained. We will focus on proving the equivalence of the problem with the complete integrability of the dynamical system induced by the map F, and related issues; on the use of the local linearization given by the Bochner Theorem; and on the use the Normal Form theory. We also present some open questions in this setting.

Keywords Globally periodic maps · Integrable discrete systems · Lie Symmetries · Linearizations · Periodic difference equations · Reversible maps

1 Introduction

This paper deals with *globally periodic* maps and difference equations. A map $F : \mathcal{U} \to \mathcal{U}$, defined on an open set $\mathcal{U} \subseteq \mathbb{K}^n$ (where $\mathbb{K}$ is either $\mathbb{R}$ or $\mathbb{C}$) is *globally*

A. Cima · A. Gasull (✉) · F. Mañosas
Departament de Matemàtiques, Universitat Autònoma de Barcelona, Barcelona, Spain
e-mail: gasull@mat.uab.cat

A. Cima
e-mail: cima@mat.uab.cat

F. Mañosas
e-mail: manyosas@mat.uab.cat

V. Mañosa
Departament de Matemàtiques, Universitat Politècnica de Catalunya, Barcelona, Spain
e-mail: victor.manosa@upc.edu

L. Alsedà i Soler et al. (eds.), *Difference Equations, Discrete Dynamical Systems and Applications*, Springer Proceedings in Mathematics & Statistics 180, DOI 10.1007/978-3-662-52927-0_7

periodic if there exists some $p \in \mathbb{N}$ such that $F^p(x) = x$ for all $x \in \mathscr{U}$, where $F^k = F \circ F^{k-1}$, $k \geq 2$. The minimal p satisfying this property is called the *period* of F.

Observe that if F is globally periodic, then all the orbits of the discrete dynamical system generated by F are periodic.

This paper, rather of being a survey on global periodicity, wants to show the techniques that we have developed to detect globally periodic maps inside a parametric families of maps. The reader is addressed to [1–3], and the references therein, to have a general overview on the problem.

We start by showing some simple and well-known examples of globally periodic maps:

Example 1 Some linear fractional maps defined in suitable open subsets of $\mathbb{R}^2$ or $\mathbb{R}^3$:

(a) The 5-periodic map $F(x, y) = (y, (1 + y)/x)$, is a paradigmatic example known as the Lyness' map. This map is associated to the celebrated difference equation $x_{n+2} = (1 + x_{n+1})/x_n$.
(b) The 6-periodic map $F(x, y) = (y, y/x)$.
(c) The 8-periodic maps $F^{\pm}(x, y, z) = (y, z, (\pm 1 + y \pm z)/x)$. The plus case is known as the Todd's map. The minus one appears in [4].

Example 2 The piecewise linear map $F(x, y) = (y, \max(0, y) - x)$ is 5-periodic, see [4]. Its associated globally 5-periodic recurrence is

$$x_{n+2} = \max(0, x_{n+1}) - x_n. \tag{1}$$

This recurrence is the *ultradiscrete* version of the Lyness recurrence; see [5] for further details on ultradiscrete systems.

Example 3 The Coxeter's maps (see [6, 7] and also Sect. 3.3), which are $(n + 3)$-periodic:

$$F_n(x_1, \ldots, x_n) = (x_2, \ldots, x_n, 1 - \cfrac{x_n}{1 - \cfrac{x_{n-1}}{1 - \cfrac{x_{n-2}}{1 - \cdots \cfrac{x_2}{1 - x_1}}}}) \tag{2}$$

All of the above maps are associated to some difference equation. Recall that, in general, it is equivalent to study a map of the form

$$F(x_1, \ldots, x_n) = (x_2, \ldots, x_{n-1}, f(x_1, \ldots, x_n)), \tag{3}$$

or the difference equation

$$x_{j+n} = f(x_j, x_{j+1}, \ldots, x_{j+n-1}). \tag{4}$$

Example 4 The maps

$$F(x,y)=\left(\frac{y}{1+x+y},\frac{-x}{1+x+y}\right)$$

and

$$F(x,y)=\left(\frac{-y}{\sqrt[3]{1+4y^3-4x^3}},\frac{x}{\sqrt[3]{1+4y^3-4x^3}}\right)$$

are 4-periodic. These maps are given by the flow at time $2\pi/4$ of the planar 2π-*isochronous centers* $\dot{x}=-y+x^2$, $\dot{y}=x(1+y)$, and $\dot{x}=-y+4x^2y^2$, $\dot{y}=x+4xy^3$, respectively. Recall that a differential system, or a vector field, is called *isochronous* in an invariant open set $\mathscr{U}\subset\mathbb{R}^n$, if there exists $T>0$ such that $\varphi(T,\mathbf{x})=\mathbf{x}$ for all $\mathbf{x}\in\mathscr{U}$, where φ is the flow associated to the differential system. In this case the flow is also called isochronous. It is clear that for each $p\in\mathbb{N}^+$, every T-isochronous flow φ gives rise to several globally p-periodic maps, via the *stroboscopic maps* $F_j(\mathbf{x})=\varphi\,(jT/p,\mathbf{x})$, where and $j\in\mathbb{N}^+, j<p$ and $(j,p)=1$.

Given a m-dimensional parametric family of maps, say F_λ, the goal of this work is to determine all the values of λ such that F_λ is globally periodic, together with their corresponding periods. A simple example is given by the family of maps $F(x,y)=(y,(\lambda+y)/x)$, $\lambda\in\mathbb{C}$, which are globally periodic if and only if $\lambda\in\{0,1\}$.

In this paper we will summarize the tools that we are using to face this question. More concretely, the techniques that we will present here are:

- Detect some special properties of the dynamical system induced by globally periodic maps; see Sect. 2.
- Use the local linearization given by the Bochner Theorem; see Sect. 3.
- Find integrability type results and relate globally periodic maps with isochronous flows; see Sect. 4.
- Use of the Normal Forms Theory to compute some *periodicity conditions*; see Sect. 5.

We will pay a special attention to those families of maps coming from globally periodic difference equations.

2 General Properties of Globally Periodic Maps

2.1 *Some Properties*

From now on, $\mathscr{U}$ will denote an open set of $\mathbb{K}^n$, where $\mathbb{K}$ can be $\mathbb{R}$ or $\mathbb{C}$, depending on the context. We list some well-known properties of globally periodic maps that can be used to detect them in some given parametric families of maps.

(i) If a map F is p-periodic, then for any integer k the map F^k is also periodic.
(ii) If a map $F : \mathscr{U} \to \mathscr{U}$ is periodic then it has to be bijective in $\mathscr{U}$. Indeed $F^{-1} = F^{p-1}$.
(iii) Any *pointwise periodic* homeomorphism in $\mathscr{U}$ must be globally periodic; see [8]. Recall that F is called pointwise periodic in $\mathscr{U}$ if for every $\mathbf{x} \in \mathscr{U}$ there exists an integer $k(\mathbf{x})$ such that $F^{k(\mathbf{x})}(\mathbf{x}) = \mathbf{x}$.
(iv) If a *rational* map is p-periodic in an open subset of $\mathbb{K}^n$ then it has to be periodic, also of period p, in the whole real or complex space, except at the points where F or its iterates are not well defined.
(v) If $F : \mathscr{U} \to \mathscr{U}$ is a p-periodic map and $\mathbf{x} \in \mathscr{U}$ is a fixed point of F then $((DF)_{\mathbf{x}})^p = \mathrm{Id}$. Thus the eigenvalues of $(DF)_{\mathbf{x}}$ have to be p-roots of the unity. Moreover, the matrix $(DF)_{\mathbf{x}}$ diagonalices in $\mathbb{C}$; see [9].
(vi) The fixed points of a periodic map are neither attractors nor repellers.
(vii) *Rational* globally periodic maps have zero algebraic entropy; see [10, 11].

2.2 Some Applications of the Above Properties

In this section we will consider the family of nth order rational difference equation

$$x_{j+n} = \frac{A_1 x_j + A_2 x_{j+1} + \cdots + A_n x_{j+n-1} + A_0}{B_1 x_j + B_2 x_{j+1} + \cdots + B_n j_{j+n-1} + B_0}, \tag{5}$$

with initial condition $(x_1, x_2, \ldots, x_n) \in (0, \infty)^n$, and $\sum_{i=0}^n A_i > 0$, $\sum_{i=0}^n B_i > 0$, $A_i \geq 0$, $B_i \geq 0$, and $A_1^2 + B_1^2 \neq 0$.

The following well-known globally periodic difference equations, see Example 1, belong to the above family:

$$x_{j+2} = \frac{x_{j+1}}{x_j}, \quad x_{j+2} = \frac{x_{j+1}+1}{x_j}, \quad x_{j+3} = \frac{x_{j+1}+x_{j+2}+1}{x_j}$$
$$x_{j+1} = x_j, \quad x_{j+1} = \frac{1}{x_j}. \tag{6}$$

The two last ones are known as trivial ones.

Observe that each p-periodic n-th order difference equation produces in a natural way periodic difference equations of higher order, see [12]. For instance, the following ones of second order:

$$x_{j+2} = \frac{x_{j+1}}{x_j}, \quad x_{j+2} = \frac{x_{j+1}+1}{x_j}, \tag{7}$$

give rise to

$$x_{j+2\ell} = \frac{x_{j+\ell}}{x_j}, \qquad x_{j+2\ell} = \frac{x_{j+\ell}+1}{x_j},$$

for any positive integer ℓ, which are also periodic. Moreover taking $x_n = \alpha y_n, \alpha \neq 0$ they can be written as

$$y_{j+2\ell} = \frac{\alpha y_{j+\ell}}{y_j}, \quad y_{j+2\ell} = \frac{\alpha y_{j+\ell} + \alpha^2}{y_j}.$$

We will say that these difference equations are *equivalent* to (7).

Next result is obtained by using almost all the properties listed in Sect. 2.1.

Theorem 1 ([12]) *Consider the nth order rational difference equation (5) with the above mentioned hypotheses. Then, for $n \in \{1, 2, 3, 4, 5, 7, 9, 11\}$, any globally periodic recurrence in the family is equivalent to one of the difference equations given in (6).*

Open question: Is the above result true for any n?

Remark 2 If A_j and B_j are no more non-negative and $n = 1$, then there are globally periodic recurrences of type (5) of all periods: the Möbius transformations.

Remark 3 When $n = 2$, that is, for difference equations of the form,

$$x_{j+2} = \frac{A_1 x_j + A_2 x_{j+1} + A_0}{B_1 x_j + B_2 x_{j+1} + B_0}, \quad A_i, B_i \in \mathbb{C},$$

by using the property that the globally periodic maps have zero algebraic entropy, it has been proved in [13, 14] that the only possible periods for the globally periodic cases are 6, 5, 8, 12, 18, and 30. Moreover all them are realizable; see also [15, 16]. Again new cases, non-equivalent to the ones of the list given in (6), appear.

2.3 A Remark About Reversibility

Traditionally a map F is said to be reversible if $F = \sigma_1 \circ \sigma_2$ where σ_1, σ_2 are involutions. Curiously all explicit known globally periodic recurrences have an associated map $F : U \to U$, where U is a open subset of $\mathbb{R}^n$, that satisfies that $\sigma \circ F$ is an involution where $\sigma : \mathbb{R}^n \to \mathbb{R}^n$ is defined by $\sigma(x_1, x_2, \ldots, x_{n-1}, x_n) = (x_n, x_{n-1}, \ldots, x_2, x_1)$. Then for these periodic recurrences we get $F = \sigma \circ (\sigma \circ F)$ and they are reversible.

Due to this fact some authors claimed that this is a general property for the real recurrences and tried to prove it. The following implicit example shows that there are recurrences F such that $\sigma \circ F$ is not an involution.

Proposition 4 ([17]) *There exists a second order globally 3-periodic recurrence which associated map F is such that $\sigma \circ F$ is not an involution.*

Proof Consider the map $F = \Phi \circ L \circ \Phi^{-1}$, where

$$\Phi(x, y) = (x + g(y), y + g(-x - y))$$

with

$$g(z) = -z - z^2 - z^3 \quad \text{and} \quad L(x, y) = (y, -x - y).$$

It can be seen that F writes as $F(x, y) = (y, f(x, y))$ for a suitable f. Since F is 3-periodic we get that $F^{-1}(x, y) = (f(x, y), x)$. On the other hand $(\sigma \circ F \circ \sigma)(x, y) = (f(y, x), x)$. Then the equality $\sigma \circ F \circ \sigma = F^{-1}$ which is equivalent to $(\sigma \circ F)^2 = \text{Id}$ is also equivalent to $f(x, y) = f(y, x)$. Computing the Taylor series of f we obtain $f(x, y) \neq f(y, x)$. Hence $\sigma \circ F$ is not an involution as we wanted to prove. ∎

Nevertheless observe that for the above map $F = \Phi \circ L \circ \Phi^{-1}$, if we consider $\widetilde{\sigma} := \Phi \circ \sigma \circ \Phi^{-1}$ it holds that:

- $\widetilde{\sigma}$ is an involution.
- The map $\widetilde{\sigma} \circ F$ is also an involution.

So the following natural question arises:

Open question: Is any globally periodic recurrence reversible?

3 Linearization of Globally Periodic Maps and Recurrences

3.1 The Bochner Theorem

Let $\mathscr{V}$ be an open subset homeomorphic to $\mathbb{K}^n$. It is said that a map $F : \mathscr{V} \to \mathscr{V}$, $\mathscr{C}^r$-linearizes (globally) if there exists a $\mathscr{C}^r$-diffeomorphism, ψ, with $r \in \{0, 1, \ldots, \infty, \omega\}$, such that $\psi \circ F = L \circ \psi$ for all points in $\mathscr{V}$, where L is a linear map defined on a suitable domain.

An adaptation of a classical result due S. Bochner states that every globally periodic map with a fixed point locally linearizes:

Theorem 5 (Bochner [18] and [19, Chap. V]) *Let $F : \mathscr{U} \to \mathscr{U}$ be a p-periodic $\mathscr{C}^r$-diffeomorphism, where $\mathscr{U}$ is an open set of $\mathbb{K}^k$ and $r \neq 0$. Let $\mathbf{x}_0 \in \mathscr{U}$ be a fixed point of F. Then, there exists a neighbourhood of $\mathbf{x}_0$ where F is $\mathscr{C}^r$-conjugated with the linear map $L(\mathbf{x}) = (DF)_{\mathbf{x}_0}\mathbf{x}$. Moreover the linearization is given by the local $\mathscr{C}^r$-diffeomorphism*

$$\psi(\mathbf{x}) = \frac{1}{p}\sum_{i=0}^{p-1}(DF)_{\mathbf{x}_0}^{-i}\left(F^i(\mathbf{x})\right).$$

Proof Since F is p-periodic $((DF)_{\mathbf{x}_0})^p = \mathrm{Id}$. So $(\det((DF)_{\mathbf{x}_0})^p = 1$ and $(DF)_{\mathbf{x}_0}$ is invertible. Consider ψ as in the statement. By the inverse function theorem it is clear that the map ψ is a local $\mathscr{C}^r$-diffeomophism because $D\psi(\mathbf{x}_0) = \mathrm{Id}$. Using the p-periodicity of F and $(DF)_{\mathbf{x}_0}$ we get that for any $\mathbf{x}$ in a certain open neighborhood of $\mathbf{x}_0$,

$$\begin{aligned}\psi(F(\mathbf{x})) &= \tfrac{1}{p}\left[F(\mathbf{x}) + (DF)_{\mathbf{x}_0}^{-1}F^2(\mathbf{x}) + \cdots + (DF)_{\mathbf{x}_0}^{-(p-1)}F^p(\mathbf{x})\right]\\ &= \tfrac{1}{p}(DF)_{\mathbf{x}_0}\left[(DF)_{\mathbf{x}_0}^{-1}F(\mathbf{x}) + \cdots + (DF)_{\mathbf{x}_0}^{-p}F^p(\mathbf{x})\right]\\ &= (DF)_{\mathbf{x}_0}\cdot\psi(\mathbf{x}) = L(\psi(\mathbf{x}))\end{aligned}$$

and the result follows. ∎

An easy, but useful, consequence of the Bochner Theorem is:

Proposition 6 *Let F be a analytic map having a fixed point $\mathbf{x}_0$. Assume that F is globally p-periodic and let k be the minimum positive k such that $((DF)_{\mathbf{x}_0}))^k = \mathrm{Id}$, then $k = p$.*

For instance, a simple corollary of the above result says that if an analytic map $F(x, y) = (-x + O_2(x, y), -y + O_2(x, y))$ is globally p-periodic then it has to be an involution, that is $p = 2$. Here $O_2(x, y)$ denotes second order terms at the origin.

In $\mathbb{R}$ and $\mathbb{R}^2$ every p-periodic map globally linearize. The last case is a consequence of Kerékjártó's Theorem.

Theorem 7 (Kerékjártó 1919, [20]) *Let $\mathscr{U} \subseteq \mathbb{R}^2$ be homeomorphic to $\mathbb{R}^2$, and let F be a p-periodic map in $\mathscr{U}$. Then F is $\mathscr{C}^0$-linearizable.*

As a consequence of the above result it is easy to prove that every planar orientation preserving globally periodic map is given by the stroboscopic map of $\mathscr{C}^0$ flow.

Corollary 8 *For every preserving orientation globally periodic map F defined in an open set $\mathscr{U}$, homeomorphic to $\mathbb{R}^2$, there exists a continuous isochronous periodic flow φ such that $F(\mathbf{x}) = \varphi(T, \mathbf{x})$ for all $\mathbf{x} \in \mathscr{U}$.*

Proof Kerékjártó's Theorem ensures the existence of a homeomorphism ψ such that

$$\psi \circ F = L \circ \psi$$

where L is a periodic linear map. On the other hand, we can consider that

$$L(\mathbf{x}) = A\,\mathbf{x}, \text{ with } A = \begin{pmatrix}\cos T & -\sin T\\ \sin T & \cos T\end{pmatrix},$$

and where $T = 2k\pi/n$, $(k, n) = 1$ and n is the period of F. Hence L is the "time-T" stroboscopic map of the isochronous flow φ_L associated to the harmonic oscillator $\dot{u} = -v,\ \dot{v} = u$. Therefore

$$F(x) = \psi^{-1} \circ L \circ \psi(x) = \psi^{-1} \circ \varphi_L(T, \psi(x)) =: \varphi(T, x).$$

∎

As we will see in next sections, sometimes the Bochner Theorem can be used to construct global linearizations of smooth globally periodic maps. However there are examples of globally periodic homeomorphisms which do not linearize. The first one was given by Bing in 1952, [21, p. 361]; see also [22]. In general, even in a differentiable scenario the search of linearizations for globally periodic maps needs stronger tools than the Bochner Theorem because although any periodic $\mathscr{C}^1$-diffeomorphism in $\mathbb{R}^k$, with $k \leq 6$ must have a fixed point, there are periodic $\mathscr{C}^1$-diffeomorphisms in $\mathbb{R}^k$, $k \geq 7$ without fixed points; see [23, 24].

3.2 *Applications of the Bochner Theorem*

Some results that can be proved using Bochner Theorem are the following:

Theorem 9 ([25]) *Any $(n+1)$-periodic recurrence of class $\mathscr{C}^k$ defined in an open connected subset of $\mathbb{R}^n$ can be $\mathscr{C}^k$-linearized.*

The proof consists in showing that the Bochner linearization is in this case globally invertible. For other globally p-periodic difference equations the same idea also works.

Unfortunately, recently we have proved that there are simple explicit involutions $(p = 2)$, not coming from a difference equation, for which the linearization given by the Bochner Theorem is not globally invertible; see [26]. Non explicit examples appear in [27].

Proposition 10 ([26]) *Let $F : \mathbb{R}^2 \longrightarrow \mathbb{R}^2$ be given by*

$$F(x, y) = \big(x + 4xy + f(x, y), -y + 2(x^2 + y^2) - f(x, y)\big), \tag{8}$$

where $f(x, y) = 4(x + y)^2(y - x) - 4(x + y)^4$. Then F is an involution, has $(0, 0)$ as a fixed point and its associated Bochner linearization $\psi = \dfrac{1}{2}\Big(\mathrm{Id} + (DF)^{-1}_{(0,0)} \circ F\Big)$ is not a global diffeomorphism.

Despite this example, the global invertibility of the Bochner linearizations for maps coming from difference equations is still an open question.

Open questions:

Let $\mathscr{U}$ be an open set, homeomorphic to $\mathbb{R}^n$.

- Assume that $\mathscr{U}$ is invariant by the map associated to a globally periodic $\mathscr{C}^1$-difference equation. Is the Bochner linearization associated to a fixed point in $\mathscr{U}$ globally invertible?
- Is any globally periodic $\mathscr{C}^1$-map, $F : \mathscr{U} \to \mathscr{U}$ globally linearizable?

3.3 A Fascinating Example: The Coxeter Recurrence

Notice that the global linearizations are not only associated to a given globally periodic difference equation but also to a given invariant open set, homeomorphic to $\mathbb{R}^n$. For instance, consider the 5-periodic Lyness' map $F(x, y) = (y, (1 + y)/x)$. It can be seen that F and its iterates are defined in $\mathbb{R}^2 \setminus \mathscr{L}$ where $\mathscr{L}$ is the union of the straight lines $x = 0, x = -1, y = 0, y = -1, x + y = -1$. Clearly $\mathbb{R}^2 \setminus \mathscr{L}$ has twelve connected components and F fixes two of them and permutes the rest. We denote by $\mathscr{U}_1 = \{(x, y) \in \mathbb{R}^2 : x > 0, y > 0\}$ and by $\mathscr{U}_2$ the interior of the triangle with vertices $(-1, 0)$, $(-1, -1)$ and $(0, -1)$ which are the two invariant components of F. By using the Bochner theorem we prove:

Proposition 11 ([25]) *The maps $F|_{\mathscr{U}_1}$ and $F|_{\mathscr{U}_2}$ are globally $\mathscr{C}^\omega$-linearizable.*

The Coxeter recurrence, which is an extension of the Lyness one, is much more complicated; see [6]. For instance, Coxeter maps have many fixed points and on a neighborhood of each of them they are locally conjugate with *different* linear models.

In the statement of the two following results [ ] denotes the integer part function.

Lemma 12 ([6]) *There are exactly $[(n + 2)/2]$ linear globally periodic recurrences of order n, and period $n + 3$ without a line of fixed points (that is without eigenvalue $\lambda = 1$).*

Theorem 13 ([6]) *The Coxeter map F_n associated to the recurrence (2), has exactly $[(n + 2)/2]$ fixed points. Moreover at each of these fixed points F_n is locally conjugated to a different linear map.*

In other words all linear $(n + 3)$-periodic recurrence of order n without the eigenvalue 1, is present in the Coxeter map F_n.

It would be interesting to know if the corresponding local conjugations given by the Bochner Theorem are global in the corresponding invariant open sets.

4 Integrability Properties of Globally Periodic Maps

4.1 Existence and Construction of Complete Set of First Integrals

A non-constant function $V : \mathcal{U} \subseteq \mathbb{R}^n \to \mathbb{R}$ is called a *first integral* or *invariant* of the dynamical system generated by F if

$$V(F(\mathbf{x})) = V(\mathbf{x}).$$

If a map F possesses a first integral V, then all the orbits lie in some level set of V, or in other words, the level sets of V are invariant by F.

A map $F : \mathcal{U} \subseteq \mathbb{R}^n \to \mathbb{R}^n$ is *integrable* if it has $n-1$ functionally independent first integrals in $\mathcal{U}$, and *completely integrable* if it has n functionally independent first integrals in $\mathcal{U}$. In [28], it was proved that there is a strong relation between globally periodic maps and complete integrability.

Theorem 14 ([28]) *Let $F : \mathcal{U} \subseteq \mathbb{R}^n \to \mathcal{U}$ be an injective analytic map defined in an open set $\mathcal{U}$. The following statements hold:*

(i) If F is globally periodic, then it is completely integrable.
(ii) If F is completely integrable and it has n functionally independent first integrals $V_1, V_2, \cdots, V_n$ such that

$$\operatorname{Card}\left(\bigcap_{i=1}^{n}\{V_i = c_i\} \cap \mathcal{U}\right) \leq K, \text{ for all } c_i \in \mathbb{R}, \tag{9}$$

being K a given positive integer, then F is globally periodic.

Some remarks:

1. The proof of this result is constructive, so given a globally periodic map F, we are able to obtain a complete set of n first integrals; see Proposition 15.
2. Notice that the statement is given for analytic maps because we use strongly this condition to prove that the system of first integrals that we construct is formed by n functionally independent functions. However in many cases the method of construction gives functionally independent integrals even if the regularity is relaxed. For instance in [28, Prop. 18] two functionally independent integrals are given for the continuous map associated to the recurrence (1).
3. Condition (9) cannot be removed as the following example shows: Consider the map $F : \mathbb{R}^2 \to \mathbb{R}^2$,
$$F(x, y) = (x + 2\pi, y).$$
This map F is bijective on the whole $\mathbb{R}^2$ and it is clearly not p-periodic for all $p \in \mathbb{N}$. Moreover, the dynamical system generated by it, is completely integrable. For

instance, $V_1(x, y) = y - \sin(x)$ and $V_2(x, y) = y$ are two functionally independent first integrals for F. On the other hand, it is clear that for many c_1 and c_2, the finiteness condition (9) is not satisfied.

Example 5 It is well known that the Lyness map $F(x, y) = (y, (a + y)/x)$ has, for all value $a \in \mathbb{C}$, the first integral

$$V_{1,a}(x, y) = \frac{(x + 1)(y + 1)(x + y + a)}{xy}.$$

When $a = 0$ it has also the first integral

$$V_{2,0}(x, y) = \frac{x^4y^2 + x^2y^4 + y^4 + y^2 + x^2 + x^4}{x^2y^2}$$

and for $a = 1$, the first integral $V_{2,1}(x, y) = P(x, y)/(x^2y^2)$, where

$$\begin{aligned} P(x, y) =& xy^4 + \left(x^3 + x^2 + 2x + 1\right) y^3 + \left(x^3 + 5x^2 + 3x + 2\right) y^2 \\ &+ \left(x^4 + 2x^3 + 3x^2 + 3x + 1\right) y + x^3 + 2x^2 + x. \end{aligned}$$

It can be easily seen that $V_{1,i}$ and $V_{2,i}$ for $i = 0$ or $i = 1$ are couples of functionally independent functions. So, as Theorem 14 predicts, globally periodic cases are completely integrable.

We also recall how to construct many independent first integrals for globally periodic maps.

Proposition 15 ([28]) *Let $F : \mathcal{U} \subset \mathbb{R}^n \to \mathcal{U}$ be a globally p-periodic map on U. Let*

$$\Phi : \mathcal{U}^p = \overbrace{\mathcal{U} \times \mathcal{U} \times \cdots \times \mathcal{U}}^{p} \longrightarrow \mathbb{K}$$

be a symmetric function. Then, whenever it is not a constant function,

$$V_\phi(\mathbf{x}) = \Phi(\mathbf{x}, F(\mathbf{x}), \ldots, F^{p-1}(\mathbf{x}))$$

is a first integral of the dynamical system generated by F.

Using the above result with some families of symmetric functions, for instance symmetric polynomials, it is possible to obtain complete sets of first integrals, see again [28].

Example 6 Consider the map $F(x, y) = (y, c/(xy))$, corresponding to the difference equation $x_{n+2} = c/(x_n x_{n+1})$. It holds that

$$(x, y) \to \left(y, \frac{c}{xy}\right) \to \left(\frac{c}{xy}, x\right) \to (x, y),$$

so F is globally 3-periodic. In this case, then $n = 2$ and $p = 3$.

A complete set of first integrals can be constructed using two suitable symmetric functions. Taking $\sigma_1(a, b, c, d, e, f) = a + b + c + d + e + f$, we get:

$$V_1(x, y) = 2\left(x + y + \frac{c}{xy}\right).$$

Taking now $\sigma_2(a, b, c, d, e, f) = a^2 + b^2 + c^2 + d^2 + e^2 + f^2$, we obtain

$$V_2(x, y) = 2\left(x^2 + y^2 + \frac{c^2}{x^2y^2}\right).$$

Note that the symmetric function $\sigma_3(a, b, c, d, e, f) = abcdef$ gives a constant function and is not a first integral. It is easy to prove that V_1 and V_2 are functionally independent first integrals.

Example 7 Using the symmetric function $S_2(a_1, \ldots, a_{np}) = a_1^2 + a_2^2 + \ldots + a_{np}^2$, we get that any globally p-periodic map $F : \mathcal{U} \subset \mathbb{R}^n \to \mathcal{U}$ has always the first integral

$$V(\mathbf{x}) = \sum_{i=0}^{p-1} \|F^i(\mathbf{x})\|^2.$$

It is easy to see that the non-empty level sets of this first integral are compact. This will be usefull in the proof of Proposition 22.

4.2 A Remark on Invariant Curves of Rational Maps and Global Periodicity

Globally periodic rational maps in $\mathbb{K}^2$ are in fact birational (i.e. its inverse is also rational), because $F^{-1} = F^{p-1}$. By Theorem 14 and Proposition 15 in this case a complete set of rational first integrals can be chosen. Consequently, they preserve a couple of fibrations of the plane, which are generically transversal, and given by algebraic curves.

A theorem of Hurwitz of 1893, states that any compact Riemann surface with genus $g > 1$ admits at most $84(g-1)$ conformal automorphisms, that is, homeomorphisms of the surface onto itself which preserve the local structure; see [29, 30]. In our context, Hurwitz theorem can be stated as (see [31]):

Theorem 16 (Hurwitz, 1893) *The group of birational maps on a non-singular algebraic curve of genus $g > 1$ is finite of order at most* $84(g-1)$.

The following result, is a consequence of the above one and properties (*iii*) and (*iv*) given in Sect. 2.1.

Corollary 17 *A birational map in $\mathcal{U} \subseteq \mathbb{K}^2$ with a rational first integral V, such that the curves $\{V = c\}$ have generically genus $g > 1$, and are non-singular, must be globally periodic.*

Proof If the curves $\{V = c\}$ have generically genus $g > 1$ and are non-singular, then there exists an open set $\mathcal{V} \subseteq \mathcal{U}$ foliated by curves of these type. By Hurwitz Theorem on each of these curves the map must be periodic, so F is pointwise periodic on $\mathcal{V}$, and hence by the Montgomery Theorem ([8], see also property (iii) in Sect. 2.1) F must be globally periodic on $\mathcal{V}$. Since F is rational then it must be periodic on the whole $\mathbb{K}^2$ except at the points where its iterates are not well defined, as property (*iv*) in Sect. 2.1, states. ∎

Example 8 It is now common knowledge that for the Lyness map $F(x, y) = (y, (a + y)/x)$, and for all value $a \in \mathbb{C}$ the first integrals $V_{1,a}$ given in Example 5 produce fibrations of the plane given by curves that have generically genus 1 (of course, for some particular level sets the curves are straight lines so they have genus 0).

It is interesting to notice that when $a = 0$ the integral $V_{2,0}$ produces fibrations of the plane given by curves that have generically genus 7. Finally, when $a = 1$ the integral $V_{2,1}$ generically fibers the plane by curves of genus 6.

4.3 Measure Preservation

Another interesting property is that any globally periodic map has associated a natural invariant measure. Recall that a measure m is invariant for a map F if $m(F^{-1}(B)) = m(B)$ for any measurable set B.

Lemma 18 *Let $\mathcal{U}$ be an open subset of $\mathbb{R}^n$ and let $F : \mathcal{U} \to \mathcal{U}$ be a $\mathcal{C}^k$-globally periodic map, $k \geq 1$. Then F has the invariant measure*

$$m(B) = \sum_{i=0}^{p-1} \int_{F^i(B)} d\mathbf{x}.$$

Moreover m is absolutely continuous with respect the Lebesgue measure with $\mathcal{C}^{k-1}$-density $\nu(\mathbf{x}) = \sum_{i=0}^{p-1} \left|\det\left(DF^i(\mathbf{x})\right)\right|$.

Proof Let B be a Lebesgue measurable subset of $\mathcal{U}$. Since $F^{-1} = F^{p-1}$,

$$m(F^{-1}(B)) = \sum_{i=0}^{p-2} \int_{F^i(B)} d\mathbf{x} + \int_{F^{-1}(B)} d\mathbf{x} = \sum_{i=0}^{p-1} \int_{F^i(B)} d\mathbf{x} = m(B).$$

So m is an invariant measure.

By using the change of variables formula we obtain

$$m(B)=\sum_{i=0}^{p-1}\int_{F^i(B)}d\mathbf{x}=\int_B\sum_{i=0}^{p-1}\left|\det\left(DF^i(\mathbf{x})\right)\right|d\mathbf{x}=\int_B\nu(\mathbf{x})\,d\mathbf{x},$$

giving the desired result. Notice that ν is a $\mathscr{C}^{k-1}$ function because the global periodicity of F implies that $\det\left(DF^i(\mathbf{x})\right)\neq 0$. ∎

4.4 Existence and Construction of Lie Symmetries

A *Lie Symmetry* of a map F is a *vector field* X such that F maps any orbit of the differential system

$$\dot{\mathbf{x}}=X(\mathbf{x}), \tag{10}$$

into another orbit of the system. An equivalent definition is the following: It is a vector field $\mathbf{X}$ such that the differential equation (10) is invariant by the change of variables $\mathbf{u}=F(\mathbf{x})$. This property is characterized by

$$X(F(\mathbf{x}))=DF(\mathbf{x})\,X(\mathbf{x}),\quad \text{for all}\quad \mathbf{x}\in\mathscr{U}. \tag{11}$$

The existence of Lie Symmetries is an important issue in the theory of discrete integrability; see for instance [32]. From a dynamic viewpoint this importance is clear in the integrable case, where the dynamics are in fact one dimensional. Next results illustrate this fact.

Theorem 19 ([33]) *Let $F:\mathscr{U}\subseteq\mathbb{R}^n\to\mathscr{U}$ be a diffeomorphism having a Lie symmetry* $\mathbf{X}$*, and let γ be an orbit of X, preserved by F, i.e. $F:\gamma\to\gamma$. Then the dynamics of F restricted to γ is either conjugated to a rotation, conjugated to a translation of the line, or constant, according whether γ is homeomorphic to $\mathbb{S}^1$, $\mathbb{R}$, or a point, respectively.*

Theorem 20 ([33]) *Let $F:\mathscr{U}\subseteq\mathbb{R}^n\to\mathscr{U}$ be an integrable map, with functionally independent first integrals $V_1,\ldots,V_{n-1}$ and such that it preserves a measure $m(B)=\int_B\nu(\mathbf{x})\,d\mathbf{x}$, with $\nu\in\mathscr{C}^1(\mathscr{U})$ and non-vanishing in $\mathscr{U}$. Then F has a Lie Symmetry X sharing with F the same set of functionally independent first integrals. Moreover*

$$\begin{aligned}X(\mathbf{x})&=\left(-\partial V_1(\mathbf{x})/\partial y,\,\partial V_1(\mathbf{x})/\partial x\right)/\nu(\mathbf{x})\text{ if } n=2,\text{ and}\\ X(\mathbf{x})&=\left(\nabla V_1(\mathbf{x})\times\nabla V_2(\mathbf{x})\times\cdots\times\nabla V_{n-1}(\mathbf{x})\right)/\nu(\mathbf{x})\text{ if } n>2,\end{aligned} \tag{12}$$

where $\times$ denotes the cross product in $\mathbb{R}^n$.

Summarizing the results of this section we have:

1. Analytic globally periodic maps have n functionally independent first integrals.
2. Every differentiable globally periodic map has an invariant measure with density $\nu(\mathbf{x})=\sum_{i=0}^{p-1}\left|\det\left(DF^i(\mathbf{x})\right)\right|$.

3. An integrable map with an invariant measure $m(B) = \int_B \nu(\mathbf{x})\,d\mathbf{x}$, has a Lie Symmetry of the form (12).

Collecting all the facts above, we have that every globally periodic map has n linearly independent Lie Symmetries, thus obtaining the following result:

Corollary 21 *Let F be a differentiable globally p-periodic map defined on an open set $\mathcal{U} \subseteq \mathbb{R}^n$, with a complete set of differentiable first integrals $V_1, \ldots, V_n$. Then there exist n Lie Symmetries of F given by*

$$X_k(\mathbf{x}) = \left(-\partial V_k(\mathbf{x})/\partial y, \partial V_k(\mathbf{x})/\partial x\right)/\nu(\mathbf{x}),$$

for $k = 1, 2$, if $n = 2$ and

$$X_k(\mathbf{x}) = \left(\nabla V_1(\mathbf{x}) \times \nabla V_2(\mathbf{x}) \times \cdots \times \widehat{\nabla V_k(\mathbf{x})} \times \cdots \times \nabla V_n(\mathbf{x})\right)/\nu(\mathbf{x}),$$

for all $k = 1, \ldots, n$, if $n > 2$. In all cases ν can be chosen as in Lemma 18.

We will say that a flow of a differential equation is *periodic* if all their trajectories are either periodic or critical points. Next result shows that at least $n - 1$ of the above Lie Symmetries give rise to periodic flows as a consequence of the result below.

Proposition 22 *Let $F : \mathcal{U} \subseteq \mathbb{R}^n \to \mathbb{R}^n$ be a p-periodic map defined in an open set $\mathcal{U}$, having $n - 1$ functionally independent first integrals, $V_1, \ldots, V_{n-1}$ and being*

$$V_1(\mathbf{x}) = \sum_{i=0}^{p-1} \|F^i(\mathbf{x})\|^2.$$

the first integral given in Example 7. Then, the induced foliation of $\mathcal{U}$ given by the transversal levels sets $L_{h_1,\ldots,h_{n-1}} := \pitchfork_{i=1}^{n-1} \{V_i = h_i\}$ is diffeomorphic to a finite disjoint union of $\mathbb{S}^1$.

Proof Is easy to prove that the nonempty level sets $L_{h_1} = \{V_1 = h_1\}$ are compact. Now, since $V_1 \ldots V_{n-1}$ are functionally independent, the locus where they intersect transversally

$$L_{h_1,\ldots,h_{n-1}} := \left\{\mathbf{x} \in \bigcap_{i=1}^{n-1} \{V_i = h_i\} \text{ such that Rank}\left(\nabla V_1(\mathbf{x}), \ldots, \nabla V_{n-1}(\mathbf{x})\right) = n - 1\right\}$$

is a compact 1-dimensional manifold, so it must be diffeomorphic to a finite disjoint union of $\mathbb{S}^1$; see for instance [34, p. 208].

Therefore, the associated Lie symmetries give rise to periodic orbits on each connected component of the (non-empty) sets $L_{h_1,\ldots,h_{n-1}}$ which are, of course, generic. Also observe that the locus of non-transversality of the energy levels will correspond to singular points of the associated Lie Symmetries. Therefore there exist at least $n - 1$ linearly independent *periodic* Lie symmetries. ∎

Open question:

- Is every *orientation preserving* globally periodic map of class $\mathscr{C}^1$ associated to a periodic flow φ? That is: Is it true that

$$F(\mathbf{x}) = \varphi(\tau(\mathbf{x}), \mathbf{x})$$

for some periodic flow $\varphi(t, \mathbf{x})$?
- Furthermore, is every orientation preserving globally periodic map of class $\mathscr{C}^1$ the stroboscopic map of an isochronous flow? That is: Is it true that

$$F(\mathbf{x}) = \varphi(\tau, \mathbf{x})$$

for some continuous periodic flow $\varphi(t, \mathbf{x})$ and some $\tau \in \mathbb{R}^+$?

Observe that the above questions are not equivalent. Indeed, if we have a periodic vector field $X : \mathscr{U} \subseteq \mathbb{R}^n \to \mathbb{R}^n$, it can be reparameterized by

$$Y(\mathbf{x}) = T(\mathbf{x})X(\mathbf{x}): \tag{13}$$

where $T(\mathbf{x})$ is the function that assigns to each point the period of the periodic orbit passing through it. If T were a smooth function in $\mathscr{U}$ then it would be an 1-isochronous vector field. But sometimes, T is not regular in $\mathscr{U}$ because it tends to infinity. A characterization of when T is regular has been given recently in [35]. Notice that indeed, there are examples of periodic flows with some periodic orbits such that the periods of the neighbouring orbits tends to infinity as they approach to the *singular* periodic orbit. These examples were constructed by G. Reeb (1952), D. Epstein (1972), D. Sullivan (1976) and E. Voght (1977); see [36–39] respectively. All these classical examples are not polynomial. Recently we have learned about the existence of a polynomial one due to D. Peralta–Salas [40].

Observe also that the obstruction to give a positive answer to the first question in the above open problem is that, although a globally periodic map has at least $n-1$ Lie Symmetries that are periodic vector fields, it is difficult to guarantee that there exists at least one of them such that F preserves any solution γ of the differential equation $\dot{\mathbf{x}} = X(\mathbf{x})$. The following example shows that sometimes this obstruction can be avoided.

Example 9 The map

$$F(x, y, z) = \left(z, \frac{1+y+z}{x}, \frac{1+x+y+z+xz}{xy}\right)$$

is globally 4-periodic because it corresponds to $G \circ G$, where G is the Todd's map given in Example 1; see also [41]. In the first octant it has the Lie symmetry given by

$$X(x, y, z) = xyz\,(\nabla V_1(x, y, z) \times \nabla V_2(x, y, z))$$

where

$$V_1 = \frac{(x+1)(y+1)(z+1)(1+x+y+z)}{xyz}, \quad V_2 = \frac{(x+1)(z+1)}{y}.$$

Since F preserves any solution γ of the differential equation $\dot{\mathbf{x}} = X(\mathbf{x})$, this implies that in the first quadrant F can be seen as the stroboscopic map of the isochronous flow of the vector field $Y(\mathbf{x}) = T(\mathbf{x})X(\mathbf{x})$ on $\mathbb{R}^3$.

5 A Normal Form Theory Approach

The theory of Normal forms for dynamical systems, both discrete and continuous, goes back to Poincaré and Lyapunov, and it is a well-known tool for the study of bifurcations of dynamical systems; see for instance [42–44].

In this section we outline a method, introduced in [45], which is the discrete analog of the way in that Normal Form Theory is applied in the theory of planar differential equations in order to obtain center and isochronicity conditions [46–48]. First, let us recall briefly how Normal Form theory works for these two problems. Given a planar analytic differential equation

$$\begin{aligned} \dot{x} &= -y + f_2(x, y) + f_3(x, y) + O_4(x, y), \\ \dot{y} &= x + g_2(x, y) + g_3(x, y) + O_4(x, y), \end{aligned} \tag{14}$$

it is well known that the origin is either a center or a focus. In this later case the stability of the origin is given by the sign of the first non-null number of a list of polynomial expressions in the coefficients of f_j and g_j, called the *Lyapunov quantities* $V_3, V_5, \ldots$ obtained from the coefficients of f_j and g_j. In particular, the first Lyapunov quantity only depends on f_2, f_3, g_2, and g_3, that is $V_3 = V_3(f_2, f_3, g_2, g_3)$.

The normal form of (14) is

$$\begin{aligned} \dot{x} &= -y - (x^2 + y^2)(V_3 y + T_2 x) + O_4(x, y), \\ \dot{y} &= x + (x^2 + y^2)(V_3 x - T_2 y) + O_4(x, y), \end{aligned} \tag{15}$$

see [43, p. 154] or [44, Sect. 3.1B]. So only the term V_3 (which gives the stability) and the term T_2 (that gives information about the period of the orbits) remain.

Following the same idea, and since the Bochner Theorem ensures that globally periodic maps with a fixed point locally linearize, we can apply the Normal Form theory to analytic maps with a fixed point in order to obtain some explicit expressions of the obstructions to be locally linearized around the point. The vanishing of these expressions will be necessary conditions for global periodicity.

5.1 Normal Form Theory and Global Periodicity

In this section we briefly recall some issues of Normal Form Theory. We also indicate how to use it to obtain some necessary conditions for global periodicity.

Let $F := F^{(1)} : \mathbb{K}^2 \to \mathbb{K}^2$, be a family of smooth maps satisfying $F^{(1)}(0) = 0$. Let

$$F^{(1)}(\mathbf{x}) = F_1^{(1)}(\mathbf{x}) + F_2^{(1)}(\mathbf{x}) + \cdots + F_k^{(1)}(\mathbf{x}) + O(|\mathbf{x}|^{k+1})$$

be the Taylor expansion of F at $\mathbf{0}$, where $F_r^{(1)} \in \mathscr{H}_r$, the real vector space of maps whose components are homogeneous polynomials of degree r.

A sequence of transformations Φ_n can be constructed, in such a way that at each step, Φ_n removes (if it is possible), the terms of the corresponding homogeneous part of degree n, obtaining a sequence of maps of the form

$$F^{(n)}(\mathbf{y}) = L\,\mathbf{y} + F_{n+1}^{(n)}(\mathbf{y}) + O(|\mathbf{y}|^{n+2}),$$

where $L\mathbf{y} := F_1^{(1)}(\mathbf{y}) = DF^{(1)}(\mathbf{0})\,\mathbf{y}$.

For instance, the transformation

$$\mathbf{x} = \Phi_3(\mathbf{y}) := \mathbf{y} + \phi_3(\mathbf{y}),$$

with $\phi_3 \in \mathscr{H}_3$, conjugates the map $F^{(2)}$ with a new map $F^{(3)}$, via

$$F^{(2)}(\Phi_3) = \Phi_3(F^{(3)}).$$

From the above equation, it can be easily seen that the term $F_3^{(3)} = L\,\phi_3(\mathbf{y}) - \phi_3(L\mathbf{y}) + F_3^{(2)}(\mathbf{y})$. In other words,

$$F^{(3)}(\mathbf{y}) = L\,\mathbf{y} + L\,\phi_3(\mathbf{y}) - \phi_3(L\mathbf{y}) + F_3^{(2)}(\mathbf{y}) + O(|\mathbf{y}|^4).$$

If $\phi_3(\mathbf{y})$ can be chosen in such a way that

$$M_L\,(\phi_3(\mathbf{y})) := L\,\phi_3(\mathbf{y}) - \phi_3(L\mathbf{y}) = -F_3^{(2)}(\mathbf{y}), \tag{16}$$

then the order-3 terms of $F^{(3)}$ can be removed and it is transformed into

$$F^{(3)}(\mathbf{y}) = L\,\mathbf{y} + O(|\mathbf{y}|^4).$$

The vectorial equation (16), whose unknowns are the coefficients of ϕ_3, is called the order 3 *homological equation* associated with L, and the existence of solutions of it, is the necessary and sufficient condition to be able to remove the homogeneous terms of degree 3.

If there is any obstruction to solve the homological equation (16), this can be interpreted as an obstruction in the linearization process. These obstructions are polynomial expressions in the coefficients of $F^{(2)}$, that must vanish in order to continue with the linearization process. The equations obtained by equaling these polynomial expressions to zero are called *periodicity conditions*; see [45].

For instance, when $L = \operatorname{diag}(\alpha, 1/\alpha)$, there are no obstructions for the order 2 homological equation. But there appear some ones in the order 3 homological equation. The associated periodicity conditions in this case, are given by

$$f_{2,1}^{(2)} = 0 \text{ and } g_{1,2}^{(2)} = 0. \tag{17}$$

Using the above method, and computing the explicit expressions of equations (17), we prove:

Theorem 23 ([45]) *Consider a smooth complex map of the form*

$$F(x, y) = \left(\alpha x + \sum_{i+j\geq 2} f_{i,j}x^i y^j, \frac{1}{\alpha} y + \sum_{i+j\geq 2} g_{i,j}x^i y^j\right), \tag{18}$$

where α is a primitive p-root of unity, $p \geq 5$. Then the conditions

$$\left(f_{2,1} + f_{1,1}g_{1,1}\right)\alpha^4 - f_{1,1}\left(2f_{2,0} - g_{1,1}\right)\alpha^3 + \\ \left(2g_{2,0}f_{0,2} - f_{1,1}f_{2,0} + f_{1,1}g_{1,1}\right)\alpha^2 - \left(f_{2,1} + f_{1,1}f_{2,0}\right)\alpha + f_{1,1}f_{2,0} = 0,$$

$$g_{0,2}g_{1,1}\alpha^4 - \left(g_{1,2} + g_{0,2}g_{1,1}\right)\alpha^3 + \left(f_{1,1}g_{1,1} + 2g_{2,0}f_{0,2} - g_{0,2}g_{1,1}\right)\alpha^2 \\ + g_{1,1}\left(-2g_{0,2} + f_{1,1}\right)\alpha + f_{1,1}g_{1,1} + g_{1,2} = 0$$

are necessary for F to be p-periodic.

5.2 An Application of Theorem 23

As an application of an extension of Theorem 23 presented in [45], where an additional third condition is given, we study the global periodicity for the 2-periodic Lyness recurrence

$$x_{n+2} = \frac{a_n + x_{n+1}}{x_n}, \quad \text{where} \quad a_n = \begin{cases} a \text{ for } n = 2\ell + 1, \\ b \text{ for } n = 2\ell, \end{cases} \tag{19}$$

and $a, b \in \mathbb{C}$. We obtain:

Theorem 24 ([45]) *The only globally periodic recurrences in* (19) *are:*

(i) *The cases $a = b = 0$ which is 6-periodic, and $a = b = 1$, which is 5-periodic.*

(ii) The cases $a = (-1 \pm i\sqrt{3})/2$ *and* $b = \overline{a} = 1/a$*, which is* 10*-periodic.*

Notice that the cases given in (i) correspond to the well-known autonomous globally periodic Lyness recurrences given in Example 1. The ones corresponding to (ii) are new non-autonomous globally periodic recurrences.

Observe that the sequence $\{x_n\}$ can be reobtained as

$$(x_1, x_2) \xrightarrow{G_a} (x_2, x_3) \xrightarrow{G_b} (x_3, x_4) \xrightarrow{G_a} (x_4, x_5) \xrightarrow{G_b} (x_5, x_6) \xrightarrow{G_a} \cdots$$

where $G_\alpha(x, y) = (y, (\alpha + y)/x)$, with $\alpha \in \{a, b\}$. So the behavior of (19) is determined by the dynamical system generated by the map:

$$F(x, y) := G_b \circ G_a(x, y) = \left(\frac{a+y}{x}, \frac{a+bx+y}{xy}\right)$$

which always has some fixed point. In [45], the extension of Theorem 23 is applied to F for proving Theorem 24.

Acknowledgments The authors are partially supported by Spanish Ministry of Economy and Competitiveness through grants MTM2008-03437 (first and second authors); DPI2011-25822 and DPI2016-77407-P (third author); and MTM2011-26995-C02-01 (fourth author). Both CoDALab and GSD-UAB groups are supported by Generalitat de Catalunya through the SGR program.

References

1. Balibrea F., Linero-Bas A.: Some new results and open problems on periodicity of difference equations. In: Iteration theory (ECIT 04), Grazer Math. Ber. **350**, 15–38 (2006)
2. Grove, E.A., Ladas, G.: Periodicities in Nonlinear Difference Equations. Advances in discrete mathematics and applications, vol. 4. Chapman & Hall/CRC Press, Boca Raton (2005)
3. Linero-Bas, A.: Some results on periodicity of difference equations. In: Liz, E., Mañosa, V. (eds.) Proceedings of the Workshop Future Directions in Difference Equations, pp. 121–143. Universidade de Vigo, Vigo (2011)
4. Csörnyei, M., Laczkovich, M.: Some periodic and non-periodic recursions. Monatshefte für Mathematik **132**, 215–236 (2001)
5. Ochiai, T., Nacher, J.C.: Inversible Max–Plus algebras and integrable systems. J. Math. Phys. **46**, 063507 (2005)
6. Cima, A., Gasull, A., Mañosas, F.: On coxeter recurrences. J. Differ. Equ. Appl. **18**, 1457–1465 (2012)
7. Coxeter, H.S.M.: Frieze patterns. Acta Arith. **18**, 297–310 (1971)
8. Montgomery, D.: Pointwise periodic homeomorphisms. Am. J. Math. **59**, 118–120 (1937)
9. Kurshan, R.P., Gopinath, B.: Recursively generated periodic sequences. Canad. J. Math. **26**, 1356–1371 (1974)
10. Bellon, M.P.: Algebraic entropy of birational maps with invariant curves. Lett. Math. Phys. **50**, 79–90 (1999)
11. Diller, J., Favre, C.: Dynamics of bimeromorphic maps of surfaces. Am. J. Math. **123**, 1135–1169 (2001)
12. Cima, A., Gasull, A., Mañosas, F.: On periodicity rational difference equations of order k. J. Differ. Equ. Appl. **10**, 549–559 (2004)

13. Bedford, E., Kim, K.: Periodicities in linear fractional recurrences: degree growth of birational surface maps. Mich. Math. J. **54**, 647–670 (2006)
14. Bedford, E., Kim, K.: Linear fractional recurrences: periodicities and integrability. Ann. Fac. Sci. Toulouse Math. **20**, 33–56 (2011)
15. Rubió-Massegú, J.: On the global periodicity of discrete dynamical systems and application to rational difference equations. J. Math. Anal. Appl. **343**, 182–189 (2008)
16. Rubió-Massegú, J., Mañosa, V.: Normal forms for rational difference equations with applications to the global periodicity problem. J. Math. Anal. Appl. **332**, 896–918 (2007)
17. Cima, A., Gasull, A., Mañosas, F.: New periodic recurrences with applications. J. Math. Anal. Appl. **382**, 418–425 (2011)
18. Bochner, S.: Compact groups of differentiable transformations. Ann. Math. **46**, 372–381 (1945)
19. Montgomery, D., Zippin, L.: Topological Transformation Groups. Interscience, New York (1955)
20. Constantin, A., Kolev, B.: The theorem of Kerékjártó on periodic homeomorphisms of the disc and the sphere. Enseign. Math. **40**, 373–413 (1994)
21. Bing, R.H.: A homeomorphism between the 3-sphere and the sum of two solid horned spheres. Ann. Math. **80**, 78–93 (1964)
22. Bing, R.H.: Inequivalent families of periodic homeomorphisms of E3. Ann. Math. **56**, 354–362 (1952)
23. Hayes, R., Kwasik, S., Mast, J., Schultz, R.: Periodic maps $\mathbb{R}^7$ without fixed points. Math. Proc. Camb. Philos. Soc. **132**, 131–136 (2002)
24. Kister, J.M.: Differentiable periodic actions on E^8 without fixed points. Am. J. Math. **85**, 316–319 (1963)
25. Cima, A., Gasull, A., Mañosas, F.: Global linearization of periodic difference equations. Discret. Contin. Dyn. Syst. **32**, 1575–1595 (2012)
26. Cima, A., Gasull, A., Mañosas, F.: Simple examples of planar involutions with non-global Montgomery–Bochner linearizations. Appl. Math. Lett. **25**, 2086–2088 (2012)
27. Pires B., Teixeira M.A.: On global linearization of planar involutions. Bull. Braz. Math. Soc. (N.S.) **43**, 637–653 (2012)
28. Cima, A., Gasull, A., Mañosa, V.: Global periodicity and complete integrability of discrete dynamical systems. J. Differ. Equ. Appl. **12**, 697–716 (2006)
29. Conder, M.: Hurwitz groups with given centre. Bull. Lond. Math. Soc. **34**, 725–728 (2002)
30. Conder, M.: An update on Hurwitz groups. Groups Complex. Cryptol. **2**, 35–49 (2010)
31. Jogia, D., Roberts, J.A.G., Vivaldi, F.: An algebraic geometric approach to integrable maps of the plane. J. Phys. A: Math. Gen. **39**, 1133–1149 (2006)
32. Haggar, F., Byrnes, G.B., Quispel, G.R.W., Capel, H.W.: k-integrals and k-Lie symmetries in discrete dynamical systems. Phys. A **233**, 379–394 (1996)
33. Cima, A., Gasull, A., Mañosa, V.: Studying discrete dynamical systems through differential equations. J. Differ. Equ. **244**, 630–648 (2008)
34. Guillemin, V., Pollack, A.: Differential Topology. Prentice Hall, New Jersey (1974)
35. Maksymenko, S.I.: Period functions for $\mathcal{C}^0$ – and $\mathcal{C}^1$ flows. Ukr. Math. J. **62**, 1109–1125 (2010)
36. Epstein, D.B.A.: Periodic flows on three-manifolds. Ann. Math. **95**, 66–82 (1972)
37. Reeb G.: Sur certaines proprietes topologiques des varietes feuilletees. Actual. Sci. Ind. **1183**, Hermann, Paris (1952)
38. Sullivan, D.: A counterexample to the periodic orbit conjecture. Inst. Hautes Etudes Sci. Publ. Math. **46**, 5–14 (1976)
39. Vogt, E.: A periodic flow with infinite Epstein hierarchy. Manuscr. Math. **22**, 403–412 (1977)
40. Peralta-Salas D.: Private Communication (2009)
41. Cima A., Gasull A., Mañosa V.: Some properties of the k–dimensional Lyness' map. J. Phys. A: Math. Theor. **41**, 285205, 18 pp. (2008)
42. Arrowsmith, D.K., Place, C.M.: An Introduction to Dynamical Systems. Cambridge University Press, Cambridge (1990)

43. Guckenheimer, J., Holmes, P.: Nonlinear Oscillations, Dynamical Systems, and Bifurcations of Vector Fields. Springer, New York (1983)
44. Wiggins, S.: Introduction to Applied Nonlinear Dynamical Systems and Chaos. Springer, New York (1990)
45. Cima A., Gasull A., Mañosa V.: Global periodicity conditions for maps and recurrences via Normal Forms. Int. J. Bifur. Chaos Appl. Sci. Eng. **23**, 1350182, 18 pp. (2013)
46. Chow, S.-N., Li, C., Wang, D.: Normal Forms and Bifurcation of Planar Vector Fields. Cambridge University Press, Cambridge (1994)
47. Farr, W.W., Li, C., Labouriau, I.S., Langford, W.F.: Degenerate Hopf bifurcation formulas and Hilbert's 16th problem. SIAM J. Math. Anal. **20**, 13–30 (1989)
48. Hassard, B., Kazarinoff, N.D., Wan, Y.H.: Theory and Applications of the Hopf Bifurcations. Cambridge University Press, Cambridge (1980)

On Bifurcations of Homoclinic Tangencies in Area-Preserving Maps on Non-orientable Manifolds

Amadeu Delshams, Marina Gonchenko and Sergey Gonchenko

Abstract We study bifurcations of non-orientable area-preserving maps with quadratic homoclinic tangencies. We study the case when the maps are given on non-orientable two-dimensional manifolds. We consider one and two parameter general unfoldings and establish results related to the appearance of elliptic periodic orbits.

Keywords Bifurcation · Homoclinic tangency · Area-preserving map · Non-orientable · Elliptic orbit

1 Introduction

The main goal of the paper is the studying bifurcations of area-preserving maps (APMs) given in non-orientable manifolds and having homoclinic tangencies. In the case of dissipative systems, the corresponding problems are quite traditional and many results obtained here have a fundamental value for the theory of dynamical chaos. One of such results, known as *theorem on cascade of periodic sinks (sources)*, goes back as early as to the paper [1] of Gavrilov and Shilnikov. Note that in [1] the general case was considered, i.e. it was assumed that (i) the initial diffeomorphism has a saddle fixed (periodic) point O whose the invariant manifolds are quadratically tangent at the points of some homoclinic orbit; (ii) multipliers λ and γ of O are such that $0 < |\lambda| < 1 < |\gamma|$ and the saddle value $\sigma \equiv |\lambda||\gamma|$ is not equal to 1. In this case,

A. Delshams · M. Gonchenko (✉)
Departament de Matemàtica Aplicada I, Universitat Politècnica de Catalunya, Barcelona, Spain
e-mail: marina.gonchenko@upc.edu

A. Delshams
e-mail: amadeu.delshams@upc.edu

S. Gonchenko
Institute of Applied Mathematics and Cybernetics, Nizhny Novgorod University, Nizhny Novgorod, Russia
e-mail: gonchenko@pochta.ru

L. Alsedà i Soler et al. (eds.), *Difference Equations, Discrete Dynamical Systems and Applications*, Springer Proceedings in Mathematics & Statistics 180, DOI 10.1007/978-3-662-52927-0_8

bifurcations of the homoclinic tangency lead to the appearance of asymptotically stable (if $\sigma < 1$) or completely unstable (if $\sigma > 1$) periodic orbits. Moreover, in any one parameter general unfolding such orbits are observed for values of the parameter belonging to an infinite sequence (cascade) of intervals that do not mutually intersect and accumulate to the value of the parameter corresponding to the initial homoclinic tangency.

This theorem admits a huge circle of extensions. For example, in the multidimensional case, criteria of the appearance of periodic sinks under homoclinic bifurcations were established in [2–6]. In the dimension two, main bifurcations in the case $\sigma = 1$ were studied in [7, 8] where it was shown that both stable and unstable periodic orbits and invariant circles are born under such homoclinic bifurcations. Another generalization is related to the case of two-dimensional diffeomorphisms having nontransversal heteroclinic cycles, i.e. there are several saddle fixed (periodic) points which form a cycle due to intersections of their invariant manifolds and some of these intersections are nontransversal. If the saddle values of all these points are less than 1 (or greater than 1), then results look to be quite similar to the homoclinic case with $\sigma < 1$ (or $\sigma > 1$), although intervals of the existence of periodic sinks (sources) can intersect here [9]. However, new very interesting dynamical phenomena occur when the heteroclinic cycle contains at least two points such that $\sigma < 1$ for one of them and $\sigma > 1$ for the other point. It was shown in [10] that bifurcations of such heteroclinic cycles lead to the appearance of infinitely many coexisting attracting and repelling periodic orbits (and also to stable and unstable invariant circles [11, 12]). Note that in the reversible case, as it was shown in [13, 14], infinitely many coexisting attracting, repelling and elliptic periodic orbits can be born under bifurcations of such heteroclinic cycles. Moreover, the so-called *phenomenon of mixed dynamics*, [10, 11, 14], is observed here, i.e. when these orbits can be not separated in common, i.e. the closures of the sets of orbits of different types have nonempty intersections. In particular, as follows from [6, 15], it is generically when the closures of periodic sinks and sources (also elliptic points in the reversible case) contain nontrivial uniformly hyperbolic set (e.g. Smale horseshoes).

Concerning related results in the conservative case, we mention, above all, the well-known result of S. Newhouse, [16], on the appearance of so-called 1-elliptic periodic orbits (there is only one pair multipliers, $e^{i\varphi}$ and $e^{-i\varphi}$ with $\varphi \neq 0, \pi$, on the unit circle) under bifurcations of homoclinic tangencies of multidimensional symplectic maps. However, the Newhouse theorem from [16] does not give answer whether these 1-elliptic points are generic.[1] In turn, this fact is principally important in the two-dimensional case where an 1-elliptic point is elliptic and the genericity means the KAM-stability. The birth of generic elliptic points under homoclinic bifurcations in symplectic two-dimensional maps was established in [19, 20]. Note that the analogous problem was considered in [21, 22] when studying bifurcations of three-dimensional conservative flows with a homoclinic loop of a saddle-focus equilibrium. In papers [23, 24], it was deduced conditions at which two-dimensional

[1] The birth of 2-elliptic generic periodic orbits was proved in [17, 18] for the case of four-dimensional symplectic maps with homoclinic tangencies to saddle-focus fixed points.

symplectic maps with nontransversal heteroclinic cycles have infinitely many generic elliptic periodic orbits.

In the paper we consider area preserving (but not symplectic) maps given in two-dimensional manifold $\mathcal{M}_2$ and study their homoclinic bifurcations. The same as in [20], we construct bifurcation diagrams for the so-called single-round periodic orbits (see Definition 1). Above of all, we establish the theorem on one parameter cascade of elliptic periodic points, see Theorem 1. However, we note that, unlike the dissipative case, intervals of the existence of elliptic periodic orbits can intersect and, thus, for the analysis of the corresponding bifurcation phenomena, it is required to consider at least two parameter general unfoldings. As a result, we establish the new (two parameter) version of the theorem on cascade of elliptic points, see Theorem 2. We show that at certain (codimension 2) conditions, the so-called *phenomenon of global resonance*, see [20, 25, 26], can occur here which shows itself in the fact that area-preserving maps on $\mathcal{M}_2$ with quadratic homoclinic tangencies can possess infinitely many coexisting elliptic periodic orbits of all successive *even* periods $2k_0, 2(k_0+1), \dots,$ beginning from some even integer $2k_0$, see Theorem 3.

2 Statement of the Problem and Main Results

2.1 *Definition of the Homoclinic Tangency*

Consider a C^r-smooth ($r \geq 3$) area-preserving map f_0 given on a non-orientable manifold $\mathcal{M}_2$ and satisfying the following conditions.

A. f_0 has a saddle fixed point O with multipliers λ and λ^{-1}, where $|\lambda| < 1$.
B. f_0 has a homoclinic orbit Γ_0 at whose points the stable and unstable invariant manifolds of the saddle O have a quadratic tangency.

Let $\mathcal{H}_0$ be a (codimension one) bifurcation surface composed of area-preserving C^r-maps of $\mathcal{M}_2$ close to f_0 and such that every map of $\mathcal{H}_0$ has a nontransversal homoclinic orbit close to Γ_0. Let f_ε be a family of area-preserving C^r-maps that contains the map f_0 at $\varepsilon = 0$. We suppose that the family depends smoothly on parameters $\varepsilon = (\varepsilon_1, \dots, \varepsilon_m)$ and satisfies the following condition.

C. The family f_ε is transverse to $\mathcal{H}$ at $\varepsilon = 0$.

Let U be a small neighborhood of $O \cup \Gamma_0$. It consists of a small disk U_0 containing O and a number of small disks surrounding those points of Γ_0 that do not lie in U_0 (see Fig. 1).

Definition 1 A periodic or homoclinic orbit entirely lying in U is called *p-round* if it has exactly p intersection points with any disk of the set $U \backslash U_0$.

In the work we study bifurcations of *single-round* ($p = 1$) *periodic orbits* in the families f_ε. Note that every point of such an orbit can be considered as a fixed point

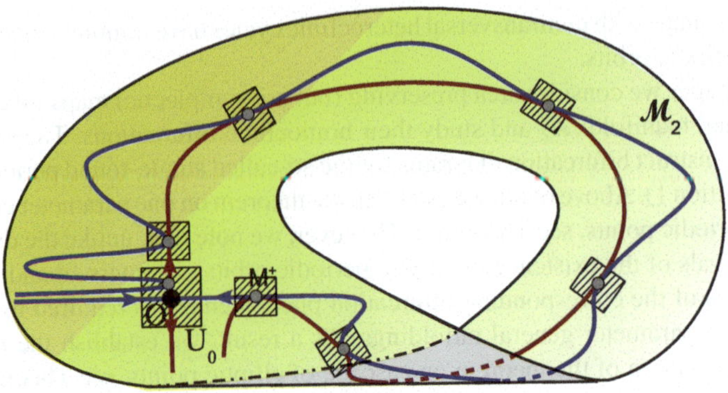

Fig. 1 An example of non-orientable area-preserving map (on the Möbius band) having a quadratic homoclinic tangency at the points of a homoclinic orbit Γ_0. Some of these homoclinic points are shown as *grey circles*. Also a small neighborhood of the set $O \cup \Gamma_0$ is shown to be the union of a number of "squares"

of the corresponding *first return map*. Such a map is usually constructed as a superposition $T_k = T_1 T_0^k$ of two maps $T_0(\varepsilon) \equiv f_\varepsilon\big|_{U_0}$, called *local map*, and $T_1 \equiv T_1(\varepsilon)$, called *global map*. The map T_1 is defined as $T_1 \equiv f_\varepsilon^q$ and acts from a small neighbourhood $\Pi^- \subset U_0$ of some point $M^- \in W^u_{loc}(O)$ of the orbit Γ_0 into a neighbourhood $\Pi^+ \subset U_0$ of another point $M^+ \in W^s_{loc}(O)$ of Γ_0, where q is an integer such that $f_0^q(M^-) = M^+$. Thus, any fixed point of T_k is a point of a single-round periodic orbit for f_ε with period $k+q$. We study maps T_k for all any sufficiently large integer k.

Both the maps T_0 and T_1 are area-preserving maps. In the paper we assume that

D. The map T_0 is symplectic in U_0, whereas, the map T_1 is area-preserving and non-orientable.

The condition D means that one can introduce in U_0 some *canonical coordinates* (x, y) such that $J(T_0) \equiv +1$ and $J(T_1) \equiv -1$, where $J(T)$ is the Jacobian of map T. Moreover, since $J(T_1) = -1$, it follows that the homoclinic orbit Γ_0 is not "homotopic to zero".[2] Thus we have a homoclinic situation as in Fig. 1.

Remark 1 In distinct of planar symplectic maps, see an example in Fig. 2, area-preserving maps on non-orientable manifolds can have only one (nontransversal) homoclinic to O orbit. It is a consequence of the fact that any global symplectic structure can not be defined here, [27]. In particular, as it is seen from Fig. 1, the curve $\ell = \{W^u(\text{from } O \text{ to } M^+) \cup W^s(\text{from } M^+ \text{ to } O) \cup O+\}$ is not Jordan one in $\mathcal{M}_2$.

[2] In other words, any "homoclinic loop" (e.g. the curve $\ell(O, M^+, O)$ composed from a connected pieces of manifolds W^u and W^s with border points O and M^+) is not contractible in $\mathcal{M}$, see Fig. 1.

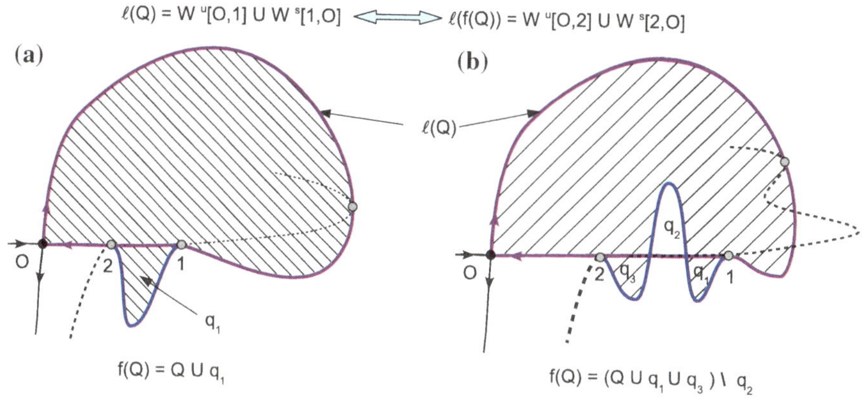

Fig. 2 In both cases, the domain Q has the boundary $\ell = \{$segment of W^u from O to $1\} \cup \{$segment of W^s from 1 to $O\}$ and Q is transformed under a map f into the domain $f(Q)$ (dashed one) having the boundary $f(\ell) = \{$segment of W^u from O to $2\} \cup \{$segment of W^s from 2 to $O\}$. **a** This case with only one (nontransversal) homoclinic orbit impossible for planar APMs, since $S(f(Q)) = S(Q) + S(q_1) > S(Q)$. **b** This case of homoclinic tangency with the existence of other (transverse) homoclinic orbits is possible for planar APMs, since the domains Q and $f(Q)$ can have the same area in case (**b**), then $S(q_1) + S(q_2) = S(q_3)$

2.2 *The Local and Global Maps T_0 and T_1*

Concerning the local canonical coordinates in U_0, we can introduce such coordinates (x, y) in which the local map T_0 can be written in in the following form, see [20],

$$\begin{aligned} \bar{x} &= \lambda x \left(1 + \beta_1 \cdot xy + \ldots + \beta_n \cdot (xy)^n\right) + xO\left(|xy|^n(|x| + |y|)\right), \\ \bar{y} &= \lambda^{-1} y \left(1 + \beta_1 \cdot xy + \ldots + \beta_n \cdot (xy)^n\right)^{-1} + yO\left(|xy|^n(|x| + |y|)\right), \end{aligned} \tag{1}$$

which is called the finitely smooth normal form of n-th order. The map T_0 can be brought to this form by a canonical change of coordinates, which is either C^r-smooth for $n = 1$ or C^{r-2n} for $n \geq 2$ [15, 20].

The normal forms (1) are very suitable for effective calculation of maps T_0^k : $(x_0, y_0) \to (x_k, y_k)$ with sufficiently large integer k. Indeed, as was shown [20], if T_0 is given by (1), then the map T_0^k can be written, for any integer k, as follows

$$\begin{aligned} x_k &= \lambda^k x_0 \cdot R_n^{(k)}(x_0 y_k, \varepsilon) + \lambda^{(n+1)k} P_n^{(k)}(x_0, y_k, \varepsilon), \\ y_0 &= \lambda^k y_k \cdot R_n^{(k)}(x_0 y_k, \varepsilon) + \lambda^{(n+1)k} Q_n^{(k)}(x_0, y_k, \varepsilon), \end{aligned} \tag{2}$$

where

$$R_n^{(k)} \equiv 1 + \tilde{\beta}_1(k)\lambda^k x_0 y_k + \cdots + \tilde{\beta}_n(k)\lambda^{nk}(x_0 y_k)^n, \tag{3}$$

$\tilde{\beta}_i(k), i = 1, \ldots, n$, are some polynomials (of degree i) with respect to k (e.g. $\tilde{\beta}_1 = \beta_1 k$, $\tilde{\beta}_2 = \beta_2 k - \frac{1}{2}\beta_1^2 k^2, \ldots$) and the functions $P_n^{(k)}$, $Q_n^{(k)}$ are uniformly bounded[3] in k.

Remark 2 (1) The normal form of the first order ($n = 1$) for T_0

$$\begin{aligned} \bar{x} &= \lambda x\,(1 + \beta_1 \cdot xy) + xO\,(|xy|(|x| + |y|))\,, \\ \bar{y} &= \lambda^{-1} y\,(1 - \beta_1 \cdot xy) + yO\,(|xy|(|x| + |y|)) \end{aligned} \tag{4}$$

is well-known from [19, 28] where it was proved the existence of normalizing C^{r-1}-coordinates. The existence of C^r-smooth canonical changes of coordinates (which are C^{r-2}-smooth with respect to parameters) bringing a symplectic saddle map to form (4) was proved in [15].

(2) Note that form (1) can be considered as a certain finitely smooth approximation of the following analytical Moser normal form

$$\bar{x} = \lambda(\varepsilon)x \cdot B(xy, \varepsilon), \quad \bar{y} = \lambda^{-1}(\varepsilon)y \cdot B^{-1}(xy, \varepsilon), \tag{5}$$

taking place for $\lambda > 0$ [30], where $B(xy, \varepsilon) = 1 + \beta_1 \cdot xy + \cdots + \beta_n \cdot (xy)^n + \ldots$. Note that the form (5) is integrable: it has integral xy; whereas, $\bar{x}\bar{y} = xy + O\left((xy)^{n+1}(|x| + |y|)\right)$ for the case of form (1).

In coordinates (1), the equations of $W^s_{loc} \cap U_0$ and $W^u_{loc} \cap U_0$ are $y = 0$ and $x = 0$, respectively. Therefore, we can write that $M^+ = (x^+, 0)$, $M^- = (0, y^-)$. Without loss of generality, we assume that $x^+ > 0$ and $y^- > 0$. Let the neighbourhoods Π^+ and Π^- of the homoclinic points M^+ and M^-, respectively, be sufficiently small and such that $T_0(\Pi^+) \cap \Pi^+ = \emptyset$, $T_0^{-1}(\Pi^-) \cap \Pi^- = \emptyset$. Then, as usually (see e.g. [1, 29]), the successor map from Π^+ into Π^- by orbits of T_0 is defined, for all sufficiently small ε, on the set consisting of infinitely many strips $\sigma_k^0 \equiv \Pi^+ \cap T_0^{-k}\Pi^-$, $k = \bar{k}, \bar{k}+1, \ldots$. The image of σ_k^0 under T_0^k is the strip $\sigma_k^1 = T_0^k(\sigma_k^0) \equiv \Pi^- \cap T_0^k \Pi^+$. As $k \to \infty$, the strips σ_k^0 and σ_k^1 accumulate on W^s_{loc} and W^u_{loc}, respectively.

We can write the global map $T_1(\varepsilon) : \Pi^- \to \Pi^+$ as follows

$$\overline{x} - x^+ = F(x, y - y^-, \varepsilon), \quad \overline{y} = G(x, y - y^-, \varepsilon), \tag{6}$$

where $F(0) = 0, G(0) = 0$. Besides, we have that $G_y(0) = 0, G_{yy}(0) = 2d \neq 0$ which follows from the fact (condition B) that at $\varepsilon = 0$ the curve $T_1(W^u_{loc})$: $\{\overline{x} - x^+ = F(0, y - y^-, 0), \overline{y} = G(0, y - y^-, 0)\}$ has a quadratic tangency with $W^s_{loc} : \{\bar{y} = 0\}$ at M^+. When parameters vary this tangency can split and, moreover, we can introduce the corresponding splitting parameter as $\mu \equiv G(0, 0, \varepsilon)$. By condition C, the parameter μ must belong to the set of parameters ε; and, without loss of generality, we assume that $\varepsilon_1 \equiv \mu$. Accordingly, we can write the following Taylor expansions

[3] along with all derivatives by coordinates and parameters up to order either $(r - 2)$ for $n = 1$ or $(r - 2n - 2)$ for $n \geq 2$.

$$\begin{aligned} F(x, y-y^-, \varepsilon) &= ax + b(y-y^-) + e_{20}x^2 + e_{11}x(y-y^-) + e_{02}(y-y^-)^2 + \text{h.o.t.}, \\ G(x, y-y^-, \varepsilon) &= \mu + cx + d(y-y^-)^2 + f_{20}x^2 + f_{11}x(y-y^-) + f_{30}x^3 \\ &\quad + f_{21}x^2(y-y^-) + f_{12}x(y-y^-)^2 + f_{03}(y-y^-)^3 + \text{h.o.t.}, \end{aligned} \quad (7)$$

where the coefficients $a, b, \ldots, f_{03}$ (as well as x^+ and y^-) depend smoothly on ε. Note also that

$$J(T_1) = \det \begin{pmatrix} F_x & F_y \\ G_x & G_y \end{pmatrix} \equiv -1 \quad (8)$$

since T_1 is non-orientable map by condition D. In particular, we have

$$\begin{aligned} bc &\equiv +1, \\ \tilde{R} &= (2a + 2e_{02}/bd - bf_{11}/d) \equiv 0. \end{aligned} \quad (9)$$

2.3 On Geometrical and Topological Properties of APMs on $\mathcal{M}_2$ with Homoclinic Tangencies

It is easy to see from (6), (7) that μ is the parameter of splitting of manifolds $W^s(O_\varepsilon)$ and $W^u(O_\varepsilon)$ with respect to the homoclinic point M^+. Indeed, the curve $l_u = T_1(W^u_{loc} \cap \Pi^-)$ has the equation $l_u \ : \ \bar{y} = \mu + \dfrac{d}{b^2}(\bar{x} - x^+)^2(1 + O(\bar{x} - x^+))$. Since the equation of W^s_{loc} is $y = 0$ for all (small) ε, it implies that the manifolds $T_1(W^u_{loc})$ and W^s_{loc} do not intersect for $\mu d > 0$, intersect transversally at two points for $\mu d < 0$, and have a quadratic tangency (at M^+) for $\mu = 0$. In turn, since the strips σ^1_k accumulate on the segment $W^u_{loc} \cap \Pi^-$ as $k \to \infty$, it follows that the images $T_1(\sigma^1_k)$ of σ^1_k under T_1 have a horseshoe form and, moreover, horseshoes $T_1(\sigma^1_k)$ accumulate on l_u as $k \to \infty$. Therefore, the first return maps $T_k = T_1 T_0^k : \sigma^0_k \to \sigma^0_k$ are, in fact, conservative horseshoe maps with the Jacobian -1.

When μ varies near zero value infinitely many bifurcations of horseshoes creation (destruction) occur. In the paper we study these bifurcations and show that they include birth (disappearance) of *elliptic periodic points*.

However, we can also see these horseshoe bifurcations must have different scenarios depending on a type of the initial homoclinic tangency. Indeed, at $\mu = 0$ a character of reciprocal position of the strips σ^0_k and their horseshoes $T_1(\sigma^1_k)$ is essentially defined by the signs of the parameters λ, c and d. Moreover, by this feature, we can select 6 different cases of area-preserving maps with quadratic homoclinic tangencies. The corresponding examples are shown in Figs. 3 and 4. Note that in the cases with $\lambda < 0$ we can always consider d to be positive: if d is negative for the given pair of homoclinic points, M^+ and M^-, we can take another pair of points, like $\{T_0(M^+), M^-\}$ or $\{M^+, T_0^{-1}(M^-)\}$, for which the corresponding d' becomes positive.

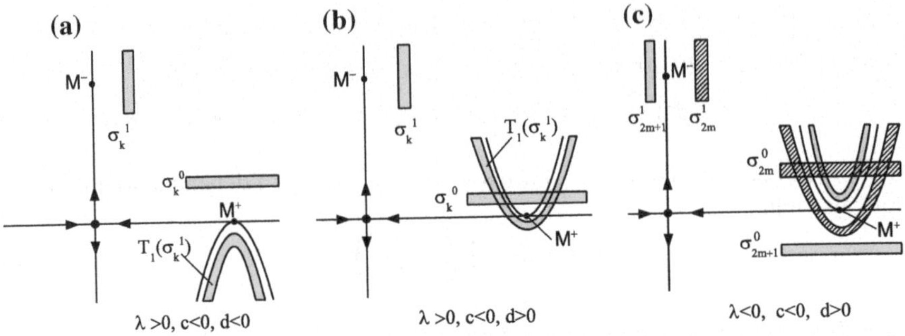

Fig. 3 Types of APMs with a homoclinic tangency for $c < 0$

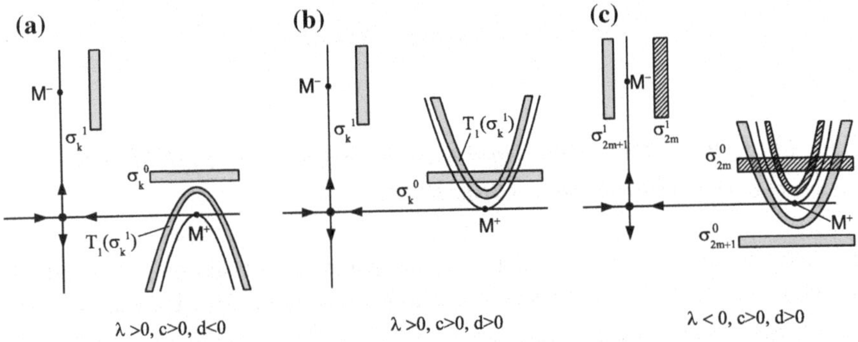

Fig. 4 Types of APMs with a homoclinic tangency

Note that in the cases with $c < 0$, see Fig. 3, a reciprocal position of all the strips σ_j^0 and their horseshoes $T_1(\sigma_j^1)$ at $\mu = 0$ is defined quite simply: $\sigma_j^0 \cap T_1(\sigma_j^1) = \emptyset$ if $\lambda > 0, d < 0$; the strips σ_j^0 and horseshoes $T_1(\sigma_j^1)$ have regular intersections if $\lambda > 0, d > 0$; the corresponding intersections are either regular for even j or empty for odd j if $\lambda < 0, d > 0$. Recall that the regular intersection means here (by [31, 32]) that the set $\sigma_j^0 \cap T_1(\sigma_j^1)$ consists of two connected components and, moreover, the first return map $T_j \equiv T_1 T_0^j : \sigma_j^0 \mapsto \sigma_j^0$ is *the Smale horseshoe map*: its nonwandering set Ω_j is hyperbolic and $T_j\big|_{\Omega_j}$ is topologically conjugate to the Bernoulli shift with two symbols (for more details see [20, 32]). Therefore, we can say that every map f_0 in the case $c < 0, d > 0$ has infinitely many horseshoes Ω_j, where j runs all sufficiently large positive integers (resp., even positive integers) in the case $\lambda > 0$ (resp., in the case $\lambda < 0$). On the other hand, every map f_0 with $\lambda > 0, c < 0, d < 0$ has no horseshoes at all (in a small neighbourhood U).

In the cases of homoclinic tangencies with $c > 0$, see Fig. 4, a reciprocal position of the strips σ_j^0 and horseshoes $T_1(\sigma_j^1)$ depends also on other invariants of the homoclinic structure. The principal such invariant is, [25, 26, 31],

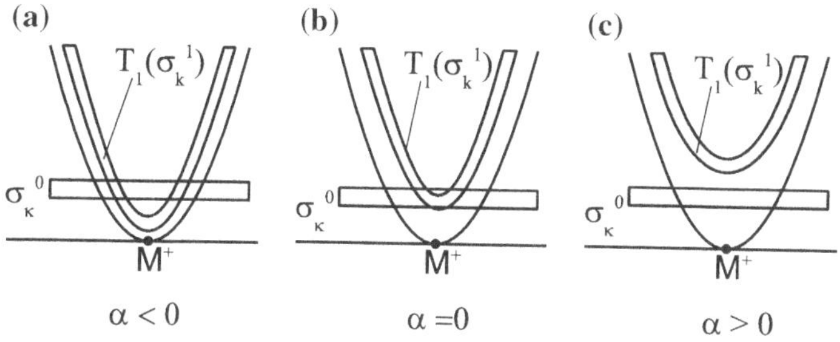

Fig. 5 A horseshoe geometry of symplectic maps with a homoclinic tangency in the case $\lambda > 0, c > 0, d > 0$ for various α

$$\alpha = \frac{cx^+}{y^-} - 1. \tag{10}$$

Note that, first of all, the sign of α is important here. For example, in Fig. 5 it is shown a reciprocal position of the strips σ_j^0 and horseshoes $T_1(\sigma_j^1)$ (with sufficiently large j) for various values of α for the case $\lambda > 0, c > 0, d > 0$. It follows from (2) that the strip σ_k^0 is posed from the line $y = 0$ (i.e. W^s_{loc} on a distance of order $\lambda^k y^-$; it follows from (7) and (2) that the top of the horseshoe $T_1(\sigma_k^1)$ is posed from the line $y = 0$ on a distance of order $c\lambda^k x^+$, see Fig. 5c. Thus, if $c\lambda^k x^+ > \lambda^k y^-$, i.e. $\alpha > 0$, the top of the horseshoe $T_1(\sigma_k^1)$ is above the strip σ_k^0. However, if $c\lambda^k x^+ < \lambda^k y^-$, i.e. $\alpha < 0$, the top of the horseshoe $T_1(\sigma_k^1)$ is below the strip σ_k^0. In fact, this geometrical observation can be extended to the following result.[4]

- if $\alpha < 0$, then f_0 has infinitely many horseshoes Ω_j;
- if $\alpha > 0$, then there exists such a neighbourhood $U(O \cap \Gamma_0)$ in which dynamics of f_0 is trivial: only orbits O and Γ_0 do not leave U under iterations of f_0.

The case $\alpha = 0$ is "bifurcational", since infinitely many horseshoes appear (disappear) when varying α near zero (even without splitting the initial tangency).

Thus, we can draw the following conclusions:

(1) the cases of homoclinic tangencies with $c < 0$ or with $\alpha \neq 0$ at $c > 0$ are "ordinary" and it is sufficient to study bifurcations of single-round periodic orbits within framework of one parameter general families only (with the parameter μ);
2) the cases of homoclinic tangencies with $\alpha = 0$ at $c > 0$ are "special" and it is necessary to consider at least two parameter general unfoldings (for example, with parameters μ and α).

In the paper we adhere to this approach and present in Sect. 2.4 our main results in the case under consideration (APMs on non-orientable manifolds).

---

[4] The rigorous proof requires quite delicate analytical considerations which are not presented here, see e.g. [20, 23, 24, 26].

2.4 Main Results

The following theorem can be considered as a new type of theorems on cascades of elliptic points in APMs close to those with homoclinic tangencies which takes place for APMs on non-orientable manifolds, and it is complementary to the symplectic variants established in [19, 20].

Theorem 1 (One parameter cascades of elliptic points in APMs on non-orientable manifolds)

Let f_0 be APM satisfying conditions A, B and D and f_μ be a one parameter family that unfolds generally (under condition C) at $\mu = 0$ the quadratic homoclinic tangency. For any interval $(-\mu_0, \mu_0)$ values of μ, there exists such integer and positive $\bar{k}$ that the following holds:

1. *All close to f_0 APMs have no single-round elliptic periodic orbits.*
2. *There exist intervals $\mathrm{e}_k^2 \subset I_\varepsilon$, $k = \bar{k}, \bar{k}+1, \ldots,$ which accumulate at $\mu = 0$ as $k \to \infty$ and such that the map f_μ has at $\mu \in \mathrm{e}_k^2$ a double-round elliptic orbit of period $2(k+q)$.*
3. *If $c < 0$ or $\alpha \neq 0$ in the case $c > 0$, then intervals e_k^2 accumulate at $\mu = 0$ as $k \to \infty$ and do not intersect for sufficiently large and different integer k.*
4. *Any interval e_k^2 has border points $\mu = \mu_k^-$ and $\mu = \mu_k^{2-}$ such that the map f_μ has a single-round periodic orbit with multipliers $+1$ and -1 for $\mu = \mu_k^-$ and a double-round periodic orbit with double multiplier -1 for $\mu = \mu_k^{2-}$.*
5. *The angular argument φ of the multipliers $e^{\pm i\varphi}$ of the elliptic points at $\mu \in \mathrm{e}_k^2$ depends monotonically on μ and the elliptic point is generic (KAM-stable) for all such μ, except for those where $\varphi(\mu) = \frac{\pi}{2}, \frac{2\pi}{3}, \arccos(-1/4)$.*

Note that Theorem 1 does not give answer on the question on a mutual position of the intervals e_k^2 in the critical case $\alpha = 0$. But this moment is quite important, since relates to the coexistence of elliptic orbits of different periods. Therefore, we assume now that f_0 is a map satisfying conditions A, B and D with $\alpha = 0$ and consider a two parameter family $\{f_{\mu,\alpha}\}$, a general unfolding for the initial tangency with $\alpha = 0$. Note that, as it follows from formula (10), α can vanish only in the case $c > 0$. Let D_ε be sufficiently small neighbourhoods (of diameter $\epsilon > 0$) of the origin in the parameter planes (μ, α).

It follows from Theorem 1 that in the case $c > 0$, there exist infinitely many open domains E_k^2 in D_ϵ such that the following holds.

- If $(\mu, \alpha) \in E_k^2$, then the map $f_{\mu,\alpha}$ has a double-round elliptic orbit of period $2(k+q)$.
- The domains E_k^2 accumulate at the axis $\mu = 0$ as $k \to \infty$.
- Any domain E_k^2 has two boundaries, bifurcation curves L_k^- and L_k^{2-}, which correspond, respectively, to the existence of a single-round nondegenerate periodic orbit with double multipliers $+1$ and a double-round nondegenerate periodic orbit with double multipliers -1.

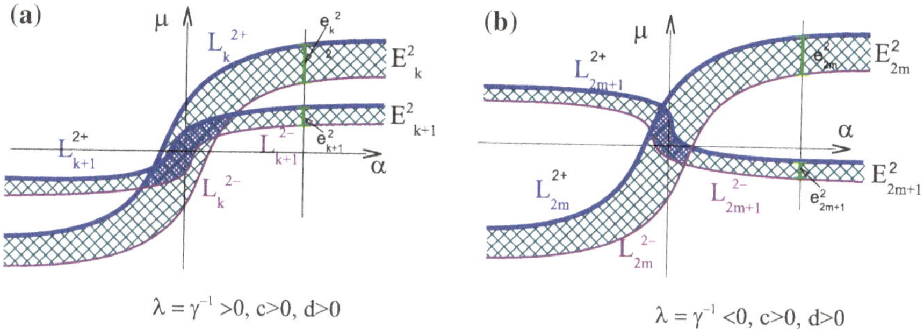

Fig. 6 Elements of the bifurcation diagrams for the families $f_{\mu,\alpha}$

Moreover, the following result takes place

Theorem 2 *All domains E_k^2 are mutually crossed and intersect the axis $\mu = 0$.*

In Fig. 6 some qualitative illustrations to Theorem 2 are shown for the cases where (a) $\lambda > 0, c > 0, d > 0$ and (b) $\lambda < 0, c > 0, d > 0$.

Introduce now the following quantity

$$s_0 = dx^+(ac + f_{20}x^+) - \frac{1}{4}(f_{11}x^+)^2 , \qquad (11)$$

which is calculated by coefficients of the global map T_1, see formula (7), and plays an important role in global dynamics of the map f_0 with $\alpha = 0$.

Theorem 2 shows that elliptic (double-round) periodic orbits of different periods can coexist when values of μ and α vary near zero. Moreover, infinitely many such orbits can coexist, in principle, at the global resonance $\mu = 0, \alpha = 0$. The following theorem give us sufficient conditions for this event.

Theorem 3 (On global resonance in APMs on non-orientable manifolds)
Let f_0 be APM satisfying conditions A, B and D. We assume also that the resonant condition $\alpha = 0$ takes place for f_0. Then, if s_0 is such that $-1 < s_0 < 0$, then there exists such integer $\bar{k}$ that f_0 has infinitely many double-round elliptic periodic orbits of all successive even periods $2(k + q)$, where $k \geq \bar{k}$. Moreover, if $s_0 \neq -\frac{1}{2}; -\frac{3}{4}; -\frac{5}{8}$, then all these elliptic orbits are generic (KAM-stable).

In the rest part of the paper we prove these and related results.

3 Rescaling Lemma

In principle, for the study of bifurcations in the first return map $T_k = T_1 T_0^k$ we could write it in the initial coordinates using formulas (7) and (2) for the maps T_1 and T_0^k, respectively, and, after, work with the obtained formulas. However, there is a more effective way for studying bifurcations. Namely, we can bring maps T_k to some unified form for all large k using the so-called rescaling method as it has been done in many papers.[5] After this, we can study (one time) bifurcations in the unified map and "project" obtained results onto the first return maps T_k for various k. This "universal" map is deduced in the following lemma.

Lemma 1 (Rescaling lemma for APMs with homoclinic tangencies on non-orientable manifolds)

Let a family f_ε under consideration satisfy condition A–D. Then, for every sufficiently large k, the first return map $T_k : \sigma_k^0 \to \sigma_k^0$ can be brought, by a linear transformation of coordinates and parameters, to the following form

$$\begin{aligned} \bar{X} &= Y + k\lambda^{2k}\varepsilon_k^1, \\ \bar{Y} &= M + X - Y^2 + \frac{f_{03}}{d^2}\lambda^k Y^3 + k\lambda^{2k}\varepsilon_k^2, \end{aligned} \tag{12}$$

where functions $\varepsilon_k^{1,2}(X, Y, M)$ are defined on some asymptotically big domain covering in the limit $k \to +\infty$ all finite values of X, Y and M, and these functions are uniformly bounded in k along with all derivatives up to order $(r-4)$); and the following formula takes place for M:

$$M = -d(1+\nu_k^1)\lambda^{-2k}\left(\mu + \lambda^k(cx^+ - y^-)(1 + k\beta_1\lambda^k x^+ y^-)\right) - s_0 + \nu_k^2\,; \tag{13}$$

the coefficient s_0 satisfies (11) and $\nu_k^{1,3} = O(\lambda^k)$, $\nu_k^{2,4} = O(k\lambda^k)$ are some asymptotically small coefficients.

Proof We will use the representation of the symplectic map T_0 in the "second normal form", i.e. in form (1) for $n = 2$.[6] Then the map $T_0^k : \sigma_k^0 \to \sigma_k^1$, for all sufficiently large k, can be written in the following form

$$x_k = \lambda^k x_0(1 + \beta_1 k\lambda^k x_0 y_k) + O(k^2\lambda^{3k}), \quad y_0 = \lambda^k y_k(1 + \beta_1 k\lambda^k x_0 y_k) + O(k^2\lambda^{3k}). \tag{14}$$

Then, using formulae (7) and (14), we can write the first return map $T_k : \sigma_k^0 \to \sigma_k^0$ in the following form

[5] see e.g. the papers [17, 19, 22–24, 26, 33, 34] in which the rescaling method was applied for the conservative case.

[6] Of course, we lose, a little, in a smoothness, since the second order normal form is C^{r-2} only. However, we get more principally important information on form of the first return maps. On the other hand, our considerations cover also C^∞ and real analytical cases.

$$\begin{aligned}\bar{x} - x^+ &= a\lambda^k x + b(y - y^-) + e_{02}(y - y^-)^2 + \\ &+ O(k|\lambda|^{2k}|x| + |y - y^-|^3 + |\lambda|^k |x||y - y^-|),\end{aligned}$$

$$\begin{aligned}&\lambda^k \bar{y}\left(1 + k\lambda^k \beta_1 \bar{x}\bar{y}\right) + k\lambda^{3k} O(|\bar{x}| + |\bar{y}|) = \\ &= \mu + c\lambda^k x\left(1 + k\lambda^k \beta_1 xy\right) + d(y - y^-)^2 + \lambda^{2k} f_{02}\, x^2 + \\ &+ \lambda^k f_{11}(1 + k\lambda^k \beta_1 xy)\, x(y - y^-) + \lambda^k f_{12}\, x(y - y^-)^2 + f_{03}(y - y^-)^3 + \\ &+ O\left((y - y^-)^4 + \lambda^{2k}|x||y - y^-| + k|\lambda|^{3k}|x| + k\lambda^{2k}|x||y - y^-|^2\right), \quad (15)\end{aligned}$$

where $x = x_0$, $y = y_k$.

Below, we will denote by α_k^i, $i = 1, 2, \ldots$, some asymptotically small in k coefficients such that $\alpha_k^i = O(k\lambda^k)$. Now we shift the coordinates

$$\eta = y - y^-, \quad \xi = x - x^+ - \lambda^k x^+(a + \alpha_k^1),$$

in order to nullify the constant term (independent of coordinates) in the first equation of (15). Thus, (15) is recast as follows

$$\begin{aligned}&\bar{\xi} = a\lambda^k \xi + b\eta + e_{02}\eta^2 + O\left(k\lambda^{2k}|\xi| + |\eta|^3 + |\lambda|^k O(|\xi||\eta|)\right), \\ &\lambda^k \bar{\eta}(1 + \alpha_k^2) + k\lambda^{2k} O(|\bar{\xi}| + \bar{\eta}^2) + k\lambda^{3k} O(|\bar{\eta}|) = \\ &\quad = M_1 + c\lambda^k \xi(1 + \alpha_k^3) + \eta^2(d + \lambda^k f_{12}x^+) + \lambda^k \eta(f_{11}x^+ + \alpha_k^4) + \lambda^k f_{11}\xi\eta + f_{03}\eta^3 + \\ &\quad + O\left(\eta^4 + k|\lambda|^{3k}|\xi| + k\lambda^{2k}(\xi^2 + \eta^2) + \lambda^k|\xi|\eta^2\right), \qquad (16)\end{aligned}$$

where

$$M_1 = \mu + \lambda^k(cx^+ - y^-)(1 + k\lambda^k \beta_1 x^+ y^-) + \lambda^{2k} x^+(ac + f_{02}x^+) + O(k\lambda^{3k}). \qquad (17)$$

Now, we rescale the variables:

$$\xi = -\frac{b(1 + \alpha_k^2)}{d + \lambda^k f_{12}x^+}\lambda^k u\,, \quad \eta = -\frac{1 + \alpha_k^2}{d + \lambda^k f_{12}x^+}\lambda^k v. \qquad (18)$$

System (16) in coordinates (u, v) is rewritten in the following form

$$\begin{aligned}\bar{u} &= v + a\lambda^k u - \frac{e_{02}}{bd}\lambda^k v^2 + O(k\lambda^{2k}), \\ \bar{v} &= M_2 + u(1 + \alpha_k^5) - v^2 + \\ &+ v(f_{11}x^+ + \alpha_k^6) - \frac{f_{11}b}{d}\lambda^k uv + \frac{f_{03}}{d^2}\lambda^k v^3 + O(k\lambda^{2k}), \qquad (19)\end{aligned}$$

where

$$M_2 = -\frac{d + \lambda^k f_{12} x^+}{1 + \alpha_k^2} \lambda^{-2k} M_1.$$

The following shift of coordinates (we remove the linear in v terms from the second equation)

$$u_{new} = u - \frac{1}{2}(f_{11}x^+ + \alpha_k^6), \; v_{new} = v - \frac{1}{2}(f_{11}x^+ + \alpha_k^6),$$

brings map (19) to the following form

$$\begin{aligned} \bar{u} &= v + a\lambda^k u - \frac{e_{02}}{bd}\lambda^k v^2 + O(k\lambda^{2k}), \\ \bar{v} &= M_3 + u - v^2 - \frac{f_{11}b}{d}\lambda^k uv + \frac{f_{03}}{d^2}\lambda^k v^3 + O(k\lambda^{2k}), \end{aligned} \tag{20}$$

where

$$M_3 = M_2 + \frac{(f_{11}x^+)^2}{4}.$$

Now, we make the following linear change of coordinates

$$x \;=\; u + \tilde{\nu}_k^1 v \;\; , \;\; y \;=\; v - \tilde{\nu}_k^2 u \; , \tag{21}$$

where

$$\tilde{\nu}_k^1 = -\frac{e_{02}}{bd}\lambda^k, \; \tilde{\nu}_k^2 = -\frac{e_{02}}{bd}\lambda^k - a\lambda^k. \tag{22}$$

Then, system (20) is rewritten as

$$\begin{aligned} \bar{x} &= y + M_3\tilde{\nu}_k^1 + O(k\lambda^{2k}), \\ \bar{y} &= M_3 + x - y^2 + a\lambda^k y - \tilde{R}\lambda^k xy + \frac{f_{03}}{d^2}\lambda^k y^3 + O(k\lambda^{2k}), \end{aligned} \tag{23}$$

where $\tilde{R} = (2a + 2e_{02}/bd - bf_{11}/d) \equiv 0$ by (9). Hence, map (23) has the following form

$$\begin{aligned} \bar{x} &= y + M_3\tilde{\nu}_k^1 + O(k\lambda^{2k}), \\ \bar{y} &= M_3 + x - y^2 + a\lambda^k y + \frac{f_{03}}{d^2}\lambda^k y^3 + O(k\lambda^{2k}), \end{aligned} \tag{24}$$

Finally, make one more shift of coordinates

$$X = x - \frac{1}{2}a\lambda^k - \tilde{\nu}_k^1 M_3, \;\; Y = y - \frac{1}{2}a\lambda^k,$$

in order to nullify in (24) the constant term in the first equation and the linear in y term in the second equation. After this, we obtain the final form (12) of map T_k in the rescale coordinates where formula (13) takes place for the parameter M.

4 Proofs of the Main Results

Bifurcations in first return maps T_k can be studied now with using their normal forms deduced by the rescaling Lemma 1. Since these normal forms coincide up to asymptotically small as $k \to \infty$ terms with the non-orientable conservative Hénon map, we recall in the next section some necessary results on bifurcations of fixed points in one parameter families of conservative Hénon map in non-orientable case.

4.1 *On Bifurcations of Fixed Points in the Conservative Non-orientable Hénon Maps*

Thus, the Rescaling Lemma 1 shows that the unified limit form for the first return maps T_k is the non-orientable and conservative Hénon map

$$\bar{x} = y, \ \bar{y} = M + x - y^2, \tag{25}$$

with the Jacobian $J = -1$. Bifurcations of fixed points in the conservative Hénon family are well-known.

Since the Hénon map (25) is not orientable, it can not have elliptic fixed points. However, elliptic points of period 2 exist for $M \in (0, 1)$. The map has no fixed points for $M < 0$, it has one fixed point $\bar{O}(0, 0)$ with multipliers $\nu_1 = +1, \nu_2 = -1$ at $M = 0$ and two saddle fixed points $(\bar{O}_1(-\sqrt{M}, -\sqrt{M})$ and $\bar{O}_2(\sqrt{M}, \sqrt{M}))$ at $M > 0$. Besides, an elliptic orbit of period 2 exists for $0 < M < 1$, it consists of two points $(p_1(-\sqrt{M}, \sqrt{M})$ and $p_2(\sqrt{M}, -\sqrt{M}))$; the value $M = +1$ corresponds to the period doubling bifurcation of this orbit. See Fig. 7 for an illustration. Note that the elliptic orbit of period 2 is generic for all $M \in (0, 1)$ except for $M = \frac{1}{2}$, $M = \frac{3}{4}$ which correspond to the strong resonances $1:4$ and $1:3$, respectively, and $M = \frac{5}{8}$ which corresponds to vanishing the first Birkhoff coefficient at the cycle $\{p_1, p_2\}$, see [14].

4.2 *Proof of Theorem 1*

The proof is deduced from the rescaling Lemma 1. Indeed, since bifurcations of fixed points of the Hénon map (25) are known, we can use this information directly for

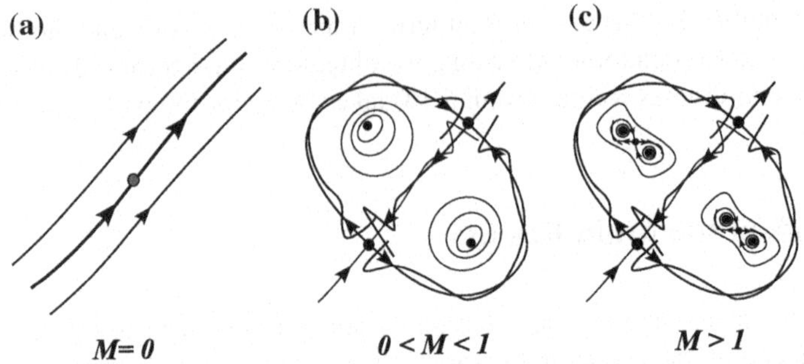

Fig. 7 The main bifurcation scenario in the non-orientable conservative Hénon map

recovering bifurcations of single-round periodic orbits in the family f_μ. We need only to know relations between the parameters of the rescaled map (12) and the initial parameters (i.e., in fact, between M and μ).

In the case under consideration, the relations between M and μ are given by formula (13) from which we find μ as follows

$$\mu = -\lambda^k y^- \alpha(1 + k\beta_1 \lambda^k x^+ y^-) - \frac{1}{d}(M + s_0 + \hat{\rho}_k^1)\lambda^{2k}, \tag{26}$$

where $\hat{\rho}_k^1 = O(k\lambda^k)$ is some small coefficient and $\alpha = \dfrac{cx^+}{y^-} - 1$ (see formula (10)).

As it follows from Lemma 1, the conservative non-orientable Hénon map $\bar{x} = y, \ \bar{y} = M + x - y^2$, where M satisfies (13), is normal (rescaled) form for the first return maps T_k with all sufficiently large k. This Hénon map has no elliptic fixed points, however, period 2 elliptic points exists for $0 < M < 1$. Thus, we obtain, by (13), that the first return map T_k has a fixed point with multipliers $\nu_1 = +1, \nu_2 = -1$ (i.e. when $M = 0$) if

$$\mu = \mu_k^{\pm} = -\lambda^k y^- \alpha(1 + k\beta_1 \lambda^k x^+ y^-) - \frac{1}{d}(s_0 + \hat{\rho}_k)\lambda^{2k}, \tag{27}$$

and a period 2 point with multipliers $\nu_1 = \nu_2 = -1$ (i.e. when $M = 1$) if

$$\mu = \mu_k^{2-} = -\lambda^k y^- \alpha(1 + k\beta_1 \lambda^k x^+ y^-) - \frac{1}{d}(s_0 + 1 + \hat{\rho}_k)\lambda^{2k}, \tag{28}$$

Thus, the first return map T_k has in this case a period two elliptic periodic point when $\mu \in \mathsf{e}_k^2$, where e_k^2 is the interval of values of μ with the border points $\mu = \mu_k^{\pm}$ and $\mu = \mu_k^{2-}$. Evidently, if $\alpha \neq 0$, the intervals e_k^2 with sufficiently large and different k do not intersect.

4.3 Proof of Theorem 2

By (27) and (28), the equations of bifurcation curves L_k^- and L_k^{2-}, which are boundaries of the domain E_k^2, can be written as follows

$$L_k^- : \ \mu = -\lambda^k y^- \left(\frac{cx^+}{y^-} - 1\right)(1 + k\beta_1 \lambda^k x^+ y^-) - \frac{s_0 + \dots}{d}\lambda^{2k}, \tag{29}$$

$$L_k^{2-} : \ \mu - \lambda^k y^- \left(\frac{cx^+}{y^-} - 1\right)(1 + k\beta_1 \lambda^k x^+ y^-) - \frac{1 + s_0 + \dots}{d}\lambda^{2k}. \tag{30}$$

Since $\lambda^{2k} \ll \lambda^k$, it means that the domains E_k^2 with sufficiently large k are not mutually intersect and do not intersect the axis $\mu = 0$, if $cx^+ \neq y^-$. Thus, the domains do not intersect always in the cases with $c < 0$. However, at the global resonance $\alpha = (cx^+/y^- - 1) = 0$ (which is possible only when $c > 0$), as it follows from (29) and (30), all domains E_k^2 with sufficiently large k are mutually intersect and all of them intersect the axis $\mu = 0$ (as in Fig. 6).

4.4 Proof of Theorem 3

Assume, for more definiteness, that $d > 0$ for all cases under consideration. The case $d < 0$ is treated in the same way. Let f_0 satisfies conditions A, B and D and has $\alpha = 0$. Then, for the one parameter family f_μ with fixed $\alpha = 0$, the intervals e_k^2 have, by (27) and (28), a form

$$\mathsf{e}_k^2 = (-1 - s_0, -s_0)\frac{\lambda^{2k}}{d}.$$

Evidently, if $-1 < s_0 < 0$, these intervals will be nested and containing $\mu = 0$. It implies that the diffeomorphism f_0 has infinitely many double-round elliptic periodic orbits.

As follows from Lemma 1, all the first return maps T_k (with sufficiently large k) are reduced to the same rescaled normal form–the non-orientable Hénon map $\bar{x} = y, \ \bar{y} = -s_0 + x - y^2$. It is well-known that, for $-1 < s_0 < 0$, the period 2 elliptic point of this map is generic if $s_0 \neq -\frac{1}{2}; -\frac{3}{4}; -\frac{5}{8}$. The exceptional cases relates, respectively, to resonances 1 : 4, 1 : 3 and such elliptic point (at $s_0 = -\frac{5}{8}$) whose the first Birkhoff coefficient is zero.

Acknowledgments The authors thank D.Turaev, L.Lerman and R.Ortega for fruitful discussions and remarks. This work has been partially supported by the Russian Scientific Foundation Grant 14-41-00044. S.G. has been supported partially by the grants of RFBR No.16-01-00324 and 14-01-00344. A.D. and M.G. have been partially supported by the Spanish MINECO-FEDER Grant MTM2012- 31714 and the Catalan Grant 2014SGR504.

References

1. Gavrilov, N.K., Shilnikov, L.P.: "On three-dimensional dynamical systems close to systems with a structurally unstable homoclinic curve".- I, Math. USSR Sbornik, **17**, 467–485 (1972); II, **19** 139–156 (1973)
2. Gonchenko, S.V.: "On stable periodic motions in systems close to a system with a nontransversal homoclinic curve".- Russian. Math. Notes **33**(5), 745–755 (1983)
3. Gonchenko, S.V., Shilnikov, L.P., Turaev, D.V.: "Dynamical phenomena in systems with structurally unstable Poincare homoclinic orbits".- Russian Acad. Sci. Dokl. Math. **47**(3), 410–415 (1993)
4. Palis, J., Viana, M.: "High dimension diffeomorphisms displaying infinitely many periodic attractors".-. Ann. Math. **140**, 207–250 (1994)
5. Gonchenko, S.V., Shilnikov, L.P., Turaev, D.V.: "Dynamical phenomena in systems with structurally unstable Poincare homoclinic orbits".- Interdisc. J. Chaos **6**(1), 15–31 (1996)
6. Gonchenko, S.V., Shilnikov, L.P., Turaev, D.V.: "On dynamical properties of multidimensional diffeomorphisms from Newhouse regions".-. Nonlinearity **20**, 923–972 (2008)
7. Gonchenko, S.V., Gonchenko, V.S.: "On bifurcations of birth of closed invariant curves in the case of two-dimensional diffeomorphisms with homoclinic tangencies". In: Proceedings of Math. Steklov Inst., vol. 244, Moscow (2004)
8. Gonchenko, S.V., Gonchenko, V.S.: "On Andronov-Hopf bifurcations of two-dimensional diffeomorphisms with homoclinic tangencies".- WIAS-preprint No.556, pp. 1–27. Berlin (2000)
9. Gonchenko, S.V., Markichev, A.S., Shatalin, A.E.: "On stable periodic orbits of two-dimensional diffeomorphisms close to a diffeomorphism with a non-transversal heteroclinic cycle". Rus. Diff. Eq. **37**, 205–215 (2001)
10. Gonchenko, S.V., Turaev, D.V., Shilnikov, L.P.: "On Newhouse regions of two-dimensional diffeomorphisms close to a diffeomorphism with a nontransversal heteroclinic cycle".-. Proc. Steklov Inst. Math. **216**, 70–118 (1997)
11. Gonchenko, S.V., Shilnikov, L.P., Sten'kin, O.V.: "On Newhouse regions with infinitely many stable and unstable invariant tori".- Proc. Int. Conf. Progress in Nonlinear Science. Nizhni Novgorod **1**, 80–102 (2002)
12. Gonchenko, S.V., Sten'kin, O.V., Shilnikov, L.P.: "On the existence of infinitely stable and unstable invariant tori for systems from Newhouse regions with heteroclinic tangencies".- Rus. Nonlinear Dyn. **2**(1), 3–25 (2006)
13. Lamb, J.S.W., Sten'kin, O.V.: "Newhouse regions for reversible systems with infinitely many stable, unstable and elliptic periodic orbits".-. Nonlinearity **17**, 1217–1244 (2004)
14. Delshams, A., Gonchenko, S.V., Gonchenko, V.S., Lazaro, J.T., Sten'kin, O.V.: "Abundance of attracting, repelling and elliptic orbits in two-dimensional reversible maps".-. Nonlinearity **26**(1), 1–35 (2013)
15. Gonchenko, S.V., Shilnikov, L.P., Turaev, D.V.: "Homoclinic tangencies of arbitrarily high orders in conservative and dissipative two-dimensional maps".-. Nonlinearity **20**, 241–275 (2007)
16. Newhouse, S.E.: "Quasi-elliptic periodic points in conservative dynamical systems". Amer. J. Math. **99**, 1061–1087 (1977)
17. Gonchenko, S.V., Shilnikov, L.P., Turaev, D.V.: "Elliptic periodic orbits near a homoclinic tangency in four-dimensional symplectic maps and Hamiltonian systems with three degrees of freedom".-. Regul. Chaotic Dyn. **3**(4), 3–26 (1998)
18. Gonchenko, S.V., Shilnikov, L.P., Turaev, D.V.: "Existence of infinitely many elliptic periodic orbits in four-dimensional symplectic maps with a homoclinic tangency". Proc. Steklov Inst. Math. **244**, 115–142 (2004)
19. Mora L., Romero, N.: "Moser's invariant curves and homoclinic bifurcations". Dyn. Sys. Appl. **6**, 29–42 (1997)
20. Gonchenko, S.V., Gonchenko, M.S.: On cascades of elliptic periodic points in two-dimensional symplectic maps with homoclinic tangencies. Regul. Chaotic Dyn. **14**(1), 116–136 (2009)

21. Biragov, V.S.: "Bifurcations in a two-parameter family of conservative mappings that are close to the Henon map". Sel. Math.Sov. **9**, 273–282 (1990). [Originally published in "Methods of qualitative theory of differential equations", Gorky State University, pp. 10–24 (1987)]
22. Biragov V.S., Shilnikov, L.P.: "On the bifurcation of a saddle-focus separatrix loop in a three-dimensional conservative system". Sel. Math. Sov. **11**, 333–340 (1992). [Originally published in "Methods of qualitative theory and theory of bifurcations", Gorky State University, pp. 25-34 (1989)]
23. Gonchenko, S.V., Shilnikov, L.P.: "On two-dimensional analytic area-preserving diffeomorphisms with infinitely many stable elliptic periodic points".-. Regul. Chaotic Dyn. **2**(3/4), 106–123 (1997)
24. Gonchenko, S.V., Shilnikov, L.P.: "On two-dimensional area-preserving diffeomorphisms with infinitely many elliptic islands". J. Stat. Phys. **101**, No.1/2, 321–356 (2000)
25. Gonchenko, S.V., Shilnikov, L.P.: "On two-dimensional area-preserving mappings with homoclinic tangencies". Dokl. Math. **63**(3), 395–399 (2001)
26. Gonchenko, S.V., Shilnikov, L.P.: "On two-dimensional area-preserving maps with homoclinic tangencies that have infinitely many generic elliptic periodic points". J. Math. Sci. **128**(2), 2767–2773 (2005). [Originally published in "Notes of Sankt-Pitersburg Steklov Math. Inst.", **300**, 155–166 (2003)]
27. Lerman, L.M.: "Homoclinic dynamics of Hamiltonian systems. Part 1: Introduction".-Moscow-Izhevsk, Inst. of Computer Investigations, p. 285 (2012)
28. Gonchenko, S.V., Shilnikov, L.P.: "Invariants of Ω-conjugacy of diffeomorphisms with a structurally unstable homoclinic trajectory". Ukr. Math. J. **42**(2), 134–140 (1990)
29. Shilnikov L.P., Shilnikov A.L., Turaev D.V., Chua L.O.: "Methods of Qualitative Theory in Nonlinear Dynamics, Part I". World Scientific, Singapore (1998)
30. Moser, J.: "The analytic invariants of an area-preserving mapping near a hyperbolic fixed point". Comm. Pure Appl. Math. **9**, 673–692 (1956)
31. Gonchenko, S.V., Shilnikov, L.P.: "Arithmetic properties of topological invariants of systems with a structurally unstable homoclinic trajectory". Ukr. Math. J. **39**(1), 21–28 (1987)
32. Gonchenko, S.V., Sten'kin, O.V., Turaev, D.V.: "Complexity of homoclinic bifurcations and Ω-moduli". Int. J. Bifurc. Chaos, **6**(6), 969–989 (1996)
33. Turaev, D., Rom-Kedar, V.: "Elliptic islands appearing in near-ergodic flows". Nonlinearity **11**(3), 575–600 (1998)
34. Rapoport, A., Rom-Kedar, V., Turaev, D.: "Approximating multi-dimensional hamiltonian flows by billiards".-. Comm. Math. Phys. **272**(3), 567–600 (2007)

A Compact Invariant Set for the Ricker Competition Model

Göran Högnäs

Abstract A compact invariant set inside the open first quadrant is constructed for the Ricker competition model (RCM)

$$F(x, y) = \begin{pmatrix} f_1(x, y) \\ f_2(x, y) \end{pmatrix} = \begin{pmatrix} xe^{r-x-by} \\ ye^{\tilde{r}-ax-y} \end{pmatrix}$$

under the condition of mutual invasibility.

Keywords Mutual invasibility · Coexistence · Time average of Ricker model

1 Introduction

The twodimensional Ricker competition model (RCM) is given by

$$F(x, y) = \begin{pmatrix} f_1(x, y) \\ f_2(x, y) \end{pmatrix} = \begin{pmatrix} xe^{r-x-by} \\ ye^{\tilde{r}-ax-y} \end{pmatrix} \tag{1}$$

where $x, y \geq 0$. (There are other equivalent formulations in the literature.) The parameters $r, \tilde{r}, a, b$ are assumed to be strictly positive, and $ab < 1$. In biological terms, x and y represent the normed densities of two species, $f_1(x, y)$ and $f_2(x, y)$ represent the densities of the two species a time unit (a generation, a year etc.) later, r and $\tilde{r}$ their carrying capacities, while a and b are competition parameters, see Luís et al. [1] for an extensive discussion with a large number of references.

The carrying capacity r of x is the equilibrium point of the x species in the absence of y. If a very small y (seen as a mutant, for example) can grow when x (the

G. Högnäs (✉)
Department of Mathematics, Center of Excellence in Optimization and Systems Engineering, Åbo Akademi University, 20500 Turku, Finland
e-mail: ghognas@abo.fi

L. Alsedà i Soler et al. (eds.), *Difference Equations, Discrete Dynamical Systems and Applications*, Springer Proceedings in Mathematics & Statistics 180, DOI 10.1007/978-3-662-52927-0_9

resident) is at equilibrium, we say that y can invade x. The condition for this is then $\tilde{r} - ar > 0$.

In this paper we will assume throughout that y can invade x and vice versa, x can invade y.

This *mutual invasibility condition*

$$r - b\tilde{r} > 0$$
$$\tilde{r} - ar > 0$$

turns out to be a coexistence condition as well: both species will survive in the long run. To be more exact: there is a compact set in the interior of the first quadrant which is F-invariant. It is noteworthy that this property does not depend on the stability properties of the fixed points or periodic points of F.

2 Preliminaries

The state space of the dynamical system generated by F is $\mathbf{R}_+^2 = \{(x, y)|x, y \geq 0\}$ but the effective state space X is

$$[0, x_{max}] \times [0, y_{max}]$$

where

$$x_{max} = \max_{x\geq 0} f_1(x, y) = f_1(1, 0) = e^{r-1}$$
$$y_{max} = \max_{y\geq 0} f_2(x, y) = f_2(0, 1) = e^{\tilde{r}-1}$$

since

$$f_1(x, y) \leq f_1(x, 0) = xe^{r-x} \text{ and } \frac{d}{dx} f_1(x, 0) = (x-1)e^{r-x}$$

and likewise for $f_2(x, y)$.

Fixed points for F are

$$(0, 0), (r, 0), (0, \tilde{r}) \text{ and } (x^*, y^*) := (\frac{r - b\tilde{r}}{1 - ab}, \frac{\tilde{r} - ar}{1 - ab}).$$

(x^*, y^*) lies in the open first quadrant by the mutual invasibility condition.

Our aim is now: To find a compact set C in the interior of $\mathbf{R}_+^2$ such that $F(C) \subset C$.

Later objective: Prove that a compact set C_1 can be chosen so that, for some natural number N, $F^N(C_1)$ is "far" inside C_1, or more precisely, $\mathbf{R}_+^2 \setminus C_1$ and $F^N(C_1)$ are at a positive distance from each other.

3 Construction of the Invariant Set

The actual construction of the compact invariant sets C and C_1 is done in several steps.

3.1 Step 1

There exists a positive number c and a triangle T of the form

$$\{(x, y) \in X \times X | x + y < c\}$$

such that, for some $\delta > 0$,

$$(x, y) \in T \Rightarrow f_1(x, y) + f_2(x, y) \geq e^{\delta}(x + y)$$

and $F(T^c) \subset T^c$, where $T^c = X \setminus T$.

3.2 Step 2

There exists a neighborhood D of the y-axis containing a strip

$$\{(x, y)|0 < x < \eta, 0 < d < y \leq y_{max}\}$$

where $0 < d < c$ and $d < f_2(0, y_{max}) = e^{\tilde{r}-1} \cdot \exp(\tilde{r} - e^{\tilde{r}-1})$ such that

$$F(D^c) \subset D^c.$$

3.3 Step 3

Similarly, there is a set E containing a strip of the form

$$\{(x, y)|0 < d' < x \leq x_{max}, 0 < y < \eta'\}$$

such that $F(E^c) \subset E^c$.

Our invariant set is then

$$(T \cup D \cup E)^c = T^c \cap D^c \cap E^c.$$

3.4 Proofs

In this subsection we prove the Steps 1–3 above.

3.4.1 Proof of Step 1

If $x + y < c_1$ for $x, y \geq 0$, then $ax + y, x + by < \max(1, a, b) \cdot c_1$. Take

$$2c_1 < \frac{\min(r, \tilde{r})}{\max(1, a, b)}.$$

Then

$$x + y < c_1 \Rightarrow r - x - by, \tilde{r} - ax - y > \frac{\min(r, \tilde{r})}{2} =: \delta > 0,$$

so

$$f_1(x, y) + f_2(x, y) = xe^{r-x-by} + ye^{\tilde{r}-ax-y} \geq xe^{\delta} + ye^{\delta} = (x + y)e^{\delta}.$$

Consider the set $\{(x, y) \in X | x + y \geq c_1\}$. The continuous function $f_1 + f_2$ has a minimum c_2 on this compact set. Let $c = \min(c_1, c_2)$. Then

$$T = \{(x, y) | x + y < c\}$$

has the property that $F(T^c) \subset T^c$:

If $c = c_2 < c_1$ then T^c consists of two parts $\{(x, y) | c_2 \leq x + y \leq c_1\}$, where the value of $f_1(x, y) + f_2(x, y)$ is at least $e^{\delta}(x + y)$, and $\{(x, y) | x + y \geq c_1\}$ where the minimum value of $f_1(x, y) + f_2(x, y)$ is c_2.

If $c = c_1 < c_2$ then $F(T^c)$ is clearly inside T^c, $F(T^c)$ is in fact at a positive distance from T.

Note. If we set $T_1 := \{(x, y) | x + y < c \cdot \mathrm{e}^{-\delta'}\}, 0 < \delta' \leq \delta$ then $F(T_1^c)$ is a subset of T^c and so it lies at a positive distance from T_1.

3.4.2 Proof of Step 2

Consider the dynamical system

$$\begin{aligned} x_{n+1} &= f_1(x_n, y_n) = x_n \exp(r - x_n - by_n) \\ y_{n+1} &= f_2(x_n, y_n) = y_n \exp(\tilde{r} - ax_n - y_n) \end{aligned}$$

where $n = 0, 1, 2, \ldots$ and (x_0, y_0) belongs to the invariant set determined in Step 1; in other words

$$(x_n, y_n) = F^n(x_0, y_0) \text{ with } (x_0, y_0) \in T^c, \tag{2}$$

where F^n is the nth iterate of F.

We have

$$\frac{x_n}{x_0} = \frac{x_n}{x_{n-1}} \cdot \frac{x_{n-1}}{x_{n-2}} \cdots \frac{x_1}{x_0} = e^{r-x_{n-1}-by_{n-1}} \cdots e^{r-x_0-by_0}$$

$$= \exp(nr - (x_0 + \cdots + x_{n-1}) - b(y_0 + \cdots + y_{n-1}))$$

and, analogously,

$$\frac{y_n}{y_0} = \exp(n\tilde{r} - a(x_0 + \cdots + x_{n-1}) - (y_0 + \cdots + y_{n-1})).$$

Further, from the second inequality,

$$-\log(\frac{y_n}{y_0}) + n\tilde{r} - a(x_0 + x_1 + \cdots + x_{n-1}) = y_0 + y_1 + \cdots + y_{n-1}. \tag{3}$$

Substituting this expression into the first equality, we get

$$\frac{x_n}{x_0} = \exp(n(r - b\tilde{r}) - (1 - ab)(x_0 + x_1 + \cdots x_{n-1}) + b\log(\frac{y_n}{y_0})) \tag{4}$$

Take a positive $\gamma < \tilde{r}$, $\gamma \ll 1$. Then we can find a d, $0 < d < \tilde{r} - \gamma$ such that

$$y_0 \in [d, y_{max}] \text{ and } 0 \leq ax_0 \leq \gamma \Rightarrow y_1 \in [d, y_{max}].$$

We will determine d below. (d must be taken smaller than the c of Step 1.)

From this we can draw the conclusion that $y_n \in [d, y_{max})$ *provided that* $a(x_0 + x_1 + \ldots x_{n-1}) \leq \gamma$. For such cases, then,

$$\frac{d}{y_{max}} \leq \frac{y_n}{y_0} \leq \frac{y_{max}}{d}$$

or

$$-\log(\frac{y_{max}}{d}) \leq \log(\frac{y_n}{y_0}) \leq \log(\frac{y_{max}}{d}).$$

Determination of d for different $\tilde{r}$ and γ.

Small $\tilde{r}$: $\tilde{r} \leq 1$, $0 < \gamma < \tilde{r}$.

We can take $d < \tilde{r} - \gamma$ because the maps $y \mapsto ye^{\tilde{r}-y}$ and $y \mapsto ye^{\tilde{r}-\gamma-y}$ are increasing on (0,1) so that

$$y_0 \in [d, 1] \Rightarrow y_0 e^{\tilde{r}-y_0} \in (d, y_{max}], y_0 e^{\tilde{r}-\gamma-y_0} \in (d, y_{max})$$

where $y_{max} = e^{\tilde{r}-1} \leq 1$.

Intermediate $\tilde{r}$: $\tilde{r} > 1, \tilde{r} - \gamma < 1$.

We can take

$$0 < d < \min(\tilde{r} - \gamma, y_{max} e^{\tilde{r}-\gamma-y_{max}}).$$

Large $\tilde{r}$: $\tilde{r} > 1, \tilde{r} - \gamma > 1$.

The maps $y \mapsto ye^{\tilde{r}-y}$ and $y \mapsto ye^{\tilde{r}-\gamma-y}$ are increasing on (0,1). $y_{max} > 1$. If we take $d \in (0, 1)$ and $d \leq y_{max} e^{\tilde{r}-\gamma-y_{max}}$ then

$$y_0 \in [d, y_{max}] \Rightarrow y_0 e^{\tilde{r}-\gamma-y_0} \in [d, y_{max}).$$

Now let $R_y := \frac{y_{max}}{d} >> 1$.

Then

$$\log R_y > \log(\frac{y_n}{y_0}) > -\log R_y$$

provided that $a(x_0 + x_1 + \cdots + x_n) < \gamma$.

Choose N such that

$$N(r - b\tilde{r}) - b \log R_y > 2\gamma \tag{5}$$

which is possible since $r - b\tilde{r} > 0$ by assumption. Also choose x_0 such that

$$x_0 + x_1 + \cdots + x_{N-1} < \min(\gamma, \frac{\gamma}{a}).$$

This can be accomplished by taking

$$x_0 < \min(\gamma, \frac{\gamma}{a})(\frac{e^{Nr}}{e^r - 1})^{-1} =: \eta_1$$

since

$$\frac{x_{i+1}}{x_i} < e^r \text{ and so } x_0 + x_1 + \cdots x_{n-1} \leq x_0(1 + e^r + \cdots e^{(n-1)r}) = x_0 \frac{e^{nr} - 1}{e^r - 1}$$

for any n.

Thus, using (3) and (4), we see that $0 < x_0 < \eta_1, d \leq y_0 \leq y_{max}$ implies that

$$\log(\frac{x_N}{x_0}) > 2\gamma - (1 - ab)(x_0 + x_1 + \cdots + x_{N-1}) > \gamma > 0$$

or

$$\frac{x_N}{x_0} > e^{\gamma} \tag{6}$$

Consider $D'' = \{(x, y) \in X | 0 < x_0 < \eta_1, d \leq y_0 \leq y_{max}\}$. Let η_2 be the minimum of the continuous function X_N for (x_0, y_0) in the compact set $(D'' \cup T)^c$. Recall that d was chosen strictly smaller than c of Step 1 so that the set does not touch the y-axis. Thus $\eta_2 > 0$. If $\eta \leq \min(\eta_1, \eta_2)$ then the strip $D' = \{(x, y) \in X | 0 < x_0 < \eta, d \leq y_0 \leq y_{max}\}$ has property that

$$(x_0, y_0) \in D' \Rightarrow \frac{x_N}{x_0} > e^{\gamma}$$

and $(x_0, y_0) \in (D' \cup T)^c$ implies $(x_N, y_N) \in (D' \cup T)^c$, i. e., $(D' \cup T)^c$ is F^N-invariant.

The compact set

$$\bigcup_{k=0}^{N-1} F^k((D' \cup T)^c)$$

is F-invariant. Let us call this set D^c.

As in Step 1 we can find a strip D_1 inside D such that $D_1 \cup T_1$ is at a positive distance from $F^N(D_1 \cup T_1)^c$. (We might need to decrease η and take $0 < \delta' \leq \min(\delta, \gamma)$ such that $d < c \cdot e^{-\delta'} - \eta$, though.)

3.4.3 The Final Steps

The proof of Step 3 is similar to that of Step 2. We obtain a set E containing a strip above the x-axis such that $(E \cup T)^c$ is F-invariant.

We also find a strip E_1 inside E such that $E_1 \cup T_1$ is at a positive distance from $F^N((E_1 \cup T_1)^c)$. (Clearly, the N can be chosen to be the same in Steps 2 and 3.)

Combining the previous steps we see that $D \cup E$ has the property that its complement $D^c \cap E^c$ is F-invariant.

Also, the complement of $D_1 \cup T_1 \cup E_1$ is mapped by F^N inside $D^c \cap E^c$. There is a positive distance between these two sets $D_1 \cup T_1 \cup E_1$, which is a neighborhood of the coordinate axes in X, and the compact $D^c \cap E^c$ in the interior of the first quadrant.

3.4.4 Note: The One-Dimensional Case

We see in our calculations in Step 2, see formula (3), that

$$\frac{y_n}{y_0} = \exp(n\tilde{r} - (y_0 + y_1 + \cdots + y_{n-1}))$$

when $x_0 = 0$. Thus

$$\frac{y_0 + y_1 + \cdots + y_{n-1}}{n} = \tilde{r} - \frac{1}{n}\log(\frac{y_n}{y_0}).$$

Since $\frac{y_n}{y_0}$ is bounded we get the following:

- The asymptotical time average of an one-dimensional Ricker process is *always* equal to its fixed point.
- In case y_0 is periodic with period n, then, of course,

$$\frac{y_0 + y_1 + \cdots + y_{n-1}}{n} = \tilde{r}.$$

Acknowledgments This research was presented at the 18th International Conference on Difference Equations and Applications (ICDEA 2012) in Barcelona. Fruitful discussions in the inspiring conference atmosphere with Rafael Luís, Eduardo Cabral Balreira and Saber Elaydi are gratefully acknowledged. I also want to thank the Magnus Ehrnrooth Foundation for travel support.

Reference

1. Luís, R., Elaydi, S., Oliveira, H.: Stability of a Ricker-type competition model and the competitive exclusion principle. J. Biolog. Dyn. **5**, 636–660 (2011)

On a Stochastic Ricker Competition Model

Göran Högnäs

Abstract We model the evolution of two populations U_t, V_t in competition by a two-dimensional size-dependent branching process. The population characteristics are assumed to be close to each other, as in a resident-mutant situation. Given that $U_t = m$ and $V_t = n$ the expected values of U_{t+1} and V_{t+1} are given by $me^{r-K(m+bn)}$ and $ne^{\tilde{r}-\tilde{K}(n+am)}$, respectively, where $r, \tilde{r}$ model the intrinsic population growth, $K, \tilde{K}$ model the force of inhibition on the population growth by the present population (such as scarcity of food), and a, b model the interaction between the two populations. For small $K, \tilde{K}$ the process typically follows the corresponding deterministic Ricker competition model closely, for a very long time. Under some conditions, notably a mutual invasibility condition, the deterministic model has a coexistence fixed point in the open first quadrant. The asymptotic behaviour is studied through the quasi-stationary distribution of the process. We initiate a study of those distributions as the inhibitive force $K, \tilde{K}$ approach 0.

Keywords Coexistence fixed point · Mutual invasibility · Demographic stochasticity · Quasi-stationary distribution (qsd) · Convergence of qsd's

1 Introduction

A general competition model between k different species put forward in [11] is

$$X_{t+1}^i = f_i(X_t, \xi_t)X_t^i, \quad 1, 2, \ldots k, \quad t = 0, 1, \ldots \tag{1}$$

where the state space $\mathbf{S}$ is a closed subset of $\Re_+^k$ and the coordinate axes form the extinction set $\mathbf{S_0}$; the ξ's represent a randomly evolving environment.

Since we concentrate on the two-dimensional case, $k = 2$, it will be more convenient to call the components X and Y, respectively. For a corresponding deterministic system we will use lower case x and y instead.

G. Högnäs (✉)
Department of Mathematics and Centre of Excellence in Optimization and Systems Engineering, Åbo Akademi University, 20500 Åbo, Finland
e-mail: ghognas@abo.fi

L. Alsedà i Soler et al. (eds.), *Difference Equations, Discrete Dynamical Systems and Applications*, Springer Proceedings in Mathematics & Statistics 180, DOI 10.1007/978-3-662-52927-0_10

There are many different f_i (growth factor of the ith species) proposed in the literature, e.g., the Leslie–Gower competition model (a generalized Beverton–Holt model) which in its deterministic form may be written

$$\begin{aligned} x_{t+1} &= b_1 \frac{1}{1 + x_t + c_1 y_t} x_t \\ y_{t+1} &= b_2 \frac{1}{1 + c_2 x_t + y_t} y_t. \end{aligned}$$

see [3]. Here we consider a *Ricker-type* function, first in a deterministic form

$$\begin{aligned} x_{t+1} &= x_t e^{r - x_t - b y_t} \\ y_{t+1} &= y_t e^{\tilde{r} - a x_t - y_t}, \end{aligned}$$

where a and b are taken close to 1, when the populations have largely the same characteristics, as in a resident-mutant situation.

In the terminology of (1), with the randomness suppressed, we have here $f_1(x, y) = e^{r-x-by}$ and $f_2(x, y) = e^{\tilde{r}-ax-y}$. Denote

$$F(x, y) = (x f_1(x, y), y f_2(x, y)) = (x e^{r-x-by}, y e^{\tilde{r}-ax-y}) \tag{2}$$

A very thorough study of the stability properties of this map F, the *Ricker competition model* RCM, is provided by [9], see also [1, 2].

The one-dimensional Ricker model, introduced in [10]

$$x_{t+1} = x_t e^{r - K x_t}, \quad t = 0, 1, 2, \ldots$$

models the evolution of the population density of a species.

- x_t is the density at time t or in the tth generation,
- r (or e^r) is the *birth rate, rate of natural increase,*
- K models the strength of an inhibitive factor, e.g., from internal competition for scarce resources (in Ricker's original paper derived from predation of the adult upon the young),
- the origin and the point $\frac{r}{K}$ are the fixed points of the system.

A key fact is: If r and x_0 are positive, the arithmetic averages of x_t of the one-dimensional Ricker model *always* approach $\frac{r}{K}$.

We note that if we pass to the normed process $K x_t$ ($K x$ being the new x) we can simplify the one-dimensional Ricker equation to

$$x_{t+1} = x_t e^{r - x_t}. \tag{3}$$

2 Demographic Stochasticity

In this section we introduce a two-dimensional competition model which is essentially the RCM as in the Introduction combined with *demographic stochasticity*.

Let q and $\tilde{q}$ be two offspring distributions with means e^r and $e^{\tilde{r}}$, respectively. We also assume that the have finite variances σ^2 and $\tilde{\sigma}^2$. Let U_t and V_t be two competing populations (modeled as branching processes) of the form

$$U_{t+1} = \sum_{j=1}^{U_t} \xi_{j,t}, \quad t = 0, 1, 2, \ldots \tag{4}$$

$$V_{t+1} = \sum_{j=1}^{V_t} \tilde{\xi}_{j,t} \quad t = 0, 1, 2, \ldots . \tag{5}$$

If $U_t = 0$ ($V_t = 0$) then we define $U_{t+1} = 0$ ($V_{t+1} = 0$). In the definitions above the ξ's and $\tilde{\xi}$'s are independent nonnegative random variables with distributions depending on the tth generation values:

$$\mathbf{P}\{\xi_{1,t} = k \mid U_t = m, V_t = n\} = \exp(-K(m + bn))q_k, \quad k = 1, 2, 3, \ldots$$
$$\mathbf{P}\{\tilde{\xi}_{1,t} = k \mid U_t = m, V_t = n\} = \exp(-\tilde{K}(am + n))\tilde{q}_k, \quad k = 1, 2, 3, \ldots$$

when m or n are positive integers. a models the relative influence of individuals of the U-species on the size of the litters $\tilde{\xi}$ of the V-species, and *vice versa* for b.

The corresponding conditional probability of $\xi_{1,t} = 0$ is given by the expression

$$1 - \exp(-K(m + bn))(1 - q_0) \tag{6}$$

and similarly for $\tilde{\xi}_{1,t}$.

(U_t, V_t) is a Markov chain on the nonnegative quadrant of $\mathbf{Z}^2$. In principle, all values are attainable in one step from any interior point. Thus the interior of $\mathbf{Z}_+^2$ constitutes one communicating aperiodic irreducible class.

The conditional means of the normed processes KU_{t+1} and $\tilde{K}V_{t+1}$ given $KU_t = x, \tilde{K}V_t = y$ are

$$x \exp\left(r - x - b\frac{K}{\tilde{K}}y\right), \quad y \exp\left(\tilde{r} - a\frac{\tilde{K}}{K}x - y\right) \tag{7}$$

showing close analogy with the deterministic cases (2, 3). Let us modify our definition of F in (2) slightly, replacing a by $a\frac{\tilde{K}}{K}$ and b by $b\frac{K}{\tilde{K}}$:

$$F(x, y) = \left(x \exp\left(r - x - b\frac{K}{\tilde{K}}y\right), y \exp\left(\tilde{r} - a\frac{\tilde{K}}{K}x - y\right)\right) \tag{8}$$

The conditional variance of KU_{t+1} is

$$Kx \exp\left(-x - b\frac{K}{\tilde{K}}y\right)\{\sigma^2 + e^{2r}(1 - e^{-x-b\frac{K}{\tilde{K}}y})\} \tag{9}$$

with a similar formula for $\tilde{K}V_{t+1}$.

Let us call the normed process (X_t, Y_t); $X_t = KU_t$, $Y_t = \tilde{K}V_t$. This is the two-dimensional stochastic Ricker competition model to be investigated in this paper.

For small values of K and $\tilde{K}$ the (X_t, Y_t) chain follows the path of the deterministic process with very high probability: The formula (9) shows that the standard deviations are of the order of $\sqrt{K}, \sqrt{\tilde{K}}$. Thus the process, whose both components are sums of independent random variables, is approximately normal with small standard deviation. More concretely, suppose we start the normed process at (x, y), then the next step is approximately normal with mean given by (7) and a standard deviation which can be made very small. As shown in [5, 8] we can infer the same thing for a finite number of steps, $(X_t, Y_t) \approx F^t(X_0, Y_0)$ for $t = 1, 2, \ldots k$, when $K, \tilde{K}$ are small enough. The difference is a sum of independent random variables and approximately bi-variate normal.

We see from (6) that the extinction probability of the X-process in one step given $X_t = x$, $Y_t = y$ is of the form

$$\left(1 - \exp\left(-x - b\frac{K}{\tilde{K}}y\right)(1 - q_0)\right)^{\frac{x}{K}} \geq \delta^{\frac{1}{K}} \tag{10}$$

where the positive number $\delta = 2^{-\log 2}$ is the minimum of the function $(1 - e^{-x})^x$.

Since the Y-component behaves similarly, we can infer that the extinction probability of our bivariate stochastic process in one step is at least

$$\delta^{\frac{1}{K}+\frac{1}{\tilde{K}}} \tag{11}$$

regardless of the present state of the process. Thus it has a finite life-time almost surely.

3 Mutual Invasibility

The two-dimensional deterministic system F defined by (8) has fixed points at the origin (corresponding to the case when both species are extinct), at the point $(r, 0)$ (the y-component is extinct and the x-component at its equilibrium), at $(0, \tilde{r})$ and possibly at

$$(x^*, y^*) = \left(\frac{r - b\frac{K}{\tilde{K}}\tilde{r}}{1 - ab}, \frac{\tilde{r} - a\frac{\tilde{K}}{K}r}{1 - ab}\right). \tag{12}$$

We will assume conditions which guarantee the existence of the fixed point in the open first quadrant. (x^*, y^*) is then called the *coexistence fixed point.*

Suppose $y_0 = \tilde{r}$ (seen, e.g., as the resident population density) and x_0 (the mutant population density) is very small. Then the first component x_1 of $(x_1, y_1) = F(x_0, y_0)$ is larger than x_0 if the condition $r > b\frac{K}{\tilde{K}}\tilde{r}$ is met. The interpretation is that the mutant x can increase from small values while the resident population is at its equilibrium. We say that x can *invade* y. Conversely, y can invade x if $\tilde{r} > a\frac{\tilde{K}}{K}r$.

We will impose two conditions:

$$ab < 1 \tag{13}$$

and the *mutual invasibility condition*

$$r > b\frac{K}{\tilde{K}}\tilde{r}, \quad \tilde{r} > a\frac{\tilde{K}}{K}r. \tag{14}$$

It is immediately seen that this is sufficient to ensure the existence of a coexistence fixed point (x^*, y^*) in the open first quadrant. The fixed point may be repelling or attracting; the conditions can be found in [9].

Remark If the population characteristics are very close to each other one could well choose $a = b = 1$. In [4] we investigated such a case: there is no coexistence fixed point so one of the two populations will eventually take over completely.

Without regard to the attractivity properties of the fixed points, the conditions (13, 14) guarantee the existence of an invariant compact set C inside the open first quadrant: $F(C) \subset C$. For the detailed calculations, see [6].

In [6] it is also proved that there exists a compact set C_1 such that it is mapped "far inside" C_1 by some iterate F^N of F. To be more precise, there is a compact set C_1 inside the open first quadrant such that the Euclidean distance between the outside of C_1 and $F^N(C_1)$ is strictly positive:

$$d(C_1^c, F^N(C_1)) > 0. \tag{15}$$

Another way of putting it is to say that $F^N(C_1)$ is farther away from the coordinate axes than C_1.

4 The Quasi-Stationary Distribution

The analysis of the process (U_t, V_t) and its normed version (X_t, Y_t) follows the outline in [5].

The Markov chain (U_t, V_t) in the interior of the first quadrant $\mathbf{Z}_+^2$ was found above to have a finite lifetime a.s., see (11). For any (m, n), both components positive, the probability of moving straight to the origin is the product of expressions such as (6):

$$(1-\exp(-K(m+bn))(1-q_0))^m(1-\exp(-\tilde{K}(am+n))(1-\tilde{q}_0))^n$$

which can be minorized by an expression of the form

$$(1-\exp(-L(m+n))(1-\bar{q}_0))^{m+n} \approx 1-(m+n)\exp(-L(m+n))(1-\bar{q}_0) \tag{16}$$

where L is taken such that $L(m+n) \le \min(K(m+bn), \tilde{K}(am+n))$ and $\bar{q}_0 = \max(q_0, \tilde{q}_0)$.

Following the corresponding proof in [5] we can now draw the conclusion that the entries of the transition probability matrix Q of the Markov chain restricted to the strictly positive states (m, n) have a finite sum. The sub-markovian transition matrix has a spectral radius (the Krein–Rutman maximal eigenvalue λ) bounded above by an expression of the form

$$1-\delta^{\frac{1}{K}+\frac{1}{\tilde{K}}}$$

where δ again is the strictly positive minimum of the function $(1-e^{-x})^x$. This is a rough upper bound because the complement of the all-positive states is the union of the non-negative coordinate axes, and not just the origin $(0, 0)$.

The Markov chain restricted to the strictly positive states admits a *quasi-stationary distribution* (qsd) π. One of the characterizations of the qsd is that it is the left eigenvector corresponding to the dominant eigenvalue λ:

$$\lambda\pi = \pi Q.$$

Another is the following invariance property: If the initial distribution is π then the conditional distribution of the first step given that it is all-positive is π, too:

$$\mathbf{P}_\pi\{(U_1, V_1) = (\cdot,\cdot) \mid U_1 > 0, V_1 > 0\} = \pi(\cdot,\cdot)$$

where we use the notation $\mathbf{P}_\pi$ to indicate that the initial distribution (the distribution of (U_0, V_0)) is π.

5 Asymptotics

Our asymptotic analysis follows the reasoning in [5] closely. The quasi-stationary distribution π depends on K and $\tilde{K}$. (We assume a and b to be constant.) As K and $\tilde{K}$ go to 0 the family of probability measures $\pi_{K,\tilde{K}}$ is *tight* provided K and $\tilde{K}$ vary boundedly in such a way that the fixed point (x^*, y^*) in (12) stays in a compact set within the interior of the first quadrant. For simplicity we take $K = \tilde{K}, 0 < K < 1$, from now on.

Theorem 1 *Let* $K = \tilde{K}$. *Let (13, 14) be satisfied. Consider the Markov chains* (X_t, Y_t) *for* $K > 0$ *and their quasi-stationary distributions* π_K. *The family of probability measures* π_K, $K > 0$, *is tight.*

Proof Recall that the conditional mean of (X_{t+1}, Y_{t+1}) given that $(X_t, Y_t) = (x, y)$ is $F(x, y) \equiv (xf_1(x, y), yf_2(x, y))$, (2), and the variance v_x of its first component is

$$Kx \exp(-x - by)\{\sigma^2 + e^{2r}(1 - e^{-x-by})\} \tag{17}$$

with a similar formula for the second component, see (9).

Following [5], Sect. 3, let $S(s)$ be the moment generating function (mgf) of the offspring distribution q and $c_{(x,y)}(s)$ the mgf of the variables $\xi_{j,t}$ of (4) given that $X(t) = x$ and $Y(t) = y$. The entropy function $c^*_{(x,y)}(z)$ of $\xi_{j,t}$ is $\sup_{s \in \Re}[zs - c_{(x,y)}(s)]$. As in [5], pp. 252–253, we get, for any positive a,

$$\mathbf{P}\{X_{t+1} > a + xf_1(x, y) \mid X_t = x, Y_t = y\} \leq \exp\left(-s\left(\frac{a}{K} - \frac{x}{K} f_1(x, y)\right) + \frac{x}{K} c_{(x,y)}(s)\right)$$

which is at most

$$\exp\left(-\frac{x}{K} c^*_{(x,y)}\left(\frac{a}{x} + e^{r-x-by}\right)\right).$$

The entropy function is 0 at the conditional mean e^{r-x-by} of $\xi_{j,t}$ and at least $\frac{1}{3v_x}(\frac{a}{x})^2$ for small a.

Since the variance varies continuously with x and y we can conclude that the exponent can be bounded away from 0 when (x, y) lies in a compact set in the interior of the first quadrant.

Thus, given a small positive a, there exists a $w > 0$ such that

$$\mathbf{P}\{X_{t+1} > xf_1(x, y) + a \mid X_t = x, Y_t = y\} \leq \exp\left(-\frac{w}{K}\right) \tag{18}$$

(and similarly for Y_{t+1}) for all (x, y) in a given compact set in the open first quadrant.

Putting it in another way, we have

$$\mathbf{P}\{(X_{t+1}, Y_{t+1}) \in U(F(x, y)) \mid (X_t, Y_t) = (x, y)\} \geq 1 - \exp\left(-\frac{w}{K}\right) \tag{19}$$

where $U(F(x, y))$ is a neighbourhood of the point $F(x, y)$.

If we take as our compact set an invariant set C, such as the one constructed in [6], and as $U(C)$ an open set containing C we get

$$\mathbf{P}\{(X_{t+1}, Y_{t+1}) \in U(C) \mid (X_t, Y_t) \in C\} \geq 1 - \exp\left(-\frac{w}{K}\right). \tag{20}$$

Since F is continuous we can extend the above relation, by appropriately adjusting the quantity $w > 0$, to the Nth iterate of F:

$$\mathbf{P}\{(X_{t+N}, Y_{t+N}) \in U(C) \mid (X_t, Y_t) \in C\} \geq 1 - \exp\left(-\frac{w}{K}\right). \tag{21}$$

Now we let N and C_1 be as in [6], i.e., the interior of C_1 is an open set containing $F^N(C_1)$. Then we obtain

$$\mathbf{P}\{(X_N, Y_N) \in C_1 \mid (X_0, Y_0) \in C_1\} \geq 1 - \exp\left(-\frac{w}{K}\right) \tag{22}$$

from which we can draw the conclusion, as in [5], that the dominant eigenvalue $\lambda = \lambda(K)$ of the transition probability matrix of the Markov chain (X_t, Y_t) restricted to the positive quadrant of $\Re^2$, differs "exponentially little" from 1:

$$1 - \exp(-\frac{u}{K}) < \lambda(K) < 1 \tag{23}$$

for some $u > 0$.

Starting the Markov chain from the point $(X_0, Y_0) = (x, y)$ the distribution of (X_1, Y_1) has mean $F(x, y)$ and a uniformly bounded variance, see (17). Hence there is a compact set $B \subset \Re^2$ such that the probability of (X_1, Y_1) being outside of B is $< \varepsilon$ for all (x, y) and all K. As in [5] we conclude that $\pi_K(B^c) < 2\varepsilon$.

In a strip S close to the y-axis, the function $(1 - e^{-x})^x$ approaches 1, it is $> e^{-v}$ for some small positive v. Take v so small that $0 < v < u$ where u satisfies (23) above. The probability of immediate exit from the state space (the open first quadrant) is at least $e^{-\frac{v}{K}}$ starting from a point in S. If the initial point is chosen according to law π_K then the exit probability is $1 - \lambda(K)$ (by the definition of qsd). As in [5] we use (23) to conclude that

$$\pi_K(S) < \exp\left(-\frac{u - v}{K}\right)$$

which can be made $< \varepsilon$ for K small enough. A similar reasoning can be applied to a strip above the x-axis.

The above remarks show that, for given positive ε, there exists a compact set inside the open first quadrant whose π_K-measure is $> 1 - 4\varepsilon$ for all K small enough.

Thus the family π_K, $K > 0$, is tight.

Theorem 2 *The limit measures of π_K as $K \to 0$ are invariant with respect to the deterministic system F. If the coexistence fixed point (x^*, y^*) of (12) is attracting then the quasi-stationary distributions converge to the point mass at the fixed point:*

$$\lim_{K \to 0} \pi_K = \delta_{(x^*, y^*)}.$$

Proof (Sketch) The invariance of the limit measures is proved as in Step 5 of [5], pp. 254f. We know, see [9], that the coexistence fixed point (x^*, y^*) is attracting if $2(1-b)\tilde{r} + 2(1-a)r - 4(1-ab) \leq (r - b\tilde{r})(\tilde{r} - ar) < (1-b)\tilde{r} + (1-a)r$. Also, by [2], if this condition is satisfied it is globally attracting so that the point mass at (x^*, y^*) is the unique invariant measure for the deterministic system F inside the open first quadrant of $\Re^2$.

6 Conclusion and Outlook

Theorems 1 and 2 show that the branching process (U_t, V_t) and its normed version (X_t, Y_t), even if it eventually goes extinct, emulates the deterministic process for a long time. The dominant eigenvalue is within $e^{-\frac{u}{K}}$ of 1 so the expected lifetime

$$\frac{1}{1-\lambda(K)}$$

of the process is exponential in $\frac{1}{K}$. When the coexistence fixed point (x^*, y^*) is attracting we can say more: the quasi-stationary distribution is a "blurred version" of the point mass at (x^*, y^*). The linear autoregressive approximation of (X_t, Y_t) in the vicinity of the fixed point, which was introduced by [8] in a similar one-dimensional case, can be used as in [7] to obtain rough estimates of the quantity u in (23). This in turn gives us an estimate for the expected life-time.

If the coexistence fixed point is repelling, the picture is less clear. We know by Theorem 1 that the qsd's do have limits but their precise nature is unknown. Simulations suggests a conjecture for $r, \tilde{r}$ not too large: the limits will be supported by the attracting cycles of the deterministic system F (as in the one-dimensional Ricker case, see [5]). A successful completion of this program is likely to require a thorough understanding of the two-dimensional dynamical system generated by F.

Acknowledgments The author has worked on Ricker competition models with Henrik Fagerholm, Mats Gyllenberg, and Brita Jung. The present paper was presented at the International Conference on Difference Equations and Applications ICDEA 2012 in Barcelona. Many discussions took place in the constructive and inspiring conference atmosphere. In particular, questions and comments by Rafael Luís, Eduardo Cabral Balreira, and Saber Elaydi are gratefully acknowledged. Travel funding was provided by the Magnus Ehrnrooth Foundation.

References

1. Balreira, E.C., Elaydi, S., Luís, R.: Geometric methods for global stability in the Ricker competition model. In: ICDEA 2012 18th International Conference on Difference Equations and Applications, Abstracts Book, pp. 96–97
2. Balreira, E.C., Elaydi, S., Luís, R.: Local stability implies global stability in the Ricker competition model. In: ICDEA 2012 18th International Conference on Difference Equations and Applications, Abstracts Book, pp. 102–103

3. Cushing, J.M., Levarge, S., Chitnis, N., Henson, S.M.: Some discrete competition models and the competitive exclusion principle. J. Differ. Equ. Appl. **10**(13–15), 1139–1151 (2004)
4. Fagerholm, H., Gyllenberg, M., Högnäs, G.: Competition and invasion in stochastic population models. Åbo Akad. Reports Comput. Sci. Math., Ser. A **191**, 1–27 (2007)
5. Högnäs, G.: Quasi-stationary distributions for a stochastic Ricker model. Stoch. Process. Appl. **70**, 243–263 (1997)
6. Högnäs, G.: A compact invariant set for the Ricker competition model. In: Alsedà, Ll., Cushing, J., Elaydi, S., Pinto, A. (eds.) Discrete Dynamical Systems and Applications, Springer Proceedings in Mathematics, vol. 180, 127–134 (2016)
7. Jung, B.: Exit times for multivariate autoregressive processes. Stoch. Process. Appl. **123**, 3052–3063 (2013). arXiv:1211.2085 [math.PR]
8. Klebaner, F.C., Nerman, O.: Autoregressive approximation in branching processes with a threshold. Stoch. Process. Appl. **51**, 1–7 (1994)
9. Luís, R., Elaydi, S., Oliveira, H.: Stability of a Ricker-type competition model and the competitive exclusion principle. J. Biol. Dyn. **5**, 636–660 (2011)
10. Ricker, W.E.: Stock and recruitment. J. Fish. Res. Board Can. **11**, 559–623 (1954)
11. Schreiber, S.J.: Persistence for stochastic difference equations: a mini-review. J. Differ. Equ. Appl. **18**(8), 1381–1403 (2012)

A Repeated Strategy for Dumping

J. Martins, N. Banik and Alberto A. Pinto

Abstract In this work, we study the phenomena of dumping in a duopoly market through an infinitely repeated game. We consider two firms of different countries competing in the same country. When both firms are cooperating, if the foreign firm deviates from cooperation this can be interpreted as dumping and a period of punishment can be imposed to the foreign firm. After this, firms can play continuously the deviation-punishment game or compete à la Cournot. Previously, we observe that the repeated strategy of deviation-punishment is not adopted in the case of symmetric demand equations. Here, we observe that this strategy of repeated dumping can appear as the best repeated strategy when the demand equations are non-symmetric.

Keywords Dumping · Duopoly · Collusion · Cournot

1 Introduction

The phenomena of dumping appear when a firm practices a price for a certain good in the foreign market lower than the price charged for the same good in the domestic market [14]. Since this can be seen as selling at less than the fair value, the dumping is assumed to be an unfair practice in international trade. This type of pricing policy

J. Martins (✉)
LIAAD-INESC TEC; Department of Mathematics, School of Technology and Management, Polytechnic Institute of Leiria, Campus 2, Morro do Lena - Alto do Vieiro, 2411-901 Leiria, Portugal
e-mail: jgouveia.martins@gmail.com

N. Banik
Center for Advance Financial Studies, Institute for Financial Management and Research, Chennai 600034, India
e-mail: nilbanik@gmail.com

A.A. Pinto
LIAAD-INESC TEC; Faculty of Sciences, Department of Mathematics, University of Porto, Rua do Campo Alegre, 687, 4169-007 Porto, Portugal
e-mail: aapinto1@gmail.com

L. Alsedà i Soler et al. (eds.), *Difference Equations, Discrete Dynamical Systems and Applications*, Springer Proceedings in Mathematics & Statistics 180, DOI 10.1007/978-3-662-52927-0_11

is frequently associated with a deliberate action of large companies to eliminate competition in foreign markets, in order to consolidate as monopolies. Hence, this phenomena is highly related with companies that have a huge productive capacity. To protect the domestic industry, many governments developed anti-dumping laws to impose penalties on suspiciously low-priced imports. To govern the application of anti-dumping measures there exist the "anti-dumping agreement", created by the World Trade Organization (WTO). Te WTO aims to supervise and liberalize international trades, promoting the negotiation between members to ensure a freely and fair trade between nations.

Dumping is studied in this work as a phenomena that happens in an international trade between two nations. Hence, we consider two firms of different countries competing in quantities of a certain good at the same country [12]. The trade runs infinitely, in a discrete number of periods of a game that have three different strategies: collusion, deviation from collusion followed by punishment and Cournot. In the collusion strategy both firms cooperate in their mutual benefit and produce the quantities that maximize the joint profit. After a period of the game when both firms play collusion, the foreign firm might deviates from collusion abandoning the cooperation between both. This deviation can be understood as dumping since the price practiced by the foreign firm decreases in the period of deviation. As consequence of deviation, the foreign firm might suffer a punishment in the period after the deviation if the dumping is proved. The punishment results from a penalty imposed by the government of the home country, usually, as a tariff per unit of the good. We assume that this tariff puts the foreign firm out of the market in the punishment period and, therefore, the home firm realizes the monopoly profit in this period of the game. To impose anti-dumping duties, the home firm have to lobby its government by a certain price that will decrease its own profit in the punishment period. There are two ways in which the domestic firm can induce its government to impose a tariff. First, the domestic firm can strategically alter its behavior (trying to induce the deviation of the foreign firm) and thereby influence anti-dumping outcome in the following stage of the game (see Ethier and Fischer [5], Fischer [6], Staiger and Wolak [13] and Reitzes [11]). Second, by mounting political pressure (see Moore [9, 10], DeVault [2], and Hansen and Prusa [7, 8]). After the two periods of the game, when the foreign firm deviates from collusion and suffer the punishment, the home firm has two possible repeated strategies to adopt: allow the repetition of the strategies taken by both firms in the previous two periods, originating the repeated deviation-punishment strategy, or force a Cournot competition where the firms produce the amount of output independently of each other to maximize its own profits, originating the deviation-punishment followed by Cournot strategy. To implement this repeated game we developed an economical model in [1] and observe that the optimal strategies in the symmetric case of the model are repeated collusion and deviation-punishment followed by Cournot. Here, we consider the non-symmetric model and observe that the repeated deviation-punishment strategy should also be adopted for some parameter values.

2 The Economic Model

We consider an international trade market constituted by a duopoly, where one firm from the home country is competing in quantities of production for a certain good with another firm from a foreign country. Let F_1 denote the home firm and F_2 the foreign firm. Let q_i denote the produced quantities and p_i the selling prices for firm F_i, $i = 1, 2$. We assume a quadratic utility function that origins the linear inverse demand functions [3, 4, 15]

$$\begin{aligned} p_1 &= \alpha_1 - \beta_1 q_1 - \gamma q_2 \\ p_2 &= \alpha_2 - \gamma q_1 - \beta_2 q_2 \end{aligned}, \tag{1}$$

with $\alpha_i > 0$, $\beta_i > 0$ and $\beta_1\beta_2 \geq \gamma^2$. The parameters α_i represents the maximum price that anyone would pay for the good and β_i measures the negative relationship between the quantity demanded and the price. The parameter γ measures the degree of substitutability of the goods: these can be substitutes if $\gamma > 0$, independent if $\gamma = 0$ or complements if $\gamma < 0$. The goods are perfect substitutes when $\beta_1 = \beta_2 = \gamma$ and perfect complements when $\beta_1 = \beta_2 = -\gamma$. If $\alpha_1 = \alpha_2$ and $\beta_1 = \beta_2 = \gamma$ the model do not distinguish the goods produced by the two firms and therefore we say that they are identical. With these notations and neglecting the marginal costs, the profit for the firm F_i is given by

$$\pi_i = p_i q_i = (\alpha_i - \beta_i q_i - \gamma q_j) q_i\,, \quad i = 1, 2. \tag{2}$$

3 The Dumping Game

Now, we consider an infinitely repeated game where both firms have to choose in every periods of the game the quantities of the goods that will be produced.

The first strategy that we consider is collusion. This strategy corresponds to a cooperation in mutual benefit of both firms in order to maximize the joint profit $(\pi_1 + \pi_2)$. If both firms adopt a collusion strategy in every periods of the game, namely, the *repeated collusion strategy (COL)*, the total value of the profit realized by the home firm is given by

$$\pi_1^{COL} = \frac{\alpha_1(\alpha_1\beta_2 - \alpha_2\gamma)}{4(\beta_1\beta_2 - \gamma^2)}, \tag{3}$$

and given by

$$\pi_2^{COL} = \frac{\alpha_2(\alpha_2\beta_1 - \alpha_1\gamma)}{4(\beta_1\beta_2 - \gamma^2)}, \tag{4}$$

for the foreign firm, as derived in [1]. To simplify the expressions of the profits, Eqs. (3), (4) and the equations of the profits in the next repeated strategies are multiplied by the factor $(1-\delta)$, where $\delta \in (0;1)$ denotes the rate of discount.

Now, we assume that, after a certain period of the game when both firms play a collusion strategy, the foreign firm deviates from collusion in order to maximize its own profit. If it is proved that this deviation is dumping, a following period of punishment can be imposed to the foreign firm. The punishment results from the home firm lobby its own government to impose a prohibitive tariff on the foreign firm during the period of punishment. After these two periods of deviation and punishment of the foreign firm, we assume that the home firm might allow the repetition of this deviation-punishment strategy, resulting in the *repeated deviation-punishment strategy (DPR)*, or might force a Cournot competition, resulting in the *Deviation-Punishment followed by a Cournot strategy (DPC)*. In the case of the repeated deviation-punishment strategy, it was derived in [1] that

$$\pi_1^{DPR} = \frac{\left(2\alpha_1\beta_1\beta_2^2 - 3\beta_2\alpha_1\gamma^2 + \alpha_2\gamma^3\right)\frac{\beta_2\alpha_1-\alpha_2\gamma}{8\beta_2(\beta_1\beta_2-\gamma^2)^2} + \delta\left(\frac{\alpha_1^2}{4\beta_1} - L\right)}{1+\delta}, \tag{5}$$

is the total value of the profit for the home firm, where L is the price payed by the home firm to lobby its government. For the foreign firm, the total value of the profit is given by

$$\pi_2^{DPR} = \frac{\beta_2}{1+\delta}\left(\frac{2\alpha_2\beta_1\beta_2 - \alpha_2\gamma^2 - \gamma\beta_2\alpha_1}{4\beta_2(\beta_1\beta_2-\gamma^2)}\right)^2. \tag{6}$$

In the case of both firms play a deviation-punishment strategy in the first two periods of the game and afterwards play a Cournot competition, the total profit for the home firm is given by

$$\begin{aligned}\pi_1^{DPC} = (1-\delta)\Bigg(&\left(2\alpha_1\beta_1\beta_2^2 - 3\beta_2\alpha_1\gamma^2 + \alpha_2\gamma^3\right)\frac{\beta_2\alpha_1-\alpha_2\gamma}{8\beta_2(\beta_1\beta_2-\gamma^2)^2}\\ &+\delta\left(\frac{\alpha_1^2}{4\beta_1} - L\right)\Bigg) + \delta^2\beta_1\left(\frac{2\alpha_1\beta_2-\gamma\alpha_2}{4\beta_1\beta_2-\gamma^2}\right)^2,\end{aligned} \tag{7}$$

and the total profit for the foreign firm is given by

$$\begin{aligned}\pi_2^{DPC} = (1-\delta)\beta_2&\left(\frac{2\alpha_2\beta_1\beta_2 - \alpha_2\gamma^2 - \gamma\beta_2\alpha_1}{4\beta_2(\beta_1\beta_2-\gamma^2)}\right)^2\\ &+\delta^2\beta_2\left(\frac{2\alpha_2\beta_1-\gamma\alpha_1}{4\beta_1\beta_2-\gamma^2}\right)^2.\end{aligned} \tag{8}$$

4 Main Results and Discussion

The action of make dumping corresponds to the deviation from collusion practiced by the foreign firm. Hence, the foreign firm starts to decide between deviation-punishment strategy or maintain the collusion strategy. If the foreign firm chooses the deviation-punishment strategy, then the home firm will decide to play a repeated deviation-punishment strategy or to force a Cournot competition. To discover which strategy leads to a higher profit and therefore the best repeated strategy for both firms, we will compare the previous expressions of the profits. The case of symmetric model, parameters $\alpha_1 = \alpha_2$ and $\beta_1 = \beta_2$, was studied in [1] and the repeated deviation-punishment strategy never was the best repeated strategy. Now, we will show that this strategy appears as best repeated strategy for both firms if the parameter values do not make the model symmetric.

To decide between make dumping or maintain collusion, the foreign firm starts to observe what the home firm will choose to do in the periods after the deviation-punishment. If the home firm prefer repeat the deviation-punishment rather than Cournot competition and if the foreign firm prefer repeat the deviation-punishment strategy rather than collusion

$$\pi_2^{DPR} > \pi_2^{COL} \quad \wedge \quad \pi_1^{DPR} > \pi_1^{DPC} \tag{9}$$

then the best repeated strategy for the game is the repeated deviation-punishment strategy (DPR). To discover when this strategy should be adopted, we start to compute the values of the rate of discount δ for which $\pi_1^{DPR} = \pi_1^{DPC}$ and obtain

$$\delta_1^{DPC/DPR} = \frac{(\gamma^6 - 8\gamma^5 + 12\gamma^4 + 8\gamma^3 + 48\gamma^2 - 128\gamma + 64)\gamma}{4\gamma^7 - 96\gamma^5 + 128\gamma^4 + 336\gamma^3 - 512\gamma^2 - 320\gamma + 512}. \tag{10}$$

In this result, and throughout the following results, we fix the parameters $\alpha_1 = \alpha_2$ and $\beta_2 = 2\beta_1 = 2$. We also consider that the home firm does not pay to lobby its government, hence $L = 0$. For values of $\delta < \delta_1^{DPC/DPR}$ we observe that $\pi_1^{DPC} > \pi_1^{DPR}$ and the home firm will choose the Cournot competition after the deviation-punishment periods, and for values of $\delta > \delta_1^{DPC/DPR}$ we observe that $\pi_1^{DPR} > \pi_1^{DPC}$ and the home firm will choose to repeat the deviation-punishment after the first two periods of deviation-punishment. These choices of the home firm are illustrated in Fig. 1a. Solving the equation $\pi_2^{DPR} = \pi_2^{COL}$ for the same parameter values, we obtain the curve

$$\delta_2^{DPR/COL} = \frac{(\gamma^2 - 4\gamma + 4)\gamma^2}{8\gamma^3 - 8\gamma^2 - 16\gamma + 16}. \tag{11}$$

For values of $\delta < \delta_2^{DPR/COL}$ we observe that $\pi_2^{DPR} > \pi_2^{COL}$ and the foreign firm prefers the repeated deviation-punishment strategy. For values of $\delta > \delta_2^{DPR/COL}$ we observe that $\pi_2^{COL} > \pi_2^{DPR}$ and the foreign firm prefers the repeated collusion strategy, as

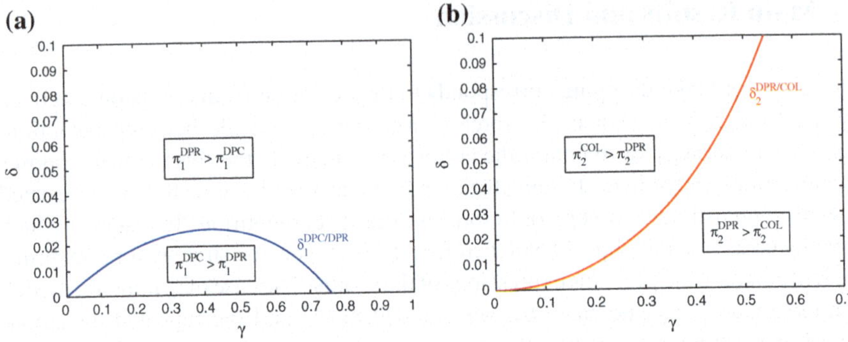

Fig. 1 The *curve* $\delta_1^{DPC/DPR}$ (*left*) where the profits of the home firm in the *DPC* and *DPR* strategies are equal. The home firm realizes a higher profit with the *DPR* strategy above the curve and with the *DPC* strategy below the *curve*. The curve $\delta_2^{DPR/COL}$ (*right*) dividing the $(\gamma; \delta)$ plane in the regions where the foreign firm prefers the *DPR* strategy and the *DPC* strategy. The values of the parameters used are: $\alpha_1 = \alpha_2$, $\beta_2 = 2\beta_1 = 2$ and $L = 0$

illustrated in Fig. 1b. Hence, by Eq. (9), we conclude that for values of δ such that

$$\delta_1^{DPC/DPR} < \delta < \delta_2^{DPR/COL} \tag{12}$$

the best repeated strategy for both firms is the repeated deviation-punishment (DPR), as illustrated in Fig. 2.

On the other hand, if the home firm will chose a Cournot competition rather than play a repeated deviation-punishment game and if the foreign firm prefer a Cournot game after the deviation-punishment periods

$$\pi_2^{DPC} > \pi_2^{COL} \quad \wedge \quad \pi_1^{DPC} > \pi_1^{DPR} \tag{13}$$

Fig. 2 The $\delta_1^{DPC/DPR}$ *curve* above which the home firm prefers the DPR strategy and the $\delta_2^{DPR/COL}$ *curve* below which the foreign firm prefers the DPR strategy. For values of δ above $\delta_1^{DPC/DPR}$ and below $\delta_2^{DPR/COL}$ the best repeated strategy for both firms is *DPR*

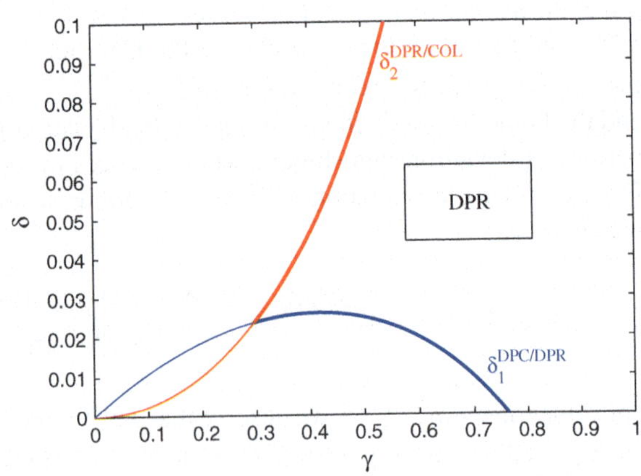

then the best repeated strategy for the game is deviation-punishment followed by Cournot (*DPC*). We start to observe that the equation $\pi_2^{DPC} = \pi_2^{COL}$ is quadratic

$$A_2\delta^2 + A_1\delta + A_0 = 0, \tag{14}$$

with

$$A_2 = -64\,\gamma^6 + 256\,\gamma^5 - 1024\,\gamma^3 + 768\,\gamma^2 + 1024\,\gamma - 1024,$$
$$A_1 = \gamma^8 + 4\,\gamma^7 - 20\,\gamma^6 - 80\,\gamma^5 + 144\,\gamma^4 + 512\,\gamma^3 - 512\,\gamma^2 - 1024\,\gamma + 1024,$$
$$A_0 = -\gamma^8 + 4\,\gamma^7 + 12\,\gamma^6 - 64\,\gamma^5 + 256\,\gamma^3 - 256\,\gamma^2.$$

The two solutions of Eq. (14) define the curve $\delta_2^{DPC/COL}$ plotted in Fig. 3a. On the right hand side of the curve $\delta_2^{DPC/COL}$ we observe that $\pi_2^{DPC} > \pi_2^{COL}$ and the foreign firm prefers a Cournot competition after the two periods of deviation-punishment. On the left hand side of the curve $\delta_2^{DPC/COL}$ we observe that $\pi_2^{COL} > \pi_2^{DPC}$ and the foreign firm prefers the repeated collusion strategy rather than deviation-punishment followed by Cournot competition. Hence, considering the solution of Eq. (14) given by $\delta_2^{DPC/COL} = \left(-A_1 - \sqrt{A_1^2 - 4A_2A_0}\right) / (2A_2)$ and the $\delta_1^{DPC/DPR}$ curve defined in Eq. (10), by Eq. (13) we conclude that for values of δ such that

$$\delta < \delta_2^{DPC/COL} \quad \wedge \quad \delta < \delta_1^{DPC/DPR} \tag{15}$$

the best repeated strategy for both firms is deviation-punishment followed by Cournot (*DPC*), as illustrated in Fig. 3b.

The best repeated strategy for the game is repeated collusion (COL) if, when the home firm will choose a Cournot competition, the foreign firm prefers collusion rather

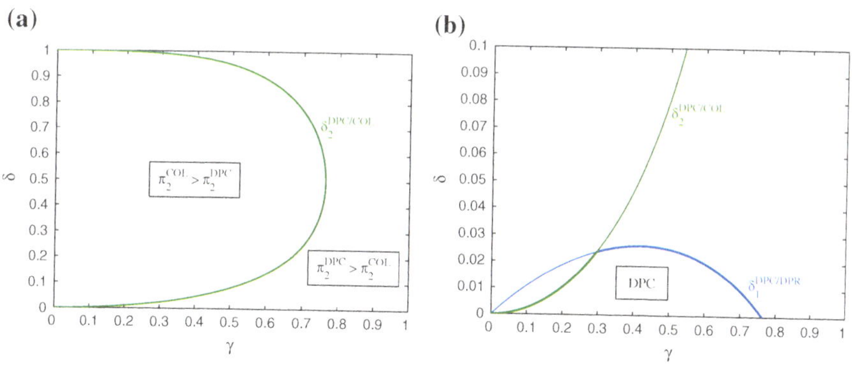

Fig. 3 The *curve* $\delta_2^{DPC/COL}$ separating the preferable strategy for the foreign firm: on the *left hand side* it prefers the *COL* strategy and on the *right hand side* it prefers the *DPC* strategy. The *curves* $\delta_2^{DPC/COL}$ and $\delta_1^{DPC/DPR}$ (*right*) defining the parameters region where the best repeated strategy for both firms is *DPC*

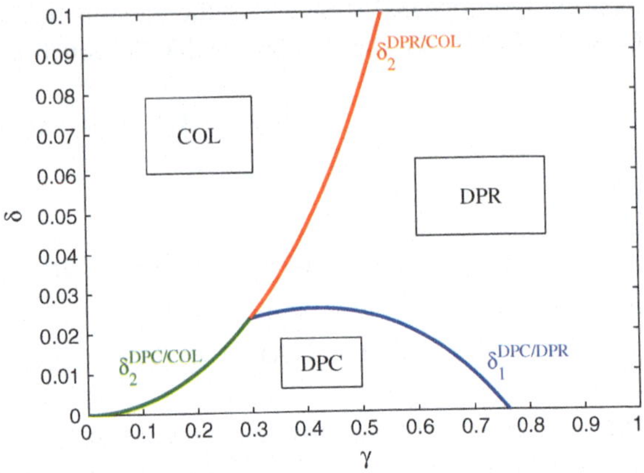

Fig. 4 The *curves* $\delta_1^{DPC/DPR}$, $\delta_2^{DPR/COL}$ and $\delta_2^{DPC/COL}$ dividing the $(\gamma;\delta)$ plane in the regions where the best repeated strategy is *COL*, *DPC* and *DPR*

than Cournot, or if, when the home firm will choose a repeated deviation-punishment strategy, the foreign firm prefers collusion rather than repeated deviation-punishment

$$\left(\pi_{T,2}^{COL} > \pi_{T,2}^{DPC} \wedge \pi_{T,1}^{DPC} > \pi_{T,1}^{DPR}\right) \quad \vee \quad \left(\pi_{T,2}^{COL} > \pi_{T,2}^{DPR} \wedge \pi_{T,1}^{DPR} > \pi_{T,1}^{DPC}\right). \quad (16)$$

Using the curves $\delta_1^{DPC/DPR}$, $\delta_2^{DPR/COL}$ and $\delta_2^{DPC/COL}$ we characterize the parameter values for which the best repeated strategy is collusion. In Fig. 4, we present this region of values in addition to the regions where the best repeated strategy is repeated deviation-punishment and deviation-punishment followed by Cournot.

For other values of the parameters $\alpha_1, \alpha_2, \beta_1$ and β_2 the curves of diagram presented in Fig. 4 are very similar. When the price payed by the home firm to lobby its own government is positive $L > 0$, we observe that the $\delta_1^{DPC/DPR}$ curve goes up and the repeated deviation-punishment tends to disappear. This is expectable since the home firm would not be much interested to pay a positive amount to lobby its government so many times.

5 Conclusions

In this work we observe that the repeated deviation-punishment strategy, that can be interpreted as a repeated strategy for dumping, appears as the best repeated strategy in an international trade if we consider non-symmetric inverse demand functions, in contrast with the symmetric case. This repeated deviation-punishment strategy is more likely to be adopted by the two firms if the goods traded are highly substitutes, which is the case of high values of γ.

Acknowledgments We thank LIAAD-INESC TEC, USP-UP project, Faculty of Sciences, University of Porto and Calouste Gulbenkian Foundation. We acknowledge the financial support received by the ERDF – European Regional Development Fund through the Operational Programme for Competitiveness and Internationalisation – COMPETE 2020 Programme within project "POCI-01-0145-FEDER-006961", and by National Funds through the FCT – Fundação para a Ciência e a

Tecnologia (Portuguese Foundation for Science and Technology) within project UID/EEA/50014/2013 and ERDF (European Regional Development Fund) through the COMPETE Program (operational program for competitiveness) and by National Funds through the FCT within Project "Dynamics, optimization and modelling", with reference PTDC/MAT-NAN/6890/2014. Alberto Pinto also acknowledges the financial support received through the Special Visiting Researcher Program (Bolsa Pesquisador Visitante Especial - PVE) "Dynamics, Games and Applications" with reference 401068/2014-5 (call: MEC/MCTI/CAPES/CNPQ/FAPS), at IMPA, Brazil. Part of this research was done during visits by the authors to IMPA (Brazil), University of São Paulo (Brazil), University of Warwick (United Kingdom), Institut Henri Poincaré (France) and SUNY (USA) who thank them for their hospitality.

References

1. Banik, N., Ferreira, F., Martins, J., Pinto, A.: An economical model for dumping by dumping in a cournot model. In: Peixoto, M.M., Pinto, A.A., Rand, D.A. (eds.) Dynamics, Games and Science II. Springer Proceedings in Mathematics. Springer, New York (2011)
2. DeVault, J.M.: Economics and the international trade commission. South. Econ. J. **60**, 463–478 (1993)
3. Dixit, A.K.: Anti-dumping and countervailing duties under oligopoly. Eur. Econ. Rev. **32**, 55–68 (1988)
4. Eaton, J., Grossman, G.M.: Optimal trade and industrial policy for the US automobile industrial policy under oligopoly. Q. J. Econ. **100**, 383–406 (1986)
5. Ethier, W.J., Fischer, R.D.: The new protectionism. J. Int. Econ. Integr. **2**, 1–11 (1987)
6. Fischer, D.R.: Endogenous probability of protection and firm behavior. J. Int. Econ. **32**, 149–163 (1992)
7. Hansen, L.W., Prusa, T.J.: Cumulation and ITC decision making: the sum of the parts is greater than the whole. Econ. Enq. **34**, 746–769 (1996)
8. Hansen, L.W., Prusa, T.J.: The economics and politics of trade policy: an empirical analysis of ITC decision making. Rev. Int. Econ. **5**, 230–245 (1997)
9. Moore, M.O.: Rules or politics? An empirical analysis of ITC antidumping decisions. Econ. Enq. **30**, 449–466 (1992)
10. Moore, M.O.: In: Krueger, A.O. (ed.) The Political Economy of Trade Protection. University of Chicago Press, Chicago (1996)
11. Reitzes, D.J.: Antidumping policy. Int. Econ. Rev. **34**, 745–763 (1993)
12. Singh, N., Vives, X.: Price and quantity competition in a differentiated duopoly. Rand J. Econ. **15**, 546–554 (1984)
13. Staiger, R.W., Wolak, F.A.: The effect of domestic antidumping law in the presence of foreign monopoly. J. Int. Econ. **32**, 265–287 (1992)
14. Viner, J.: Dumping: A Problem in International Trade. University of Chicago Press, Chicago (1923)
15. Vives, X.: Duopoly information equilibrium: Cournot and Bertrand. J. Econ. Theory **34**, 546–554 (1984)

Hypercyclic and Topologically Mixing Properties of Certain Classes of Abstract Time-Fractional Equations

Marko Kostić

Abstract In recent years, considerable effort has been directed toward the topological dynamics of abstract PDEs whose solutions are governed by various types of operator semigroups, fractional resolvent operator families and evolution systems. In this paper, we shall present the most important results about hypercyclic and topologically mixing properties of some special subclasses of the abstract time-fractional equations of the following form:

$$\begin{aligned} &\mathbf{D}_t^{\alpha_n} u(t) + c_{n-1}\mathbf{D}_t^{\alpha_{n-1}} u(t) + \cdots + c_1\mathbf{D}_t^{\alpha_1} u(t) = A\mathbf{D}_t^{\alpha} u(t), \quad t > 0, \\ &u^{(k)}(0) = u_k, \quad k = 0, \ldots, \lceil \alpha_n \rceil - 1, \end{aligned} \tag{1}$$

where $n \in \mathbb{N} \setminus \{1\}$, A is a closed linear operator acting on a separable infinite-dimensional complex Banach space E, $c_1, \ldots, c_{n-1}$ are certain complex constants, $0 \leq \alpha_1 < \cdots < \alpha_n$, $0 \leq \alpha < \alpha_n$, and $\mathbf{D}_t^{\alpha}$ denotes the Caputo fractional derivative of order α [5]. We slightly generalize results from [24] and provide several applications, including those to abstract higher order differential equations of integer order [38].

Keywords Abstract time-fractional equations · Hypercyclicity · topologically mixing property

1 Introduction and Preliminaries

The last two decades have witnessed a growing interest in fractional derivatives and their applications. In this paper, we enquire into the basic hypercyclic and topologically mixing properties of some special subclasses of the abstract time-fractional equations of the form (1), continuing in such a way the research raised in [24]. Our main result is Theorem 1, which is the kind of Desch–Schappacher–Webb and

M. Kostić (✉)
Faculty of Technical Sciences, Trg Dositeja Obradovića 6, 21125 Novi Sad, Serbia
e-mail: marco.s@verat.net

L. Alsedà i Soler et al. (eds.), *Difference Equations, Discrete Dynamical Systems and Applications*, Springer Proceedings in Mathematics & Statistics 180, DOI 10.1007/978-3-662-52927-0_12

Banasiak–Moszyński criteria for chaos of strongly continuous semigroups. For further information concerning hypercyclic and topologically mixing properties of single valued operators and abstract PDEs, we refer the reader to [2–4, 6–8, 10–22, 24, 25, 33, 36, 37]. A fairly complete information on the general theory of operator semigroups, cosine functions and abstract Volterra equations can be obtained by consulting the monographs [1, 9, 22, 35, 38].

Before going any further, it will be convenient to introduce the basic concepts used throughout the paper. We shall always assume that $(E, \|\cdot\|)$ is a separable infinite-dimensional complex Banach space, A and $A_1, \ldots, A_{n-1}$ are closed linear operators acting on E, $n \in \mathbb{N} \setminus \{1\}$, $0 \leq \alpha_1 < \cdots < \alpha_n$ and $0 \leq \alpha < \alpha_n$. By I is denoted the identity operator on E. Given $s \in \mathbb{R}$, put $\lceil s \rceil := \inf\{k \in \mathbb{Z} : s \leq k\}$. Define $m_j := \lceil \alpha_j \rceil$, $1 \leq j \leq n$, $m := m_0 := \lceil \alpha \rceil$, $A_0 := A$ and $\alpha_0 := \alpha$. The dual space of E and the space of continuous linear mappings from E into E are denoted by E^* and $L(E)$, respectively. By $D(A)$, $\mathrm{Kern}(A)$, $R(A)$, $\rho(A)$, $\sigma_p(A)$ and A^*, we denote the domain, kernel, range, resolvent set, point spectrum and adjoint operator of A, respectively. Suppose F is a closed subspace of E. Then the part of A in F, denoted by $A_{|F}$, is a linear operator defined by $D(A_{|F}) := \{x \in D(A) \cap F : Ax \in F\}$ and $A_{|F}x := Ax$, $x \in D(A_{|F})$. In the sequel, we assume that $L(E) \ni C$ is an injective operator satisfying $CA \subseteq AC$. The Gamma function is denoted by $\Gamma(\cdot)$ and the principal branch is always used to take the powers. Set $\mathbb{N}_l := \{1, \ldots, l\}$, $\mathbb{N}_l^0 := \{0, 1, \ldots, l\}$, $0^{\zeta} := 0$, $g_{\zeta}(t) := t^{\zeta-1}/\Gamma(\zeta)$ $(\zeta > 0, t > 0)$ and $g_0 :=$ the Dirac δ-distribution. If $\delta \in (0, \pi]$, then we define $\Sigma_{\delta} := \{\lambda \in \mathbb{C} : \lambda \neq 0, \ |\arg(\lambda)| < \delta\}$. Denote by $\mathcal{L}$ and $\mathcal{L}^{-1}$ the Laplace transform and its inverse transform, respectively.

It is clear that the abstract Cauchy problem (1) is a special case of the following one:

$$\begin{aligned} &\mathbf{D}_t^{\alpha_n} u(t) + A_{n-1}\mathbf{D}_t^{\alpha_{n-1}} u(t) + \cdots + A_1\mathbf{D}_t^{\alpha_1} u(t) = A\mathbf{D}_t^{\alpha} u(t), \quad t > 0, \\ &u^{(k)}(0) = u_k, \quad k = 0, \ldots, \lceil \alpha_n \rceil - 1. \end{aligned} \tag{2}$$

In what follows, we shall briefly summarize the most important facts concerning the C-wellposedness of the problem (2).

Definition 1 A function $u \in C^{m_n-1}([0,\infty) : E)$ is called a *(strong) solution* of (2) iff $A_i\mathbf{D}_t^{\alpha_i} u \in C([0,\infty) : E)$ for $0 \leq i \leq n-1$, $g_{m_n-\alpha_n} * (u - \sum_{k=0}^{m_n-1} u_k g_{k+1}) \in C^{m_n}([0,\infty) : E)$ and (2) holds. The abstract Cauchy problem (2) is said to be *C-wellposed* if:

1. For every $u_0, \ldots, u_{m_n-1} \in \bigcap_{0 \leq j \leq n-1} C(D(A_j))$, there exists a unique solution $u(t; u_0, \ldots, u_{m_n-1})$ of (2).
2. For every $T > 0$, there exists $c > 0$ such that, for every $u_0, \ldots, u_{m_n-1} \in \bigcap_{0 \leq j \leq n-1} C(D(A_j))$, the following holds:

$$\left\| u\left(t; u_0, \ldots, u_{m_n-1}\right) \right\| \leq c \sum_{k=0}^{m_n-1} \left\| C^{-1} u_k \right\|, \ t \in [0, T].$$

Although not of primary importance in our analysis, the following facts should be stated. The *Caputo fractional derivative* $\mathbf{D}_t^{\alpha_n} u$ is defined for those functions $u \in C^{m_n-1}([0,\infty) : E)$ for which $g_{m_n-\alpha_n} * (u - \sum_{k=0}^{m_n-1} u_k g_{k+1}) \in C^{m_n}([0,\infty) : E)$. If this is the case, then we have $\mathbf{D}_t^{\alpha_n} u(t) = \frac{d^{m_n}}{dt^{m_n}}[g_{m_n-\alpha_n} * (u - \sum_{k=0}^{m_n-1} u_k g_{k+1})]$. Suppose $\beta > 0$, $\gamma > 0$ and $\mathbf{D}_t^{\beta+\gamma} u$ is defined. Then the equality $\mathbf{D}_t^{\beta+\gamma} u = \mathbf{D}_t^{\beta}\mathbf{D}_t^{\gamma} u$ does not hold in general. The validity of this equality can be proved provided that any of the following conditions holds:

1. $\gamma \in \mathbb{N}$,
2. $\lceil \beta + \gamma \rceil = \lceil \gamma \rceil$, or
3. $u^{(j)}(0) = 0$ for $\lceil \gamma \rceil \leq j \leq \lceil \beta + \gamma \rceil - 1$.

Suppose $u(t) \equiv u(t; u_0, \ldots, u_{m_n-1})$, $t \geq 0$ is a strong solution of (2), with $f(t) \equiv 0$ and initial values $u_0, \ldots, u_{m_n-1} \in R(C)$. Convoluting the both sides of (2) with $g_{\alpha_n}(t)$, and making use of the equality [5, (1.21)], it readily follows that $u(t)$, $t \geq 0$ satisfies the following:

$$u(\cdot) - \sum_{k=0}^{m_n-1} u_k g_{k+1}(\cdot) + \sum_{j=1}^{n-1} g_{\alpha_n-\alpha_j} * A_j \left[u(\cdot) - \sum_{k=0}^{m_j-1} u_k g_{k+1}(\cdot) \right]$$
$$= g_{\alpha_n-\alpha} * A \left[u(\cdot) - \sum_{k=0}^{m-1} u_k g_{k+1}(\cdot) \right]. \tag{3}$$

Given $i \in \mathbb{N}_{m_n-1}^0$ in advance, set $D_i := \{ j \in \mathbb{N}_{n-1} : m_j - 1 \geq i \}$. Plugging $u_j = 0$, $0 \leq j \leq m_n - 1$, $j \neq i$, in (3), one gets:

$$\begin{aligned}
&\Big[u(\cdot; 0, \ldots, u_i, \ldots, 0) - u_i g_{i+1}(\cdot) \Big] \\
&\quad + \sum_{j \in D_i} g_{\alpha_n-\alpha_j} * A_j \Big[u(\cdot; 0, \ldots, u_i, \ldots, 0) - u_i g_{i+1}(\cdot) \Big] \\
&\quad + \sum_{j \in \mathbb{N}_{n-1} \setminus D_i} \Big[g_{\alpha_n-\alpha_j} * A_j u(\cdot; 0, \ldots, u_i, \ldots, 0) \Big] \\
&= \begin{cases} g_{\alpha_n-\alpha} * A u(\cdot; 0, \cdots, u_i, \cdots, 0), & m - 1 < i, \\ g_{\alpha_n-\alpha} * A \Big[u(\cdot; 0, \cdots, u_i, \cdots, 0) - u_i g_{i+1}(\cdot) \Big], & m - 1 \geq i, \end{cases}
\end{aligned} \tag{4}$$

where u_i appears in the ith place $(0 \leq i \leq m_n - 1)$ starting from 0. Suppose now $0 < \tau \leq \infty$, $0 \neq K \in L^1_{loc}([0,\tau))$ and $k(t) = \int_0^t K(s)\,ds$, $t \in [0,\tau)$. Denote $R_i(t)C^{-1}u_i = (K * u(\cdot; 0, \ldots, u_i, \cdots, 0))(t)$, $t \in [0,\tau)$, $0 \leq i \leq m - 1$. Convoluting formally the both sides of (4) with $K(t)$, $t \in [0,\tau)$, one obtains that, for $0 \leq i \leq m_n - 1$:

$$\Big[R_i(\cdot)C^{-1}u_i - \big(k * g_i\big)(\cdot)u_i\Big] + \sum_{j\in D_i} g_{\alpha_n-\alpha_j} * A_j\Big[R_i(\cdot)C^{-1}u_i - \big(k * g_i\big)(\cdot)u_i\Big]$$
$$+ \sum_{j\in \mathbb{N}_{n-1}\setminus D_i} \Big[g_{\alpha_n-\alpha_j} * A_j R_i(\cdot)C^{-1}u_i\Big]$$
$$= \begin{cases} \big(g_{\alpha_n-\alpha} * AR_i\big)(\cdot)C^{-1}u_i, & m-1<i, \\ g_{\alpha_n-\alpha} * A\Big[R_i(\cdot)C^{-1}u_i - \big(k * g_i\big)(\cdot)u_i\Big], & m-1\geq i. \end{cases}$$

Motivated by the above analysis, we introduce the following general definition.

Definition 2 Suppose $0 < \tau \leq \infty$, $k \in C([0,\tau))$, $C,\ C_1,\ C_2 \in L(E)$, C and C_2 are injective. A sequence $((R_0(t))_{t\in[0,\tau)}, \ldots, (R_{m_n-1}(t))_{t\in[0,\tau)})$ of strongly continuous operator families in $L(E)$ is called a (*local*, if $\tau < \infty$):

1. *k-regularized C_1-existence propagation family for* (2) if $R_i(0) = (k * g_i)(0)C_1$ and:

$$\Big[R_i(\cdot)x - \big(k * g_i\big)(\cdot)C_1x\Big] + \sum_{j\in D_i} A_j\Big[g_{\alpha_n-\alpha_j} * \Big(R_i(\cdot)x - \big(k * g_i\big)(\cdot)C_1x\Big)\Big]$$
$$+ \sum_{j\in \mathbb{N}_{n-1}\setminus D_i} A_j\big(g_{\alpha_n-\alpha_j} * R_i\big)(\cdot)x$$
$$= \begin{cases} A\big(g_{\alpha_n-\alpha} * R_i\big)(\cdot)x, & m-1<i, \\ A\Big[g_{\alpha_n-\alpha} * \Big(R_i(\cdot)x - \big(k * g_i\big)(\cdot)C_1x\Big)\Big](\cdot), & m-1\geq i, \end{cases}$$

   for any $i = 0, \ldots, m_n - 1$ and $x \in E$.
2. *k-regularized C_2-uniqueness propagation family for* (2) if $R_i(0) = (k * g_i)(0)C_2$ and:

$$\Big[R_i(\cdot)x - \big(k * g_i\big)(\cdot)C_2x\Big] + \sum_{j\in D_i} g_{\alpha_n-\alpha_j} * \Big[R_i(\cdot)A_jx - \big(k * g_i\big)(\cdot)C_2A_jx\Big]$$
$$+ \sum_{j\in \mathbb{N}_{n-1}\setminus D_i} \big(g_{\alpha_n-\alpha_j} * R_i(\cdot)A_jx\big)(\cdot)$$
$$= \begin{cases} \big(g_{\alpha_n-\alpha} * R_i(\cdot)Ax\big)(\cdot), & m-1<i, \\ g_{\alpha_n-\alpha} * \Big[R_i(\cdot)Ax - \big(k * g_i\big)(\cdot)C_2Ax\Big](\cdot), & m-1\geq i, \end{cases}$$

   for any $i = 0, \ldots, m_n - 1$ and $x \in \bigcap_{0\leq i\leq n-1} D(A_i)$.
3. *k-regularized C-resolvent propagation family for* (2) if $((R_0(t))_{t\in[0,\tau)}, \ldots, (R_{m_n-1}(t))_{t\in[0,\tau)})$ is a k-regularized C-uniqueness propagation family for (2), and if for every $t \in [0,\tau)$, $i \in \mathbb{N}^0_{m_n-1}$ and $j \in \mathbb{N}^0_{n-1}$, one has: $R_i(t)A_j \subseteq A_jR_i(t)$, $R_i(t)C = CR_i(t)$ and $CA_j \subseteq A_jC$.

Before proceeding further, we would like to draw the readers attention to the paper [26] for further information concerning some other types of (C_1, C_2)-existence and uniqueness resolvent families which can be useful in the analysis of (inhomogeneous) abstract Cauchy problems of the form (2). Notice also the following: If A is a subgenerator of a k-regularized C-resolvent propagation family $((R_0(t))_{t\in[0,\tau)}, \ldots, (R_{m_n-1}(t))_{t\in[0,\tau)})$ for (2), then, in general, there do not exist $a_i \in L^1_{loc}([0,\tau))$, $i \in \mathbb{N}^0_{m_n-1}$ and $k_i \in C([0,\tau))$ such that $(R_i(t))_{t\in[0,\tau)}$ is an (a_i, k_i)-regularized C-resolvent family with subgenerator A; cf. [22, 23, 26–31] for the basic properties of (a,k)-regularized C-resolvent families and their applications in the study of abstract Cauchy problem (2). The notions of exponential boundedness and analyticity of k-regularized C-resolvent propagation families will be understood in the sense of [26].

In the sequel, we shall consider only global C-resolvent propagation families for (2), i.e., global k-regularized C-resolvent propagation families for (1) with $k(t) \equiv 1$; in the case $C = I$, such a resolvent family is also called a *resolvent propagation family for* (2), or simply a *resolvent propagation family*, if there is no risk for confusion. It will be assumed that every single operator family $(R_i(t))_{t\geq 0}$ of the tuple $((R_0(t))_{t\geq 0}, \ldots, (R_{m_n-1}(t))_{t\geq 0})$ is non-degenerate, i.e., that the supposition $R_i(t)x = 0$, $t \geq 0$ implies $x = 0$. Henceforward we shall assume that there exist complex constants $c_1, \ldots, c_{n-1}$ such that $A_j = c_j I$, $j \in \mathbb{N}_{n-1}$. Then it is also said that the operator A is a *subgenerator* of $((R_0(t))_{t\geq 0}, \ldots, (R_{m_n-1}(t))_{t\geq 0})$. The *integral generator* $\hat{A}$ of $((R_0(t))_{t\geq 0}, \ldots, (R_{m_n-1}(t))_{t\geq 0})$ is defined as the set of all pairs $(x, y) \in E \times E$ such that, for every $i = 0, \ldots, m_n - 1$ and $t \geq 0$, the following holds:

$$\Big[R_i(\cdot)x - (k * g_i)(\cdot)Cx\Big] + \sum_{j=1}^{n-1} c_j g_{\alpha_n-\alpha_j} * \Big[R_i(\cdot)x - (k * g_i)(\cdot)Cx\Big]$$
$$+ \sum_{j\in\mathbb{N}_{n-1}\setminus D_i} c_j\big[g_{\alpha_n-\alpha_j+i} * k\big](\cdot)Cx$$
$$= \begin{cases} \big[g_{\alpha_n-\alpha} * R_i\big](\cdot)y, & m-1 < i, \\ g_{\alpha_n-\alpha} * \Big[R_i(\cdot)y - (k * g_i)(\cdot)Cy\Big], & m-1 \geq i. \end{cases}$$

By a *mild solution* of (3) we mean any function $u \in C([0,\infty) : E)$ such that the following holds:

$$u(\cdot) - \sum_{k=0}^{m_n-1} u_k g_{k+1}(\cdot) + \sum_{j=1}^{n-1} c_j g_{\alpha_n-\alpha_j} * \left[u(\cdot) - \sum_{k=0}^{m_j-1} u_k g_{k+1}(\cdot)\right]$$
$$= A\left(g_{\alpha_n-\alpha} * \left[u(\cdot) - \sum_{k=0}^{m-1} u_k g_{k+1}(\cdot)\right]\right);$$

a *strong solution* is any function $u \in C([0,\infty) : E)$ satisfying (3). It is clear that every strong solution of (3) is also a mild solution of the same problem; the converse statement is not true, in general. In the sequel, we shall always assume that, for every $i \in \mathbb{N}^0_{m_n-1}$ with $m-1 \geq i$, one has: $\mathbb{N}_{n-1} \setminus D_i \neq \emptyset$ and $\sum_{j \in \mathbb{N}_{n-1} \setminus D_i} |c_j|^2 > 0$. Then the problem (3) has at most one mild (strong) solution; cf. [26] for more details.

The proof of following auxiliary lemma follows from an application of [26, Theorem 2.12].

Lemma 1 *1. Suppose A generates an exponentially bounded, analytic C-regularized semigroup of angle $\beta \in (0, \pi/2]$ and A is densely defined. Then A is the integral generator of an exponentially bounded, analytic C-regularized propagation family $((R_0(t))_{t\geq 0}, \dots, (R_{m_n-1}(t))_{t\geq 0})$ of angle $\min(\frac{\frac{\pi}{2}+\beta}{\alpha_n-\alpha} - \frac{\pi}{2}, \frac{\pi}{2})$, provided that $\frac{\pi}{2}+\beta > \frac{\pi}{2}(\alpha_n-\alpha)$.*

2. *Suppose A generates an exponentially bounded C-regularized semigroup and A is densely defined. Then A is the integral generator of an exponentially bounded, analytic C-regularized propagation family $((R_0(t))_{t\geq 0}, \dots, (R_{m_n-1}(t))_{t\geq 0})$ of angle $\min(\frac{\pi}{2(\alpha_n-\alpha)} - \frac{\pi}{2}, \frac{\pi}{2})$, provided that $\frac{\pi}{2} < \frac{\pi}{2(\alpha_n-\alpha)}$.*

We refer the reader to [24, Definition 1.1] for the notion of a global α_n-times C-regularized resolvent family. If $n=2$, $c_1=0$, $\alpha=0$, and A is a subgenerator of a global C-regularized propagation family $((R_0(t))_{t\geq 0}, \dots, (R_{m_2-1}(t))_{t\geq 0})$, then it is obvious that $(R_0(t))_{t\geq 0}$ is a global α_2-times C-regularized resolvent family having A as subgenerator. In our recent paper [24], we have considered hypercyclic and topologically mixing properties of fractional C-regularized resolvent families. Therefore, the results of this paper can be viewed as generalizations of corresponding results from [24].

Suppose $\beta > 0$ and $\gamma > 0$. Then the *Mittag-Leffler function* $E_{\beta,\gamma}(z)$ is defined by $E_{\beta,\gamma}(z) := \sum_{n=0}^{\infty} z^n/\Gamma(\beta n+\gamma)$, $z \in \mathbb{C}$. Set, for short, $E_\beta(\cdot) := E_{\beta,1}(\cdot)$. The following asymptotic formulae for the Mittag-Leffler functions [5, 32] play a crucial role in our analysis:

$$E_\alpha(z) = \frac{1}{\alpha}e^{z^{1/\alpha}} + \varepsilon_\alpha(z),\ |\arg(z)| < \alpha\pi/2, \tag{5}$$

and

$$E_\alpha(z) = \varepsilon_\alpha(z),\ |\arg(-z)| < \pi - \alpha\pi/2, \tag{6}$$

where

$$\varepsilon_\alpha(z) = \sum_{n=1}^{N-1} \frac{z^{-n}}{\Gamma(1-\alpha n)} + O(|z|^{-N}),\ |z| \to \infty. \tag{7}$$

2 Hypercyclicity and Topologically Mixing Property for C-Resolvent Propagation Families

We recall the basic notations used henceforward: E is a separable infinite-dimensional complex Banach space, A is a closed linear operator on E, $n \in \mathbb{N} \setminus \{1\}$, $0 \leq \alpha_1 < \cdots < \alpha_n$, $0 \leq \alpha < \alpha_n$, $A_j = c_j I$ for certain complex constants $c_1, \ldots, c_{n-1}$, $m_j = \lceil \alpha_j \rceil$, $1 \leq j \leq n$, $m = m_0 = \lceil \alpha \rceil$, $A_0 = A$ and $\alpha_0 = \alpha$. We assume, in addition, that $C^{-1}AC = A$ is densely defined and that A is a subgenerator of a global C-resolvent propagation family $((R_0(t))_{t\geq 0}, \ldots, (R_{m_n-1}(t))_{t\geq 0})$. Then we know (see [26]) that A is, in fact, the integral generator of $((R_0(t))_{t\geq 0}, \ldots, (R_{m_n-1}(t))_{t\geq 0})$.

Let $i \in \mathbb{N}^0_{m_n-1}$. Then we denote by $Z_i(A)$ the set which consists of those vectors $x \in E$ such that $R_i(t)x \in R(C)$, $t \geq 0$ and that the mapping $t \mapsto C^{-1}R_i(t)x$, $t \geq 0$ is continuous. Then $R(C) \subseteq Z_i(A)$, and it can be simply proved with the help of [26, Theorem 2.8] that $x \in Z_i(A)$ iff there exists a unique mild solution of (3) with $u_k = \delta_{k,i}x$, $k \in \mathbb{N}^0_{m_n-1}$; if this is the case, the unique mild solution of (3) is given by $u(t; x) := u_i(t; x) := C^{-1}R_i(t)x$, $t \geq 0$. Set $\mathcal{D}_i = \{j \in \mathbb{N}^0_{n-1} : m_j - 1 \geq i\}$.

The Laplace transform can be used to prove the following extension of [24, Lemma 2.1].

Lemma 2 *Suppose $\lambda \in \mathbb{C}$, $x \in E$ and $Ax = \lambda x$. Then $x \in Z_i(A)$ and the unique strong solution of (3) is given by*

$$u_i(t; x) = \mathcal{L}^{-1}\left(\frac{z^{-i-1} + \sum_{j\in D_i} c_j z^{-\alpha_n-i-1+\alpha_j} - \chi_{\mathcal{D}_i}(0)\lambda z^{-\alpha_n-i-1+\alpha}}{1 + \sum_{j=1}^{n-1} c_j z^{\alpha_j-\alpha_n} - \lambda z^{\alpha-\alpha_n}}\right)(t)x,$$

for any $t \geq 0$ and $i \in \mathbb{N}^0_{m_n-1}$.

Set $P_\lambda := \lambda^{\alpha_n-\alpha} + \sum_{j=1}^{n-1} c_j \lambda^{\alpha_j-\alpha}$, $\lambda \in \mathbb{C} \setminus \{0\}$ and

$$F_i(\lambda, t) := \mathcal{L}^{-1}\left(\frac{z^{-i-1} + \sum_{j\in D_i} c_j z^{-\alpha_n-i-1+\alpha_j} - \chi_{\mathcal{D}_i}(0)P_\lambda z^{-\alpha_n-i-1+\alpha}}{1 + \sum_{j=1}^{n-1} c_j z^{\alpha_j-\alpha_n} - P_\lambda z^{\alpha-\alpha_n}}\right)(t),$$

for any $t \geq 0$, $i \in \mathbb{N}^0_{m_n-1}$ and $\lambda \in \mathbb{C} \setminus \{0\}$.

Definition 3 Let $i \in \mathbb{N}^0_{m_n-1}$, and let $\tilde{E}$ be a closed linear subspace of E. Then it is said that $(R_i(t))_{t\geq 0}$ is:

1. $\tilde{E}$*-hypercyclic* iff there exists $x \in Z_i(A) \cap \tilde{E}$ such that $\{C^{-1}R_i(t)x : t \geq 0\}$ is a dense subset of $\tilde{E}$; such an element is called a $\tilde{E}$*-hypercyclic vector* of $(R_i(t))_{t\geq 0}$;
2. $\tilde{E}$*-topologically transitive* iff for every y, $z \in \tilde{E}$ and for every $\varepsilon > 0$, there exist $x \in Z_i(A) \cap \tilde{E}$ and $t \geq 0$ such that $||y - x|| < \varepsilon$ and $||z - C^{-1}R_i(t)x|| < \varepsilon$;
3. $\tilde{E}$*-topologically mixing* iff for every y, $z \in \tilde{E}$ and for every $\varepsilon > 0$, there exists $t_0 \geq 0$ such that, for every $t \geq t_0$, there exists $x_t \in Z_i(A) \cap \tilde{E}$ such that $||y - x_t|| < \varepsilon$ and $||z - C^{-1}R_i(t)x_t|| < \varepsilon$.

In the case $\tilde{E} = E$, it is also said that a $\tilde{E}$-hypercyclic vector of $(R_i(t))_{t\geq 0}$ is a *hypercyclic vector* of $(R_i(t))_{t\geq 0}$ and that $(R_i(t))_{t\geq 0}$ is *topologically transitive*, resp. *topologically mixing*.

Suppose $C = I$, $\tilde{E} = E$ and $(R_i(t))_{t\geq 0}$ is topologically transitive for some $i \in \mathbb{N}^0_{m_n-1}$. Then $(R_i(t))_{t\geq 0}$ is hypercyclic and the set of all hypercyclic vectors of $(R_i(t))_{t\geq 0}$, denoted by $\mathrm{HC}(R_i)$, is a dense G_δ-subset of E [16]. Furthermore, the condition $\rho(A) \neq \emptyset$ combined with the proofs of [19, Lemma 3.1, Theorem 3.2] implies that $\mathrm{HC}(R_i) \cap D_\infty(A)$ is a dense subset of E.

The proof of following theorem follows from Lemma 2 and the argumentation used in the proof of [24, Theorem 2.3].

Theorem 1 *Suppose $i \in \mathbb{N}^0_{m_n-1}$, Ω is an open connected subset of $\mathbb{C}$, $\Omega \cap (-\infty, 0] = \emptyset$ and $P_\Omega := \{P_\lambda : \lambda \in \Omega\} \subseteq \sigma_p(A)$. Let $f : P_\Omega \to E$ be an analytic mapping such that $f(P_\lambda) \in Kern(P_\lambda - A) \setminus \{0\}$, $\lambda \in \Omega$ and let $\tilde{E} := \overline{span\{f(P_\lambda) : \lambda \in \Omega\}}$. Suppose Ω_+ and Ω_- are two open connected subsets of Ω, and each of them admits a cluster point in Ω. If*

$$\lim_{t\to+\infty} \left|F_i(\lambda, t)\right| = +\infty,\ \lambda \in \Omega_+ \text{ and } \lim_{t\to+\infty} F_i(\lambda, t) = 0,\ \lambda \in \Omega_-, \tag{8}$$

then $(R_i(t))_{t\geq 0}$ is $\tilde{E}$-topologically mixing.

Remark 1 1. Assume that $\langle x^*, f(P_\lambda)\rangle = 0$, $\lambda \in \Omega$ for some $x^* \in E^*$ implies $x^* = 0$. Then $\tilde{E} = E$.

2. It is not clear how one can prove an extension of [14, Theorem 2.1] for the most simplest time-fractional evolution equations of the form (1).
3. The previous theorem can be slightly improved in the following manner. Suppose $l \in \mathbb{N}$, $\Omega_1, \ldots, \Omega_l$ are open connected subsets $\Omega_1, \ldots, \Omega_l$ of $\mathbb{C}$, as well as $\Omega_{j,+}$ and $\Omega_{j,-}$ are open connected subsets of Ω_j which admits a cluster point in Ω_j, and satisfy (8) with Ω_+ and Ω_- replaced respectively by $\Omega_{j,+}$ and $\Omega_{j,-}$ $(1 \leq j \leq l)$. Assume, additionally, that $f_j : P_{\Omega_j} \to E$ is an analytic mapping, $\Omega_j \cap (-\infty, 0] = \emptyset$, $P_{\Omega_j} \subseteq \sigma_p(A)$, and $f_j(P_\lambda) \in \mathrm{Kern}(A - P_\lambda) \setminus \{0\}$, $\lambda \in \Omega_j$ $(1 \leq j \leq l)$. Set $\tilde{E} := \overline{span\{f_j(P_\lambda) : \lambda \in \Omega_j,\ 1 \leq j \leq n\}}$ and assume that Ω_j' is an open connected subset of Ω_j which admits a cluster point in Ω_j for $1 \leq j \leq l$. Then
$$\tilde{E} = \overline{span\{f_j(P_\lambda) : \lambda \in \Omega_j',\ 1 \leq j \leq l\}}$$
and one can repeat literally the proof of Theorem 1 in order to see that $(R_i(t))_{t\geq 0}$ is $\tilde{E}$-topologically mixing (cf. also [7]).
4. Let $Cf(P_\lambda) \in \tilde{E}$, $\lambda \in \Omega$. Then $A_{|\tilde{E}}$ is the densely defined integral generator of the $C_{|\tilde{E}}$-resolvent propagation family $((R_0(t)_{|\tilde{E}})_{t\geq 0}, \ldots, (R_{m_n-1}(t)_{|\tilde{E}})_{t\geq 0})$ in the Banach space $\tilde{E}$, $C^{-1}_{|\tilde{E}} A_{|\tilde{E}} C_{|\tilde{E}} = A_{|\tilde{E}}$ and the proof of Theorem 1 implies that $((R_0(t)_{|\tilde{E}})_{t\geq 0}, \ldots, (R_{m_n-1}(t)_{|\tilde{E}})_{t\geq 0})$ is topologically mixing in $\tilde{E}$. The additional assumption $\overline{C(\tilde{E})} = \tilde{E}$ implies that $((R_0(t)_{|\tilde{E}})_{t\geq 0}, \ldots, (R_{m_n-1}(t)_{|\tilde{E}})_{t\geq 0})$ is

hypercyclic and that the set of all hypercyclic vectors of $((R_0(t)_{|\tilde{E}})_{t\geq 0}, \ldots, (R_{m_n-1}(t)_{|\tilde{E}})_{t\geq 0})$ is dense in $\tilde{E}$.

5. The assumptions of Theorem 1 hold provided that $n = 2$, $c_1 = 0$, $\alpha_2 > 0$, $\alpha = 0$, $i = 0$ and $\Omega \cap i\mathbb{R} \neq \emptyset$ [24]. In this case, $F_0(\lambda, t) = E_{\alpha_2}(\lambda^{\alpha_2} t^{\alpha_2})$, $t \geq 0$, and there exist $\lambda_0 \in \Omega$ and $\delta > 0$ such that (8) holds with $\Omega_+ = \{\lambda \in \Omega : |\lambda - \lambda_0| < \delta,\ \arg(\lambda) \in (\frac{\pi}{2} - \delta, \frac{\pi}{2})\}$ and $\Omega_- = \{\lambda \in \Omega : |\lambda - \lambda_0| < \delta,\ \arg(\lambda) \in (\frac{\pi}{2}, \frac{\pi}{2} + \delta)\}$.
6. It is worth noting that the condition (8) of Theorem 1 does not hold in general. In order to illustrate this, suppose now $n = 4$, $\alpha_j = j - 1$, $j \in \mathbb{N}_4$, $\alpha = 1$, $i = 2$ and $c_1 \in \mathbb{C} \setminus \{0\}$. Then $\mathcal{D}_2 = \emptyset$ and, for every $t \geq 0$,

$$F_2(\lambda, t) = \frac{e^{\lambda t}}{(\lambda - \lambda_1)(\lambda - \lambda_2)} + \frac{e^{\lambda_1 t}}{(\lambda_1 - \lambda)(\lambda_1 - \lambda_2)} + \frac{e^{\lambda_2 t}}{(\lambda_2 - \lambda)(\lambda_2 - \lambda_1)},$$

where $\lambda_{1,2} := (-\lambda^2 \pm \sqrt{\lambda^4 + 4c_1\lambda})/(2\lambda)$. It is not difficult to prove that, for every $\lambda \in \mathbb{C} \setminus \{0\}$, the following relation holds: $\Re\lambda \neq \Re\lambda_1$. This implies that, for every $\lambda \in \mathbb{C}$ with $\Re\lambda > 0$, one has $\lim_{t\to+\infty} |F_2(\lambda, t)| = +\infty$. Regrettably, there does not exist an open connected subset Ω_- of $\mathbb{C}$ such that $\lim_{t\to+\infty} F_2(\lambda, t) = 0$, $\lambda \in \Omega_-$.
7. As far as we know, in the handbooks containing tables of Laplace transforms, the explicit forms of functions like $F_i(\lambda, t)$ have not been presented as known images, except for some very special cases of the coefficients α_j, c_j. In this place, we would like to point out the following fact. Suppose $\alpha_n - \alpha_j \in \mathbb{Q}$, $j \in \mathbb{N}^0_{n-1}$. By the well-known formula [5, (1.26)], we obtain that there exists a number $\zeta \in (0, 1)$, independent of λ, such that the function $F_i(\lambda, t)$ can be represented as the finite convolution products of functions like $E_{1,\zeta}(p_\lambda t)$.

We recommend for the reader the reference [25] for the basic hypercyclic and chaotic properties of fractionally integrated C-cosine functions. Notice that, in general, the notion of chaoticity makes no sense for the equations of the form (1).

We shall omit the proof of the following extension of [24, Theorem 2.4].

Theorem 2 *Suppose $R(C)$ is dense in E and there exists $i \in \mathbb{N}^0_{m_n-1}$ such that $(R_i(t))_{t\geq 0}$ is hypercyclic. Then $\sigma_p(A^*) = \emptyset$.*

We close the paper by giving some illustrative examples (for some other applications, the reader may consult the references [2–4, 10, 15, 33, 36, 37]).

Example 1 1. [12, 13, 24, 25] Let $a, b, c > 0$, $\zeta \in (0, 2)$, $c < \frac{b^2}{2a} < 1$ and

$$\Lambda := \left\{\lambda \in \mathbb{C} : \left|\lambda - \left(c - \frac{b^2}{4a}\right)\right| \leq \frac{b^2}{4a},\ \Im(\lambda) \neq 0 \text{ if } \Re(\lambda) \leq c - \frac{b^2}{4a}\right\}.$$

Consider the following abstract time-fractional equation:

$$\begin{cases} \mathbf{D}_t^{\alpha} u(t) = a u_{xx} + b u_x + c u := -Au, \\ u(0,t) = 0, \ t \geq 0, \\ u(x,0) = u_0(x), \ x \geq 0, \ \text{and } u_t(x,0) = 0, \ \text{if } \alpha \in (1,2). \end{cases}$$

As it is known, the operator $-A$ with domain $D(-A) = \{f \in W^{2,2}([0,\infty)) : f(0) = 0\}$, generates an analytic strongly continuous semigroup of angle $\frac{\pi}{2}$ in the space $E = L^2([0,\infty))$; the same assertion holds in the case that the operator $-A$ acts on $E = L^1([0,\infty))$ with domain $D(-A) = \{f \in W^{2,1}([0,\infty)) : f(0) = 0\}$. Assume first $\zeta \in [1,2)$, $\theta \in (\zeta\frac{\pi}{2} - \pi, \pi - \zeta\frac{\pi}{2})$ and $P(z) = \sum_{j=0}^{n} a_j z^j$ is a non-constant complex polynomial such that $a_n > 0$ and

$$-e^{i\theta} P(-\Lambda) \cap \left\{ t e^{\pm i\zeta\frac{\pi}{2}} : t \geq 0 \right\} \neq \emptyset. \tag{9}$$

Then it is not difficult to prove that $-e^{i\theta}P(A)$ generates an analytic C_0-semigroup of angle $\frac{\pi}{2} - |\theta|$. Taking into account [23, Theorem 2.17], one gets that the operator $-e^{i\theta}P(A)$ is the integral generator of an exponentially bounded, analytic ζ-times regularized resolvent family $(R_{\zeta,\theta,P}(t))_{t\geq 0}$ of angle $\frac{\pi-|\theta|}{\zeta} - \frac{\pi}{2}$. It is not difficult to show that the conditions of Theorem 1 are satisfied with $\tilde{E} = E$, which implies that $(R_{\zeta,\theta,P}(t))_{t\geq 0}$ is topologically mixing. Suppose now $\zeta \in (0,1)$, $\theta \in (-\frac{\pi}{2}, \frac{\pi}{2})$ and $P(z) = \sum_{j=0}^{n} a_j z^j$ is a non-constant complex polynomial such that $a_n > 0$ and (9) holds. Then $-e^{i\theta}P(A)$ is the integral generator of an exponentially bounded, analytic ζ-times regularized resolvent family $(R_{\zeta,\theta,P}(t))_{t\geq 0}$ of angle $\min((\frac{1}{\zeta} - 1)\frac{\pi}{2}, \frac{\pi}{2})$. Using the above arguments, we easily infer that $(R_{\zeta,\theta,P}(t))_{t\geq 0}$ is topologically mixing. Notice that (9) holds if $c < \frac{b^2}{4a}$; in the case $c \geq \frac{b^2}{4a}$, one can prove that (9) holds provided $a_0 = 0$ or $P(z) = \sum_{j=0}^{n} a_j (z+d)^j$, $z \in \mathbb{C}$, where $d \in \mathbb{C}$ and $0 \in \operatorname{int}(d - \Lambda)$. Consider now the Eq. (1) with $n = 2$, $\alpha_2 = 2$, $\alpha_1 = 0$, $\alpha = 1$, $c_1 > 0$, and A replaced by $-e^{i\theta}P(A)$ therein. Using Lemma 1(1), one gets that $-e^{i\theta}P(A)$ is the integral generator of an exponentially bounded, analytic resolvent propagation family $((R_0(t))_{t\geq 0}, (R_1(t))_{t\geq 0})$ of angle $\frac{\pi}{2} - |\theta|$. Moreover, $F_0(\lambda, t) = (1 + c_1(\lambda^2 - c_1)^{-1})e^{\lambda t} - c_1(\lambda^2 - c_1)^{-1}e^{c_1 t/\lambda}$, $t \geq 0$. By Theorem 1, we easily infer that the condition

$$e^{i\theta} P(-\Lambda) \cap i\mathbb{R} \neq \emptyset \tag{10}$$

mplies that $(R_0(t))_{t\geq 0}$ is topologically mixing. Finally, suppose that $n = 2$, $\alpha_2 - \alpha = 1$, $\alpha_1 - \alpha = -1$, $i = 1$, $c_1 > 0$ and $2 < \alpha_2 \leq 3$. Then $m_2 = 3$, $\mathcal{D}_1 = \emptyset$ and $F_1(\lambda, t) = \lambda^{-1}(1 + c_1(\lambda^2 - c_1)^{-1})e^{\lambda t} - \lambda(\lambda^2 - c_1)^{-1}e^{c_1 t/\lambda}$, $t \geq 0$. By Lemma 1(1), we get that $-e^{i\theta}P(A)$ is the integral generator of an exponentially bounded, analytic resolvent propagation family $((R_0(t))_{t\geq 0}, (R_1(t))_{t\geq 0}, (R_2(t))_{t\geq 0})$ of angle $\frac{\pi}{2} - |\theta|$. If the condition (10) is satisfied, then one can apply Theorem 1 in order to see that $(R_1(t))_{t\geq 0}$ is topologically mixing.

2. References [18, 24] Theorem 1 can be applied in the analysis of (subspace) topologically mixing properties of time-fractional wave equation and time-fractional heat equation on symmetric spaces of non-compact type (cf. [18, Theorem 3.1a, 3.2, Corollary 3.3]); here we shall also provide some applications to the abstract Cauchy problem (1). Consider, for example, the situation of [18, Theorem 3.1a]. Let X be a symmetric space of non-compact type and rank one, let $p > 2$, let the parabolic domain P_p and the positive real number c_p possess the same meaning as in [18] and let $P(z) = \sum_{j=0}^{n} a_j z^j$, $z \in \mathbb{C}$ be a non-constant complex polynomial with $a_n > 0$. Assume first $\zeta \in (1, 2)$, $\pi - n \arctan \frac{|p-2|}{2\sqrt{p-1}} - \zeta\frac{\pi}{2} > 0$ and $\theta \in (n \arctan \frac{|p-2|}{2\sqrt{p-1}} + \zeta\frac{\pi}{2} - \pi, \pi - n \arctan \frac{|p-2|}{2\sqrt{p-1}} - \zeta\frac{\pi}{2})$. Then it is obvious that $-e^{i\theta} P(\Delta_{X,p}^{\natural})$ is the integral generator of an exponentially bounded, analytic ζ-times regularized resolvent family $(R_{\zeta,\theta,P}(t))_{t\geq 0}$ of angle $\frac{1}{\zeta}(\pi - n \arctan \frac{|p-2|}{2\sqrt{p-1}} - \zeta\frac{\pi}{2} - |\theta|)$. Keeping in mind that $\operatorname{int}(P_p) \subseteq \sigma_p(\Delta_{X,p}^{\natural})$, the condition

$$-e^{i\theta} P\big(\operatorname{int}(P_p)\big) \cap \big\{te^{\pm i\zeta\frac{\pi}{2}} : t \geq 0\big\} \neq \emptyset \tag{11}$$

implies that $(R_{\zeta,\theta,P}(t))_{t\geq 0}$ is topologically mixing. Suppose now $n = 2$, $0 < a < 2$, $\alpha_2 = 2a$, $\alpha_1 = 0$, $\alpha = a$, $c_1 > 0$, $i = 0$ and $|\theta| < \min(\frac{\pi}{2} - n \arctan \frac{|p-2|}{2\sqrt{p-1}}, \frac{\pi}{2} - n \arctan \frac{|p-2|}{2\sqrt{p-1}} - \frac{\pi}{2}a)$. Then $\mathcal{D}_1 = \emptyset$ and, by Lemma 1(1), $-e^{i\theta} P(\Delta_{X,p}^{\natural})$ is the integral generator of an exponentially bounded, analytic resolvent propagation family $((R_{\theta,P,0}(t))_{t\geq 0}, \ldots, (R_{\theta,P,\lceil 2a\rceil-1}(t))_{t\geq 0})$ of angle $\min(\frac{\pi - n \arctan \frac{|p-2|}{2\sqrt{p-1}} - |\theta|}{a} - \frac{\pi}{2}, \frac{\pi}{2})$. Furthermore, the equality [5, (1.26)] can serve one to simply verify that:

$$\begin{aligned}
F_0(\lambda, t) =& \frac{\lambda^a t^{-a}}{\lambda^{2a} - c_1}\Big(E_{a,2-a}\big(\lambda^a t^a\big) - E_{a,2-a}\big(c_1\lambda^{-a} t^a\big)\Big) \\
&+ \frac{\lambda^a}{\lambda^{2a} - c_1}\Big[\lambda^a E_a\big(\lambda^a t^a\big) + (a-1)\lambda^a E_{a,2}\big(\lambda^a t^a\big) \\
&- c_1\lambda^{-a} E_a\big(c_1\lambda^{-a} t^a\big) - (a-1)c_1\lambda^{-a} E_{a,2}\big(c_1\lambda^{-a} t^a\big)\Big], \quad t > 0.
\end{aligned}$$

Invoking the asymptotic expansion formulae (5)–(7) and the above expression, it can be shown without any substantial difficulties that the condition

$$-e^{i\theta} P\big(\operatorname{int}(P_p)\big) \cap \big\{(it)^a + c_1(it)^{-a} : t \in \mathbb{R} \setminus \{0\}\big\} \neq \emptyset$$

implies that $(R_{\theta,P,0}(t))_{t\geq 0}$ is topologically mixing. Finally, let $\zeta \in (0, 1)$ and let

$$\theta \in \left(n \arctan \frac{|p-2|}{2\sqrt{p-1}} - \frac{\pi}{2}, \frac{\pi}{2} - n \arctan \frac{|p-2|}{2\sqrt{p-1}}\right).$$

Then the validity of (11) provides that $-e^{i\theta} P(\Delta_{X,p}^{\natural})$ is the integral generator of a topologically mixing ζ-times regularized resolvent family $(R_{\zeta,\theta,P}(t))_{t\geq 0}$

of angle $\min((\frac{1}{\zeta}-1)\frac{\pi}{2},\frac{\pi}{2})$. It is clear that (11) holds if $P(z)$ is of the form $P(z)=\sum_{j=0}^{n}a_j(z-c)^j$, $z\in\mathbb{C}$, where $c>c_P$.

3. References [7, 24, 34] Suppose $\zeta\in(0,1)$, $E:=L^2(\mathbb{R})$, $c>\frac{b}{2}>0$, $\Omega:=\{\lambda\in\mathbb{C}:\Re\lambda<c-\frac{b}{2}\}$, $\phi\in E^*=E$ and $\mathcal{A}_c u:=u''+2bxu'+cu$ is the bounded perturbation of the one-dimensional Ornstein–Uhlenbeck operator acting with domain $D(\mathcal{A}_c):=\{u\in L^2(\mathbb{R})\cap W^{2,2}_{loc}(\mathbb{R}):\mathcal{A}_c u\in L^2(\mathbb{R})\}$. Then $\mathcal{A}_c$ is the integral generator of a topologically mixing ζ-times regularized resolvent family $(R_\zeta(t))_{t\geq 0}$ which cannot be hypercyclic provided $b<0$ or $c\leq\frac{b}{2}$ [7, 24]. Notice also that the above assertions continue to hold in the case of ζ-times regularized resolvent families generated by bounded perturbations of multi-dimensional Ornstein–Uhlenbeck operators [7, Proposition 4.1,Theorem 4.2]; for the sake of simplicity, in the sequel of this example we shall consider only the hypercyclic and topologically mixing properties of resolvent propagation families generated by the operator $\mathcal{A}_c$ defined above. Suppose $\alpha_n-\alpha<1$. Then an application of Lemma 1(2) shows that $\mathcal{A}_c$ is the integral generator of an exponentially bounded, analytic resolvent propagation family $((R_0(t))_{t\geq 0},\ldots,(R_{m_n-1}(t))_{t\geq 0})$ of angle $\min(\frac{\pi}{2(\alpha_n-\alpha)}-\frac{\pi}{2},\frac{\pi}{2})$. If $b<0$, then $\sigma_p(\mathcal{A}_c^*)\neq\emptyset$ (cf. [7]) and, by Theorem 2, there does not exist $i\in\mathbb{N}^0_{m_n-1}$ such that $(R_i(t))_{t\geq 0}$ is hypercyclic (the case $c\leq\frac{b}{2}$ is more complicated in the newly arisen situation since it is not clear how one can prove the boundedness of $(R_i(t))_{t\geq 0}$, in general). Consider now the following case: $n=3$, $\frac{1}{3}<a<\frac{1}{2}$, $\alpha_3=3a$, $\alpha_2=2a$, $\alpha_1=0$, $\alpha=a$, $c_1<0$, $c_2>0$ and $i=1$. Then $\mathcal{D}_1=\emptyset$ and

$$\mathcal{L}\big(F_1(\lambda,t)\big)(z)=\frac{z^{3a-2}}{z^{3a}+c_2z^{2a}-z^a\big(\lambda^{2a}+c_1\lambda^{-a}+c_2\lambda^a\big)+c_1}.$$

Set $\lambda_{1,2}:=\frac{-c_2-\lambda^a\pm\sqrt{(c_2+\lambda^a)^2+4c_1\lambda^{-a}}}{2}$. Then one can simply prove that the set $\Upsilon=\{\lambda\in\mathbb{C}:(\lambda^a-\lambda_1)(\lambda^a-\lambda_2)(\lambda_1-\lambda_2)\neq 0\}$ is finite and that, for every $z\in\mathbb{C}\setminus\{0\}$ and $\lambda\in\mathbb{C}\setminus\Upsilon$,

$$z^{3a}+c_2z^{2a}-z^a\big(\lambda^{2a}+c_1\lambda^{-a}+c_2\lambda^a\big)+c_1=\big(z^a-\lambda^a\big)\big(z^a-\lambda_1\big)\big(z^a-\lambda_2\big).$$

Using the equality [5, (1.26)], we get that, for every $\lambda\in\mathbb{C}\setminus\Upsilon$,

$$F_1(\lambda,t)=\frac{t^{1-2a}E_{a,2-2a}\big(\lambda^a t^a\big)}{\big(\lambda^a-\lambda_1\big)\big(\lambda^a-\lambda_2\big)}+\frac{t^{1-2a}E_{a,2-2a}\big(\lambda_1 t^a\big)}{\big(\lambda_1-\lambda_2\big)\big(\lambda_1-\lambda^a\big)}+\frac{t^{1-2a}E_{a,2-2a}\big(\lambda_2 t^a\big)}{\big(\lambda_2-\lambda_1\big)\big(\lambda_2-\lambda^a\big)}. \tag{12}$$

Clearly, $P_\lambda=\lambda^{2a}+c_2\lambda^a+c_1\lambda^{-a}$, $\lambda\in\mathbb{C}\setminus\{0\}$, $\lim_{\lambda\to 0}(\lambda_1-(-\frac{c_2}{2}+\sqrt{c_1\lambda^{-a}}))=0$ and $\lim_{\lambda\to 0}(\lambda_2-(-\frac{c_2}{2}-\sqrt{c_1\lambda^{-a}}))=0$. This implies that there exists a sufficiently small number $\epsilon_1>0$ such that, for every $\lambda\in\mathbb{C}$ with $\Re\lambda>0$ and $|\lambda|\leq\epsilon_1$, the following holds: $\Re\lambda_2\leq-\frac{c_2}{4}$ and

$$\operatorname{dist}\Bigl(\lambda_1, \Bigl\{z \in \mathbb{C} : \arg\Bigl(z + \frac{c_2}{2}\Bigr) \in \Bigl[\frac{\pi}{2} - \frac{\pi a}{4}, \frac{\pi}{2}\Bigr]\Bigr\}\Bigr) < \min\Bigl(\frac{c_2}{4}, \frac{c_2}{2}\cot\frac{\pi a}{4}\Bigr). \tag{13}$$

Arguing similarly, we obtain that there exists a sufficiently small number $\epsilon_2 > 0$ such that, for every $\lambda \in \mathbb{C}$ with $\arg(\lambda) \in (\frac{\pi}{2}, \frac{\pi}{2a})$ and $|\lambda| \leq \epsilon_2$, the following holds: $\Re\lambda_2 \leq -\frac{c_2}{4}$ and

$$\operatorname{dist}\Bigl(\lambda_1, \Bigl\{z \in \mathbb{C} : \arg\Bigl(z + \frac{c_2}{2}\Bigr) \in \Bigl[\frac{\pi}{4}, \frac{\pi}{2} - \frac{\pi a}{4}\Bigr]\Bigr\}\Bigr) < \frac{c_2}{4}. \tag{14}$$

Furthermore, our assumption $c_1 < 0$ implies that there exists a sufficiently small number $\epsilon_3 > 0$ such that, for every $\lambda \in \mathbb{C} \setminus \{0\}$ with $|\arg(\lambda)| \leq \frac{\pi}{2a}$ and $|\lambda| \leq \epsilon_3$, we have $\Re(P_\lambda) = \Re(\lambda^{2a} + c_2\lambda^a + c_1\lambda^{-a}) \leq \epsilon_3^{2a} + |c_2|\epsilon_3^a < c - \frac{b}{2}$. Let $\epsilon_4 > 0$ satisfy that, for every $\lambda \in \mathbb{C} \setminus \{0\}$ and $|\lambda| \leq \epsilon_4$, one has $\lambda \in \Upsilon$. Put $\epsilon := \min(\epsilon_1, \epsilon_2, \epsilon_3, \epsilon_4)$, $\Omega := \Omega_1 := \Omega_2 := \{z \in \mathbb{C} \setminus \{0\} : |\arg(z)| \leq \frac{\pi}{2a},\ |z| < \epsilon\}$, $\Omega_+ := \Omega_{1,+} := \Omega_{2,+} := \{z \in \mathbb{C} : \Re z > 0, |z| < \epsilon\}$ and $\Omega_- := \Omega_{1,-} := \Omega_{2,-} := \{z \in \mathbb{C} \setminus \{0\} : \arg(z) \in (\frac{\pi}{2}, \frac{\pi}{2a}),\ |z| < \epsilon\}$. Then it is obvious that $P_\Omega \subseteq \sigma_p(\mathcal{A}_c)$. Define $f_1 : P_\Omega \to E$ and $f_2 : P_\Omega \to E$ by $f_1(z) := \mathcal{F}^{-1}(e^{-\frac{\xi^2}{2b}}\xi|\xi|^{-(2+\frac{z-c}{b})})(\cdot)$, $z \in P_\Omega$ and $f_2(z) := \mathcal{F}^{-1}(e^{-\frac{\xi^2}{2b}}|\xi|^{-(1+\frac{z-c}{b})})(\cdot)$, $z \in P_\Omega$, where $\mathcal{F}$ and $\mathcal{F}^{-1}$ denote the Fourier transform on the real line and its inverse transform, respectively. Exploiting (12)–(14) and (5)–(7), we easily infer that:

$$\lim_{t \to +\infty} \bigl|F_1(\lambda, t)\bigr| = +\infty,\ \lambda \in \Omega_+ \text{ and } \lim_{t \to +\infty} F_1(\lambda, t) = 0,\ \lambda \in \Omega_-.$$

By Remark 1(3) and the consideration given in [24, Example 2.5(iii)], we reveal that $(R_1(t))_{t \geq 0}$ is topologically mixing.

4. The study of qualitative properties of the abstract Basset–Boussinesq–Oseen equation:

$$u'(t) - \mathbf{A}\mathbf{D}_t^\alpha u(t) + u(t) = f(t),\quad t \geq 0,\ u(0) = 0\quad (\alpha \in (0, 1)), \tag{15}$$

describing the unsteady motion of a particle accelerating in a viscous fluid under the action of the gravity, has been initiated by C. Lizama and H. Prado in [31]. For further results concerning the C-wellposedness of (15), the references [27] and [28] are of importance. Our intention here is to clarify the most important facts about hypercyclic and topologically mixing properties of once integrated solutions of the Eq. (15) with $f(t) \equiv 0$. Clearly, $n = 2$, $\alpha_2 = 1$, $\alpha_1 = 0$, $c_1 = 1$, $\mathcal{D}_0 = \emptyset$ and the analysis is quite complicated in the general case since

$$\mathcal{L}\bigl(F_0(\lambda, t)\bigr)(z) = \frac{1}{z + 1 - z^\alpha\bigl(\lambda^{1-\alpha} + \lambda^{-\alpha}\bigr)}.$$

The cases $\alpha = \frac{1}{2}$ and $\alpha = \frac{1}{3}$ can be considered similarly as in the parts (2) and (3). Suppose now $\alpha = \frac{2}{3}$, $A \equiv \mathcal{A}_c$ and $c - \frac{b}{2} > 2^{1/3} + 2^{2/3}$ (cf. (3)). Then $\mathcal{A}_c$

is the integral generator of an exponentially bounded, analytic resolvent propagation family $(R_0(t))_{t\geq 0}$ of angle $\frac{\pi}{2}$. Put $\lambda_{1,2} := \frac{-\lambda^{1/3}\pm\sqrt{\lambda^{2/3}+4\lambda^{(-1)/3}}}{2}$. Then the sets $\Upsilon_1 := \{\lambda \in \mathbb{C} \setminus \{0, -4\} : (\lambda - \lambda_1)(\lambda - \lambda_2) = 0\}$ and $\Upsilon_2 := \{\lambda \in \mathbb{C} \setminus \{0\} : \Re\lambda = \Re(\lambda_1^3)\}$ are finite. Furthermore, for every $\lambda \in \mathbb{C} \setminus ((-\infty, 0] \cup \Upsilon_1)$, one has:

$$
\begin{aligned}
F_0(\lambda, t) =& \frac{E_{1/3,1/3}\big(\lambda^{1/3}t^{1/3}\big)}{\big(\lambda^{1/3} - \lambda_1\big)\big(\lambda^{1/3} - \lambda_2\big)} \\
&+ \frac{E_{1/3,1/3}\big(\lambda_1 t^{1/3}\big)}{\big(\lambda_1 - \lambda^{1/3}\big)\big(\lambda_1 - \lambda_2\big)} + \frac{E_{1/3,1/3}\big(\lambda_2 t^{1/3}\big)}{\big(\lambda_2 - \lambda^{1/3}\big)\big(\lambda_2 - \lambda_1\big)} \\
&- \big(\lambda^{1/3} + \lambda^{(-2)/3}\big)\Big[\frac{E_{1/3,2/3}\big(\lambda^{1/3}t^{1/3}\big)}{t^{1/3}\big(\lambda^{1/3} - \lambda_1\big)\big(\lambda^{1/3} - \lambda_2\big)} \\
&+ \frac{E_{1/3,2/3}\big(\lambda_1 t^{1/3}\big)}{t^{1/3}\big(\lambda_1 - \lambda^{1/3}\big)\big(\lambda_1 - \lambda_2\big)} + \frac{E_{1/3,2/3}\big(\lambda_2 t^{1/3}\big)}{t^{1/3}\big(\lambda_2 - \lambda^{1/3}\big)\big(\lambda_2 - \lambda_1\big)}\Big].
\end{aligned}
$$

Since the function $s \mapsto s^{1/3} + s^{(-2)/3}$, $s > 0$ attained its global minimum $2^{1/3} + 2^{2/3}$ for $s = 2$, we obtain that there exist positive real numbers ε_1 and ε_2 such that $\varepsilon_1 < 2 < \varepsilon_2$ and $\Re(P_\lambda) = \Re(\lambda^{1/3} + \lambda^{(-2)/3}) < c - \frac{b}{2}$, provided $\varepsilon_1 < |\lambda| < \varepsilon_2$. Set $\Omega := \Omega_1 =:= \Omega_2 := \{\lambda \in \mathbb{C} : \varepsilon_1 < |\lambda| < \varepsilon_2\}$ and $\Omega_+ := \Omega_{1,+} = := \Omega_{2,+} := \{\lambda \in \mathbb{C} : \Re\lambda > 0,\ \varepsilon_1 < |\lambda| < \varepsilon_2,\ \lambda \notin \Upsilon_2\}$. It is clear that $\Re\lambda_2 < 0$ for $\lambda \in \mathbb{C} \setminus \{0\}$, and that $\lim_{\lambda\to -2, \Im\lambda>0} \lambda_1 = \lim_{\lambda\to -2, \Im\lambda>0} \frac{-\lambda^{1/3}+\sqrt{\lambda^{2/3}+4\lambda^{(-1)/3}}}{2} = \frac{-(-2)^{1/3}+\sqrt{(-2)^{2/3}+4(-2)^{(-1)/3}}}{2}$. Direct calculation shows that the argument of the last written number belongs to the set $(-\frac{2\pi}{3}, -\frac{\pi}{6})$, which implies that there exists a sufficiently small number $\epsilon > 0$ such that the set $\Omega_- := \Omega_{1,-} := \Omega_{2,-} := \{\lambda \in \mathbb{C} : \Im\lambda > 0,\ |\lambda + 2| < \epsilon\}$ is a subset of Ω, and that $\arg(\lambda) \in (-\frac{2\pi}{3}, -\frac{\pi}{6})$ for $\lambda \in \Omega_-$. Using Remark 1(3) and (5)–(7), we obtain that $(R_0(t))_{t\geq 0}$ is topologically mixing.

5. References [11, 24] Let B, ω_1, ω_2, V_{ω_2,ω_1}, E, a and b possess the same meaning as in [11, Sect. 5] and let $Q(z)$ be a non-constant complex polynomial of degree n. Assume $0 < \zeta < 2$, $N \in \mathbb{N}$, $N > \frac{n}{2\zeta}$ and

$$R_\zeta(t) = \Big(E_\zeta\big(t^\zeta Q(z)\big)e^{-(-z^2)^N}\Big)(B),\ t \geq 0, \tag{16}$$

where the right hand side of (16) is defined by means of the $H_{a,b}$ functional calculus developed in [11]. Then $R((e^{-(-z^2)^N})(B))$ is dense in E, and $(R_\zeta(t))_{t\geq 0}$ is a ζ-times $(e^{-(-z^2)^N})(B)$-regularized resolvent family generated by $Q(B)$. Moreover, the condition

$$Q\big(\mathrm{int}\big(V_{\omega_2,\omega_1}\big)\big) \cap \big\{te^{\pm i\zeta\frac{\pi}{2}} : t \geq 0\big\} \neq \emptyset$$

implies that $(R_{\zeta}(t))_{t\geq 0}$ is both topologically mixing and hypercyclic (cf. also [25, Example 36(ii)] for the case $\zeta = 2$).

We leave to the interested reader the problem of finding some other applications of functional calculi in the analysis of hypercyclic and topologically mixing properties of the abstract Cauchy problem (1).

Acknowledgments This research is supported by grant 174024, Ministry of Science and Technological Development, Republic of Serbia.

References

1. Arendt, W., Batty, C.J.K., Hieber, M., Neubrander, F.: Vector-Valued Laplace Transforms and Cauchy Problems. Birkhäuser-Verlag, Basel (2001)
2. Banasiak, J., Moszyński, M.: A generalization of Desch-Schappacher-Webb criteria for chaos. Discret. Contin. Dyn. Syst. **12**, 959–972 (2005)
3. Banasiak, J., Lachowicz, M.: Topological chaos for birth-and-death-type-models with proliferation. Math. Model. Method. Appl. Sci. **12**, 755–775 (2002)
4. Banasiak, J., Lachowicz, M., Moszyński, M.: Chaotic behaviour of semigroups related to process of gene amplification-deamplification with cell proliferation. Math. Biosci. **206**, 200–215 (2007)
5. Bazhlekova, E.: Fractional evolution equations in Banach spaces. Dissertation, Eindhoven University of Technology (2001)
6. Bonilla, A., Miana, P.J.: Hypercyclic and topologically mixing cosine functions on Banach spaces. Proc. Am. Math. Soc. **136**, 519–528 (2008)
7. Conejero, J.A., Mangino, E.: Hypercyclic semigroups generated by Ornstein-Uhlenbeck operators. Mediterr. J. Math. **7**, 101–109 (2010)
8. Conejero, J.A., Peris, A.: Linear transitivity criteria. Topol. Appl. **153**, 767–773 (2005)
9. deLaubenfels, R.: Existence Families, Functional Calculi and Evolution Equations. Springer-Verlag, New York (1994)
10. deLaubenfels, R., Emamirad, H.: Chaos for functions of discrete and continuous weighted shift operators. Ergod. Th. Dyn. Syst. **21**, 1411–1427 (2001)
11. deLaubenfels, R., Emamirad, H., Grosse-Erdmann, K.-G.: Chaos for semigroups of unbounded operators. Math. Nachr. **261/262**, 47–59 (2003)
12. DeLaubenfels, R., Emamirad, H., Protopopescu, V.: Linear chaos and approximation. J. Approx. Theor. **105**, 176–187 (2000)
13. Desch, W., Schappacher, W., Webb, G.F.: Hypercyclic and chaotic semigroups of linear operators. Ergod. Th. Dyn. Syst. **17**, 1–27 (1997)
14. El Mourchid, S.: The imaginary point spectrum and hypercyclicity. Semigroup Forum **76**, 313–316 (2006)
15. El Mourchid, S., Metafune, G., Rhandi, A., Voigt, J.: On the chaotic behaviour of size structured cell population. J. Math. Anal. Appl. **339**, 918–924 (2008)
16. Grosse-Erdmann, K.-G.: Universal families and hypercyclic operators. Bulettin AMS **36**, 345–381 (1999)
17. Grosse-Erdmann, K.-G., Peris, A.: Linear Chaos. Springer, London (2011)
18. Ji, L., Weber, A.: Dynamics of the heat semigroup on symmetric spaces. Ergod. Th. Dyn. Syst. **30**, 457–468 (2010)
19. Kalmes, T.: Hypercyclic, mixing, and chaotic C_0-semigroups. Dissertation, Universität Trier (2006)
20. Kalmes, T.: Hypercyclic C_0-semigroups and evolution families generated by first order differential operators. Proc. Am. Math. Soc. **137**, 3833–3848 (2009)

21. Kalmes, T.: Hypercyclicity and mixing for cosine operator functions generated by second order partial diiferential operators. J. Math. Anal. Appl. **365**, 363–375 (2010)
22. Kostić, M.: Generalized Semigroups and Cosine Functions. Mathematical Institute Belgrade, Belgrade (2011)
23. Kostić, M.: (a, k)-regularized C-resolvent families: regularity and local properties. Abstr. Appl. Anal. **858242**, 27 (2009)
24. Kostić, M.: Hypercyclicity and topologically mixing property for abstract time-fractional equations. Dyn. Syst. Int. J. **27**, 213–221 (2012). doi:10.1080/14689367.2012.655263
25. Kostić, M.: Hypercyclic and chaotic integrated C-cosine functions. Filomat **26**(1), 1–44 (2012)
26. Kostić, M., Li, C.-G., Li, M., Piskarev, S.: On a class of abstract time-fractional equations. Fract. Calc. Appl. Anal. **27**, 639–668 (2012)
27. Kostić, M.: On a class of (a, k)-regularized C-resolvent families. Electron. J. Qual. Theory Differ. Equ. **94**, pp. 27 (2012)
28. Kostić, M.: Perturbation theory for abstract volterra equations. Abstr. Appl. Anal. **307684**, pp. 26 (2013)
29. Lizama, C., Prado, H.: Fractional relaxation equations on Banach spaces. Appl. Math. Lett. **23**, 137–142 (2010)
30. Lizama, C.: Regularized solutions for abstract Volterra equations. J. Math. Anal. Appl. **243**, 278–292 (2000)
31. Lizama, C., Prado, H.: On duality and spectral properties of (a, k)-regularized resolvents. Proc. Roy. Soc. Edinburgh Sect. A **139**, 505–517 (2009)
32. Mainardi, F., Gorenflo, R.: On Mittag-Leffler-type functions in fractional evolution processes. J. Comp. Appl. Math. **118**, 283–299 (2000)
33. Matsui, M., Takeo, F.: Chaotic semigroups generated by certain differential operators of order 1. SUT J. Math. **37**, 51–67 (2001)
34. Metafune, G.: L^p-Spectrum of Ornstein-Uhlenbeck operators. Ann. Scuola Norm. Sup. Pisa Cl. Sci. **30**, 97–124 (2001)
35. Prüss, J.: Evolutionary Integral Equations and Applications. Birkhäuser-Verlag, Basel (1993)
36. Takeo, F.: Chaotic or hypercyclic semigroups on a function space $C_0(I, \mathbb{C})$ or $L^p(I, \mathbb{C})$.. SUT J. Math. **41**, 43–61 (2005)
37. Takeo, F.: Chaos and hypercyclicity for solution semigroups to partial differential equations. Nonlinear Anal. **63**, 1943–1953 (2005)
38. Xiao, T.-J., Liang, J.: The Cauchy Problem for Higher-Order Abstract Differential Equations. Springer, Berlin (1998)

Periods of Homeomorphisms on Closed Surfaces

Juan Luis García Guirao and Jaume Llibre

Abstract The goal of this paper is to show what information on the set of periodic points of a homeomorphism on a closed surface can be obtained using the action of this homeomorphism on the homological groups of the closed surface.

Keywords Periodic point · Period · Homeomorphism · Closed surface

1 Introduction

Here a *closed surface* means a connected compact surface with or without boundary, orientable or not. More precisely, *an orientable connected compact surface without boundary of genus* $g \geq 0$, $\mathbb{M}_g$, is homeomorphic to the sphere if $g = 0$, to the torus if $g = 1$, or to the connected sum of g copies of the torus if $g \geq 2$. *An orientable connected compact surface with boundary of genus* $g \geq 0$, $\mathbb{M}_{g,b}$, is homeomorphic to $\mathbb{M}_g$ minus a finite number $b > 0$ of open discs having pairwise disjoint closure. In what follows $\mathbb{M}_{g,0} = \mathbb{M}_g$.

A non-orientable connected compact surface without boundary of genus $g \geq 1$, $\mathbb{N}_g$, is homeomorphic to the real projective plane if $g = 1$, or to the connected sum of g copies of the real projective plane if $g > 1$. *A non-orientable connected compact surface with boundary of genus* $g \geq 1$, $\mathbb{N}_{g,b}$, is homeomorphic to $\mathbb{N}_g$ minus a finite number $b > 0$ of open discs having pairwise disjoint closure. In what follows $\mathbb{N}_{g,0} = \mathbb{N}_g$.

J.L.G. Guirao
Departamento de Matemática Aplicada y Estadística, Hospital de Marina,
Universidad Politécnica de Cartagena, 30203 Cartagena, Región de Murcia, Spain
e-mail: juan.garcia@upct.es

J. Llibre (✉)
Departament de Matemàtiques, Universitat Autònoma de Barcelona,
Barcelona, Spain
e-mail: jllibre@mat.uab.cat

L. Alsedà i Soler et al. (eds.), *Difference Equations, Discrete Dynamical Systems and Applications*, Springer Proceedings in Mathematics & Statistics 180, DOI 10.1007/978-3-662-52927-0_13

Let $f : \mathbb{X} \to \mathbb{X}$ be a homeomorphism on a closed surface $\mathbb{X}$. A point $x \in \mathbb{X}$ is periodic of period n if $f^n(x) = x$ and $f^k(x) \neq x$ for $k = 1, \ldots, n-1$. We denote by $\text{Per}(f)$ the set of periods of all periodic points of f. The aim of the present paper is to provide some information on $\text{Per}(f)$. The statement of our first result is.

Theorem 1 *Let $\mathbb{X}$ be a closed surface and let f be a self-homeomorphism into $\mathbb{X}$. If $\mathbb{X} = \mathbb{M}_{g,b}$, then the following statement hold.*

(*a*) *If $(g, b) \in \{(1,0), (0,2)\}$ (i.e. the torus and the closed annulus respectively), then there is no information on the set* $\text{Per}(f)$.
(*b*) *If $(g, b) = (0,0)$ (i.e. the 2-dimensional sphere), then* $\text{Per}(f) \cap \{1, 2\} \neq \emptyset$.
(*c*) *If $(g, b) = (0,1)$ (i.e. the 2-dimensional disc), then* $1 \in \text{Per}(f)$.
(*d*) *If $g > 1$ and $b = 0$, then* $\text{Per}(f) \cap \{1, 2, \ldots, 2g\} \neq \emptyset$.
(*e*) *If $g > 1$ and $b > 0$, then* $\text{Per}(f) \cap \{1, 2, \ldots, 2g + b - 1\} \neq \emptyset$.

If $\mathbb{X} = \mathbb{N}_{g,b}$, then the following statements hold.

(*f*) *If $(g, b) \in \{(2,0), (1,1)\}$ (i.e. the Klein bottle and the Möebius band respectively), there is no information on the set* $\text{Per}(f)$.
(*g*) *If $(g, b) = (1,0)$ (i.e. the projective plane), then* $1 \in \text{Per}(f)$.
(*h*) *If (g, b) does not satisfy the assumptions of statements (f) or (g), then* $\text{Per}(f) \cap \{1, 2, \ldots, g + b - 1\} \neq \emptyset$.

The proof of Theorem 1 is done in Sect. 2. The main tool for proving it is a result due to Fuller [4].

The results of Theorem 1 restricted to the orientable closed surfaces without boundary, i.e. for the closed surfaces $\mathbb{M}_{g,0}$, where already obtained by Franks and Llibre in [3].

The objective of the rest of the paper is to improve the information provided in Theorem 1 using as a main tool the Lefschetz fixed point theory. We shall follow the ideas of Franks and Llibre in [3] when they improve the results of Theorem 1 for the homeomorphisms of the closed surfaces $\mathbb{M}_{g,0}$, see Theorems 5 and 6.

Let A be an $n \times n$ complex matrix. A $k \times k$ *principal submatrix* of A is a submatrix lying in the same set of k rows and columns, and a $k \times k$ *principal minor* is the determinant of such a principal submatrix. There are $\binom{n}{k}$ different $k \times k$ principal minors of A, and the sum of these is denoted by $E_k(A)$. In particular, $E_1(A)$ is the trace of A, and $E_n(A)$ is the determinant of A, denoted by $\det(A)$.

It is well known that the characteristic polynomial of A is given by

$$\det(tI - A) = t^n - E_1(A)t^{n-1} + E_2(A)t^{n-2} - \cdots + (-1)^n E_n(A).$$

Our main result is state in the following theorem.

Theorem 2 *Let $f : \mathbb{X} \to \mathbb{X}$ be a homeomorphism and let A be the integral matrix of the isomorphism $f_{*1} : H_1(\mathbb{X}, \mathbb{Q}) \to H_1(\mathbb{X}, \mathbb{Q})$ induced by f on the first homology group of $\mathbb{X}$. If $\mathbb{X}$ is either $\mathbb{M}_{g,b}$ with $b > 0$, or $\mathbb{N}_{g,b}$ with $b \geq 0$, then the following statements hold.*

(a) *If* $E_1(A) \neq 1$, *then* $1 \in \mathrm{Per}(f)$.
(b) *If* $E_1(A) = 1$ *and* $E_2(A) \neq 0$, *then* $\mathrm{Per}(f) \cap \{1, 2\} \neq \emptyset$.

If $\mathbb{X} = \mathbb{M}_{g,b}$ *with* $b > 0$, *then the following statement hold.*

(c) *If* $2g + b - 1 \geq 3$, $E_1(A) = 1$, $E_2(A) = 0$ *and* k *is the smallest integer of the set* $\{3, 4, \ldots, 2g + b - 1\}$ *such that* $E_k(A) \neq 0$, *then* $\mathrm{Per}(f)$ *has a periodic point of period a divisor of* k.

If $\mathbb{X} = \mathbb{N}_{g,b}$ *with* $b \geq 0$, *then the following statement hold.*

(d) *If* $g + b - 1 \geq 3$, $E_1(A) = 1$, $E_2(A) = 0$ *and* k *is the smallest integer of the set* $\{3, 4, \ldots, g + b - 1\}$ *such that* $E_k(A) \neq 0$, *then* $\mathrm{Per}(f)$ *has a periodic point of period a divisor of* k.

Theorem 2 is proven in Sect. 3.

2 Proof of Theorem 1

Let f be continuous self-map defined on $\mathbb{M}_{g,b}$ or $\mathbb{N}_{g,b}$, respectively. For a closed surface, the homological groups with coefficients in $\mathbb{Q}$ are linear vector spaces over $\mathbb{Q}$. We recall the homological spaces of $\mathbb{M}_{g,b}$ with coefficients in $\mathbb{Q}$, i.e.

$$H_k(\mathbb{M}_{g,b}, \mathbb{Q}) = \mathbb{Q} \oplus \overset{n_k}{\cdots} \oplus \mathbb{Q},$$

where $n_0 = 1$, $n_1 = 2g$ if $b = 0$, $n_1 = 2g + b - 1$ if $b > 0$, $n_2 = 1$ if $b = 0$, and $n_2 = 0$ if $b > 0$; and the induced linear maps $f_{*k} : H_k(\mathbb{M}_{g,b}, \mathbb{Q}) \to H_k(\mathbb{M}_{g,b}, \mathbb{Q})$ by f on the homological group $H_k(\mathbb{M}_{g,b}, \mathbb{Q})$ are $f_{*0} = (1)$, $f_{*2} = (d)$ where d is the *degree* of the map f if $b = 0$, $f_{*2} = 0$ if $b > 0$, and $f_{*1} = A$ where A is an $n_1 \times n_1$ integral matrix (see for additional details [6, 7]).

We recall that the homological groups of $\mathbb{N}_{g,b}$ with coefficients in $\mathbb{Q}$, i.e.

$$H_k(\mathbb{N}_{g,b}, \mathbb{Q}) = \mathbb{Q} \oplus \overset{n_k}{\cdots} \oplus \mathbb{Q},$$

where $n_0 = 1, n_1 = g + b - 1$ and $n_2 = 0$; and the induced linear maps are $f_{*0} = (1)$ and $f_{*1} = A$ where A is an $n_1 \times n_1$ integral matrix (see again for additional details [6, 7]).

The proof of Theorem 1 is a consequence of a general result from polyhedron homeomorphisms proved in [4], see also Halpern [5] and Brown [1] for more details on it.

Theorem 3 (Fuller's Theorem) *Let* f *be a homeomorphism of a compact polyhedron* X *into itself. If the Euler characteristic of* X *is not zero, then* f *has a periodic point with period not greater that the maximum of* $\sum_{k\ odd} B_k(X)$ *and* $\sum_{k\ even} B_k(X)$, *where* $B_k(X)$ *denotes the kth Betti number of* X.

Proof (*Proof of Theorem* 1) Assume $\mathbb{X} = \mathbb{M}_{g,b}$. Since for a closed surface $\mathbb{M}_{g,b}$ its homological groups with rational coefficients are $H_0(\mathbb{M}_{g,b},\mathbb{Q}) = \mathbb{Q}$, $H_1(\mathbb{M}_{g,b},\mathbb{Q}) = \mathbb{Q}\oplus \overset{2g}{\cdots} \oplus \mathbb{Q}$ and $H_2(\mathbb{M}_{g,b},\mathbb{Q}) = \mathbb{Q}$ if $b=0$, and $H_0(\mathbb{M}_{g,b},\mathbb{Q}) = \mathbb{Q}$, $H_1(\mathbb{M}_{g,b},\mathbb{Q}) = \mathbb{Q}\oplus \overset{2g+b-1}{\cdots} \oplus \mathbb{Q}$ and $H_2(\mathbb{M}_{g,b},\mathbb{Q}) = 0$ if $b>0$, then its Euler characteristic $\mathcal{X}(\mathbb{M}_{g,b}) = B_0(\mathbb{M}_{g,b}) - B_1(\mathbb{M}_{g,b}) + B_2(\mathbb{M}_{g,b})$ is equal to $2-2g$ if $b=0$, and $2-2g-b$ if $b>0$, where $B_k(\mathbb{M}_{g,b}) = \dim_{\mathbb{Q}}(H_k(\mathbb{M}_{g,b},\mathbb{Q}))$.

Since

$$\sum_{k\,even} B_k(\mathbb{M}_{g,b}) = 2 \neq 0 \text{ and } \sum_{k\,odd} B_k(\mathbb{M}_{g,b}) = 2g \text{ if } b=0,$$

$$\sum_{k\,even} B_k(\mathbb{M}_{g,b}) = 1 \neq 0 \text{ and } \sum_{k\,odd} B_k(\mathbb{M}_{g,b}) = 2g+b-1 \text{ if } b>0,$$

the orientable closed surfaces for which the Fuller's Theorem does not provide any information on the set of periods $\text{Per}(f)$ are the ones having zero Euler characteristic, i.e. when $g=1$ and $b=0$, and $g=0$ and $b=2$. Therefore statement (a) is proved.

If $(g,b)=(0,0)$ then $\max\{\sum_{k\,even} B_k(\mathbb{M}_{g,b}) = 2, \sum_{k\,odd} B_k(\mathbb{M}_{g,b}) = 0\} = 2$, then by Theorem 3 it follows that $\text{Per}(f)\cap\{1,2\} \neq \emptyset$. Hence statement (b) follows.

If $(g,b)=(0,1)$ then $\max\{\sum_{k\,even} B_k(\mathbb{M}_{g,b}) = 1, \sum_{k\,odd} B_k(\mathbb{M}_{g,b}) = 0\} = 1$, then by Theorem 3 it follows that $1\in\text{Per}(f)$. So statement (c) follows.

If $g>1$ and $b=0$ then $\max\{\sum_{k\,even} B_k(\mathbb{M}_{g,b}) = 2, \sum_{k\,odd} B_k(\mathbb{M}_{g,b}) = 2g\} = 2g$, then by Theorem 3 it follows that $\text{Per}(f)\cap\{1,2,\ldots,2g\}\neq\emptyset$. Hence statement (d) is proved.

If $g>1$ and $b>0$ then $\max\{\sum_{k\,even} B_k(\mathbb{M}_{g,b}) = 1, \sum_{k\,odd} B_k(\mathbb{M}_{g,b}) = 2g\} = 2g+b-1$, then by Theorem 3 it follows that $\text{Per}(f)\cap\{1,2,\ldots,2g+b-1\}\neq\emptyset$. Therefore statement (e) follows.

Assume $\mathbb{X} = \mathbb{N}_{g,b}$. Since for a closed surface $\mathbb{N}_{g,b}$ its homological groups with rational coefficients are $H_0(\mathbb{M}_{g,b},\mathbb{Q}) = \mathbb{Q}$, $H_1(\mathbb{M}_{g,b},\mathbb{Q}) = \mathbb{Q}\oplus \overset{g+b-1}{\cdots} \oplus \mathbb{Q}$ and $H_2(\mathbb{M}_{g,b},\mathbb{Q}) = 0$, then its Euler characteristic $\mathcal{X}(\mathbb{N}_{g,b}) = B_0(\mathbb{N}_{g,b}) - B_1(\mathbb{N}_{g,b}) + B_2(\mathbb{N}_{g,b}) = 2-g-b$, where $B_k(\mathbb{N}_{g,b}) = \dim_{\mathbb{Q}}(H_k(\mathbb{N}_{g,b},\mathbb{Q}))$.

Since

$$\sum_{k\,even} B_k(\mathbb{M}_{g,b}) = 1 \neq 0 \text{ and } \sum_{k\,odd} B_k(\mathbb{M}_{g,b}) = g+b-1,$$

the non-orientable closed surfaces for which the Fuller's Theorem does not provide any information on the set of periods $\text{Per}(f)$ are the ones having zero Euler characteristic, i.e. when $g=2$ and $b=0$, and $g=1$ and $b=1$. Therefore statement (f) is proved.

If $(g,b)=(1,0)$ then $\max\{\sum_{k\,even} B_k(\mathbb{M}_{g,b}) = 1, \sum_{k\,odd} B_k(\mathbb{M}_{g,b}) = 0\} = 1$, then by Theorem 3 it follows that $1\in\text{Per}(f)$. So statement (g) follows.

If (g, b) does not satisfy the assumptions of statements (f) or (g), then max $\{\sum_{k\,even} B_k(\mathbb{M}_{g,b}) = 1, \sum_{k\,odd} B_k(\mathbb{M}_{g,b}) = g + b - 1\} = g + b - 1$, then by Theorem 3 it follows that $\text{Per}(f) \cap \{1, 2, \ldots, g + b - 1\} \neq \emptyset$. Hence statement (h) is proved.

3 Proof of Theorems 2

Let $f : \mathbb{X} \to \mathbb{X}$ be a continuous map and let $\mathbb{X}$ be either $\mathbb{M}_{g,b}$ or $\mathbb{N}_{g,b}$. Then the *Lefschetz number* of f is defined by

$$L(f) = \text{trace}(f_{*0}) - \text{trace}(f_{*1}) + \text{trace}(f_{*2}).$$

For continuous maps and in particular for homeomorphisms f defined on $\mathbb{X}$ the Lefschetz fixed point theorem states (see for instance [1]).

Theorem 4 *If $L(f) \neq 0$ then f has a fixed point.*

With the aim of studying the periodic points of f we shall use the Lefschetz numbers of the iterates of f, i.e. $L(f^n)$. Note that if $L(f^n) \neq 0$ then f^n has a fixed point, and consequently f has a periodic point of period a divisor of n. In order to study the whole sequence $\{L(f^n)\}_{n\geq 1}$ it is defined the formal *Lefschetz zeta function* of f as

$$Z_f(t) = \exp\left(\sum_{n=1}^{\infty} \frac{L(f^n)}{n} t^n\right). \tag{1}$$

The Lefschetz zeta function is in fact a generating function for the sequence of Lefschetz numbers n. In order to study the whole sequence $\{L(f^n)\}_{n\geq 1}$.

For a continuous self-map of a closed surface the Lefschetz zeta function is the rational function

$$Z_f(t) = \frac{\det(I - tf_{*1})}{\det(I - tf_{*0})\det(I - tf_{*2})},$$

see for more details Franks [2]. Then, for an orientation-preserving homeomorphism $f : \mathbb{M}_{g,b} \to \mathbb{M}_{g,b}$ we have

$$Z_f(t) = \begin{cases} \dfrac{\det(I - tA)}{(1-t)^2} & \text{if } b = 0, \\[2ex] \dfrac{\det(I - tA)}{1-t} & \text{if } b > 0, \end{cases} \tag{2}$$

where $f_{*1} = A$. Note that $f_{*2} = (1)$ if $b = 0$, and $f_{*2} = (0)$ if $b > 0$.

For an orientation-reversing homeomorphism $f : \mathbb{M}_{g,b} \to \mathbb{M}_{g,b}$ we have

$$Z_f(t) = \begin{cases} \dfrac{\det(I - tA)}{1 - t^2} & \text{if } b = 0. \\ \dfrac{\det(I - tA)}{1 - t} & \text{if } b > 0. \end{cases} \tag{3}$$

Note that $f_{*2} = (-1)$ if $b = 0$, and $f_{*2} = (0)$ if $b > 0$.

Finally, for a homeomorphism $f : \mathbb{N}_{g,b} \to \mathbb{N}_{g,b}$ we have

$$Z_f(t) = \frac{\det(I - tA)}{1 - t}. \tag{4}$$

Using the Lefschetz zeta function the orientation preserving and reversing homeomorphisms on $\mathbb{M}_{g,0}$ were studied in [3]. The results there obtained are the following two theorems. Here we shall study the homeomorphisms on $\mathbb{M}_{g,b}$ with $b > 0$ and on $\mathbb{N}_{g,b}$ with $b \geq 0$.

Theorem 5 (Theorem 4 of [3]) *Let $f : \mathbb{M}_{g,0} \to \mathbb{M}_{g,0}$ be an orientation-preserving homeomorphism and let A be the $2g \times 2g$ integral matrix of the isomorphism $f_{*1} : H_1(\mathbb{M}_{g,0}, \mathbb{Q}) \to H_1(\mathbb{M}_{g,0}, \mathbb{Q})$ induced by f on the first homology group of $\mathbb{M}_{g,0}$. Then the following statements hold.*

(a) If $g = 0$, then $1 \in \mathrm{Per}(f)$.
(b) If $g > 0$ and $E_1(A) \neq 2$, then $1 \in \mathrm{Per}(f)$.
(c) If $g > 0$, $E_1(A) = 2$ and $E_2(A) \neq 1$, then $\mathrm{Per}(f) \cap \{1, 2\} \neq \emptyset$.
(d) If $g = 1$, $E_1(A) = 2$ and $E_2(A) = 1$, then there is no information on $\mathrm{Per}(f)$.
(e) If $g > 1$, $E_1(A) = 2$, $E_2(A) = 1$ and k is the smallest integer of the set $\{3, 4, \ldots, 2g\}$ such that $E_k(A) \neq 0$, then f has a periodic point of period a divisor of k.

Theorem 6 (Theorem 3 of [3]) *Let $f : \mathbb{M}_{g,0} \to \mathbb{M}_{g,0}$ be an orientation-reversing homeomorphism and let A be the $2g \times 2g$ integral matrix of the isomorphism $f_{*1} : H_1(\mathbb{M}_{g,0}, \mathbb{Q}) \to H_1(\mathbb{M}_{g,0}, \mathbb{Q})$ induced by f on the first homology group of $\mathbb{M}_{g,0}$. Then the following statements hold.*

(a) If $g = 0$, then $\mathrm{Per}(f) \cap \{1, 2\} \neq \emptyset$.
(b) If $g > 0$ and $E_1(A) \neq 0$, then $1 \in \mathrm{Per}(f)$.
(c) If $g > 0$, $E_1(A) = 0$ and $E_2(A) \neq -1$, then $\mathrm{Per}(f) \cap \{1, 2\} \neq \emptyset$.
(d) If $g = 1$, $E_1(A) = 0$ and $E_2(A) = -1$, then there is no information on $\mathrm{Per}(f)$.
(e) If $g > 1$, $E_1(A) = 0$, $E_2(A) = 1$ and k is the smallest integer of the set $\{3, 4, \ldots, 2g\}$ such that $E_k(A) \neq 0$, then f has a periodic point of period a divisor of k.

Unfortunately we cannot distinguish using the Lefschetz zeta function the orientation-preserving homeomorphisms from the orientation-reversing on $\mathbb{M}_{g,b}$ when $b > 0$, see (2) and (3). They can be distinguished if $b = 0$, because then they have different Lefschetz zeta functions, see again (2) and (3).

Proof (*Proof of Theorem* 2) Assume $b > 0$. Let $f : \mathbb{M}_{g,b} \to \mathbb{M}_{g,b}$ be a homeomorphism, and let A be the $(2g + b - 1) \times (2g + b - 1)$ integral matrix of the isomorphism $f_{*1} : H_1(\mathbb{X}, \mathbb{Q}) \to H_1(\mathbb{X}, \mathbb{Q})$ induced by f on the first homology group of $\mathbb{X}$. Then, combining the expressions (1), (2) with $b > 0$, (3) with $b > 0$ and (4) with $b \geq 0$ we obtain the following equalities

$$\begin{aligned}
\sum_{n=1}^{\infty} \frac{L(f^n)}{n} t^n &= \log(Z_f(t)) \\
&= \log\left(\frac{\det(I - tA)}{1-t}\right) \\
&= \log\left(\frac{1 - E_1(A)t + E_2(A)t^2 - \ldots + (-1)^m E_m(A)t^m}{1-t}\right) \\
&= \log(1 - E_1(A)t + E_2(A)t^2 - \ldots) - \log(1-t) \\
&= \left(-E_1(A)t + \left(E_2(A) - \frac{E_1(A)^2}{2}\right)t^2 - \ldots\right) - \left(-t - \frac{t^2}{2} - \ldots\right) \\
&= (1 - E_1(A))t + \left(\frac{1}{2} - \frac{E_1(A)^2}{2} + E_2(A)\right)t^2 + O(t^3).
\end{aligned}$$

Here $m = 2g + b - 1$ if $\mathbb{X} = \mathbb{M}_{g,b}$ with $b > 0$, or $m = g + b - 1$ if $\mathbb{X} = \mathbb{N}_{g,b}$ with $b \geq 0$. Therefore we have

$$L(f) = 1 - E_1(A), \quad \text{and} \quad L(f^2) = 1 - E_1(A)^2 + 2E_2(A).$$

Hence, if $E_1(A) \neq 1$ then $L(f) \neq 0$, and by Theorem 4 statement (a) follows.

If $E_1(A) = 1$ and $E_2(A) \neq 0$, then $L(f^2) = 2E_2(A) \neq 0$, and again by Theorem 4 we get that $\text{Per}(f) \cap \{1, 2\} \neq \emptyset$. So statement (b) is proved.

Assume now that $\mathbb{X} = \mathbb{M}_{g,b}$ with $b > 0$, $2g + b - 1 \geq 3$, $E_1(A) = 1$, $E_2(A) = 0$ and k is the smallest integer of the set $\{3, 4, \ldots, 2g + b - 1\}$ such that $E_k(A) \neq 0$. Therefore

$$\begin{aligned}
\sum_{n=1}^{\infty} \frac{L(f^n)}{n} t^n &= \log\left(\frac{1 - t + (-1)^k E_k(A)t^k + \cdots + (-1)^{b-1} E_{2g+b-1}(A)t^{2g+b-1}}{1-t}\right) \\
&= \log\left(1 + \frac{(-1)^k E_k(A)t^k + \cdots + (-1)^{b-1} E_{2g+b-1}(A)t^{2g+b-1}}{1-t}\right) \\
&= (-1)^k E_k(A)t^k + O(t^{k+1}).
\end{aligned}$$

Hence, $L(f) = \cdots = L(f^{k-1}) = 0$ and $L(f^k) = (-1)^k k E_k(A) \neq 0$. So, from Theorem 4, it follows the statement (c).

Suppose that $\mathbb{X} = \mathbb{N}_{g,b}$ with $b \geq 0$, $g + b - 1 \geq 3$, $E_1(A) = 1$, $E_2(A) = 0$ and k is the smallest integer of the set $\{3, 4, \ldots, g + b - 1\}$ such that $E_k(A) \neq 0$. Therefore

$$
\begin{aligned}
\sum_{n=1}^{\infty} \frac{L(f^n)}{n} t^n &= \log\left(\frac{1 - t + (-1)^k E_k(A)t^k + \cdots + (-1)^{g+b-1} E_{g+b-1}(A)t^{g+b-1}}{1-t}\right) \\
&= \log\left(1 + \frac{(-1)^k E_k(A)t^k + \cdots + (-1)^{g+b-1} E_{g+b-1}(A)t^{g+b-1}}{1-t}\right) \\
&= (-1)^k E_k(A)t^k + O(t^{k+1}).
\end{aligned}
$$

Again $L(f) = \cdots = L(f^{k-1}) = 0$ and $L(f^k) = (-1)^k k E_k(A) \neq 0$. Therefore, from Theorem 4, it follows the statement (d).

Acknowledgments The first author of this work was partially supported by MICINN/FEDER grant number MTM2011–22587, Junta de Comunidades de Castilla-La Mancha, grant number PEII09-0220-0222. The second author was partially supported by a MINECO grant MTM2013-40998-P, an AGAUR grant number 2014SGR-568, and the grants FP7-PEOPLE-2012-IRSES 318999 and 316338. The third author was partially supported by Fundación Séneca de la Región de Murcia grant number 12001/PI/09.

References

1. Brown, R.F.: The Lefschetz Fixed Point Theorem. Scott, Foresman and Company, Glenview, IL (1971)
2. Franks, J.: Homology and Dynamical Systems, CBMS Regional Conference Series, vol. 49, American Mathematical Society, Providence (1982)
3. Franks, J., Llibre, J.: Periods of surface homeomorphisms. Contemp. Math. **117**, 63–77 (1991)
4. Fuller, F.B.: The existence of periodic points. Ann. Math. **57**, 229–230 (1953)
5. Halpern, B.: Fixed point for iterates. Pacific J. Math. **25**, 255–275 (1968)
6. Munkres, J.R.: Elements of Algebraic Topology. Addison–Wesley, Boston (1984)
7. Vicks, J.W.: Homology Theory. An Introduction to Algebraic Topology. Springer, New York (1994). (Academic Press, New York, 1973)

Polygonal Billiards with Strongly Contractive Reflection Laws: A Review of Some Hyperbolic Properties

Gianluigi Del Magno, João Lopes Dias, Pedro Duarte, José Pedro Gaivão and Diogo Pinheiro

Abstract We provide an overview of recent results concerning the dynamics of polygonal billiards with strongly contractive reflection laws.

Keywords Polygonal billiards · Contractive reflection law · SRB measures

1 Introduction

A mathematical billiard is a continuous-time dynamical system describing the motion of a massless point particle, the *billiard ball*, moving freely with constant speed in the interior of a connected compact Riemannian manifold P with piecewise smooth boundary, the *billiard table*. The billiard ball moves along geodesics of P until it hits its boundary ∂P. If the collision with ∂P occurs at a smooth boundary point, the billiard ball reflects according to a prespecified rule, the *reflection law*. Otherwise, if

G. Del Magno · J.P. Gaivão
ISEG, CEMAPRE, Universidade Técnica de Lisboa,
Rua Do Quelhas 6, 1200-781 Lisboa, Portugal
e-mail: delmagno@iseg.utl.pt

J.P. Gaivão
e-mail: jpgaivao@iseg.utl.pt

J. Lopes Dias (✉)
Departamento de Matemática and ISEG, CEMAPRE, Universidade Técnica de Lisboa,
Rua Do Quelhas 6, 1200-781 Lisboa, Portugal
e-mail: jldias@iseg.utl.pt

P. Duarte
Departamento de Matemática and Faculdade de Ciências, CMAF,
Universidade de Lisboa, Campo Grande, Edifício C6, Piso 2, 1749-016 Lisboa, Portugal
e-mail: pduarte@ptmat.fc.ul.pt

D. Pinheiro
Department of Mathematics, Brooklyn College of The City University of New York,
2900 Bedford Ave, 11210 Brooklyn, NY, USA
e-mail: dpinheiro@brooklyn.cuny.edu

L. Alsedà i Soler et al. (eds.), *Difference Equations, Discrete Dynamical Systems and Applications*, Springer Proceedings in Mathematics & Statistics 180, DOI 10.1007/978-3-662-52927-0_14

the collision with ∂P occurs at a non-smooth boundary point, a *corner*, the forward dynamics may not be well-defined. The most commonly considered reflection law is the *elastic reflection*: the tangential component of the billiard ball velocity remains the same, while the normal component changes its sign. However, as we will see below, there exist equally reasonable alternative reflection laws with relevant mathematical implications.

Interesting questions concerning the dynamical properties of mathematical billiards include, among others, the existence and number of periodic orbits or more general invariant sets, the existence and number of dense orbits in phase space, the existence of non-trivial invariant measures, or the positivity of Lyapunov exponents. Note that all these properties depend solely on the geometry of ∂P, the metric on P and the precise reflection law under consideration. See the survey paper [13] for a recent overview of the field with a focus on open problems.

It is worth remarking that mathematical billiards are not just mathematical abstractions, but are also models for important problems in optics, acoustics or classical mechanics. A notable example is the Boltzmann gas of elastically colliding hard balls in a box in statistical mechanics [25, 28]. A second interesting example, which may also be reduced to the study of a billiard, is provided by a mechanical system consisting of a fixed number of point particles constrained to move with no friction on a line segment, bouncing off the segment ends and each other without loss of kinetic energy [29].

2 Planar Billiards

In this section we will restrict our attention to *planar billiards*, i.e. the special case when the billiard table P is a connected compact subset of $\mathbb{R}^2$ with piecewise smooth closed curve as boundary. Before proceeding, we note that the dynamics of a billiard flow is completely determined by the dynamics of a discrete time map – the *billiard map* Φ. Indeed, the billiard flow may be seen as the suspension flow over such map. Very roughly, the billiard map gives the next impact point $s' \in \partial P$ and angle of reflection $\theta' \in (-\pi/2, \pi/2)$ as a function of the previous impact $(s, \theta) \in \partial P \times (-\pi/2, \pi/2)$. Notice that at a point where the boundary ∂P is not smooth, the billiard map may not be defined and its domain of definition is only a subset of $\partial P \times (-\pi/2, \pi/2)$.

One of the most interesting features of billiard dynamics is that it exhibits a whole range of dynamical behaviours, from integrable to chaotic. Let us restrict our attention for the time being to the case of billiards with the standard elastic reflection law. Billiards in tables with smooth and strictly convex boundaries exhibit regular dynamics in some parts of the phase space. These are known as *Birkhoff* billiards. The corresponding billiard map is an area preserving twist map. Moreover, if the table is an ellipse, the billiard is integrable [4]. Indeed, Birkhoff's conjecture states that elliptic tables provide the only example of integrable billiards among the class of Birkhoff billiards. In what concerns the opposite side of the regularity spectrum, chaotic behaviour may be found in *hyperbolic* billiards, i.e. billiards with non-zero Lyapunov exponents. Key examples of this family of billiards include Sinai's table [26] and the Bunimovich stadium [6]. Both exhibit strong ergodic properties, but while the billiard map associated with Sinai's table is uniformly hyperbolic, the one associated with Bunimovich stadium is non-uniformly hyperbolic. This difference appears in the speed of the decay of correlations: Sinai's table has exponential decay of correlations, whereas the Bunimovich stadium has subexponential decay of correlations. These properties are due to dispersive effects caused by the boundary, formed by the disk and the square, in the case of Sinai's table, and to the interplay between boundary segments, formed by the semi-circles and the rectangle, for the Bunimovich stadium. Other billiards with hyperbolic behaviour were discovered by Wojtkowski [30], Markarian [20] and Donnay [9]. There is an extensive literature in this topic, see e.g. [7, 8, 15] for further details.

We are particularly interested in the case when the billiard table is a polygon: a *polygonal billiard*. For the standard elastic reflection law, these billiards are never hyperbolic since their topological entropy vanishes and thus, their Lyapunov exponents are zero [12, 14]. Its dynamical properties depend heavily on the angles of the table. If all angles of the table are a rational multiple of π, the polygonal billiard is called *rational*, otherwise it is called *irrational*. We provide next a quick overview of the main properties of polygonal billiards, but see the surveys [11–13, 27] for a rather complete overview of the topic.

Rational billiards are reasonably well understood: any orbit has only a finite number of different angles of reflection and their dynamics can be viewed as the geodesic flow on a translation surface. Flat two-dimensional tori are the only translation surfaces of genus one and correspond to integrable billiard tables, the only ones being rectangles, equilateral triangles, right isosceles triangles and right triangles with an angle $\pi/6$. If the translation surface has higher genus, then the phase space is foliated by invariant surfaces of genus higher than one. Moreover, Kerckhoff et al. [17] proved that the billiard flow on these translation surfaces is uniquely ergodic for almost all surfaces. Other properties worth mentioning include:

(i) periodic orbits exist and are dense in the phase space [5, 10, 22];
(ii) for all but countably many directions, all infinite semi-orbits are dense in configuration space, i.e. a rational polygonal billiard is minimal [16].

In what concerns the dynamical properties of irrational billiards, most of what is known is based on typical properties:

(i) arbitrarily close to a given polygon there is an irrational polygon, with vertices as close as desired, whose billiard map is ergodic with respect to the standard invariant measure. More precisely, for every $n \geq 3$, in the space of n-gons there is a dense G_δ-subset of ergodic tables [17].
(ii) the set of topologically transitive polygons is a dense G_δ-subset of the space of n-gons [16].
(iii) given an arbitrary polygon, any orbit on the corresponding billiard table is either periodic or its closure contains at least one vertex [14].

While rational billiards are never mixing and their periodic orbits forms a dense set in the phase space, it is not know whether irrational billiards can be mixing and have periodic orbits.

The dynamics of billiards with the standard elastic reflection law discussed so far is conservative, i.e. they admit an invariant measure that is absolutely continuous with respect to the phase space volume. Markarian, Pujals and Sambarino [21] studied a class of non-conservative billiards, which they call *pinball billiards*, with a non-elastic reflection law according to which the angle of reflection is a contraction of the angle of incidence. Under a set of mild assumptions, they proved that the billiard map associated with a large family of contractive reflection laws has dominated splitting, i.e. the tangent bundle splits into two invariant directions such that the growth rate along one direction dominates uniformly the growth rate along the other direction. A particular consequence of their result is that all pinball billiards on polygonal tables have dominated splitting. Further results for particular choices of billiard table shapes with a contractive reflection law were obtained for non-polygonal billiards [1, 2], the equilateral triangle billiard [3] and the square billiard [18].

3 Polygonal Billiards with a Contractive Reflection Law

We will now provide an overview of the setup considered in [19]. We consider planar polygonal billiard tables with a contractive general reflection law, that we will describe in a moment. Let P be a non self-intersecting n-gon, which we do not always assume to be convex. Denote by $|\partial P|$ the length of ∂P and define M to be the subset of $\mathbb{R}^2$ given by

$$M = (0, |\partial P|) \times \left(-\frac{\pi}{2}, \frac{\pi}{2}\right).$$

Let $0 = \tilde{s}_1 < \cdots < \tilde{s}_n < \tilde{s}_{n+1} = |\partial P|$ be the values of the arclength coordinate $s \in (0, |\partial P|)$ corresponding to vertices of the polygon P. Denote by V the set

$$V = \{\tilde{s}_0, \ldots, \tilde{s}_{n+1}\} \times \left(-\frac{\pi}{2}, \frac{\pi}{2}\right)$$

and by S_1^+ the closure of the set of all $x_0 \in M$ such that the billiard flow starting from x_0 reaches a corner of P in its first collision with ∂P. Finally, define the set N_1^+ to be given by $N_1^+ = \partial M \cup V \cup S_1^+$ and for $n \geq 1$, define

$$N_{n+1}^+ = N_n^+ \cup \Phi_f^{-1}(N_n^+)$$

and $N_\infty^+ = \bigcup_{n \geq 1} N_n^+$.

Denote by $\bar{\Phi}\colon M \setminus N_1^+ \to M$ the billiard map on P with the standard elastic reflection law. The billiard map Φ is given by

$$\bar{\Phi}(x_0) = (s(x_0), \theta(x_0)),$$

where $(s(x_0), \theta(x_0)) \in M$ are, respectively, the coordinate of the first collision point with ∂P for an orbit of the billiard flow starting at $x_0 \in M \setminus N_1^+$ and the corresponding angle of reflection. To define a billiard map with a general reflection law, we note that a reflection law is simply given by a function $f\colon (-\pi/2, \pi/2) \to (-\pi/2, \pi/2)$ and define $R_f : M \to M$ to be the map given by

$$R_f(s, \theta) = (s, f(\theta)).$$

Then, the billiard map on P with the general reflection law f, $\Phi_f : M \setminus N_1^+ \to M$, is given by

$$\Phi_f = R_f \circ \bar{\Phi}.$$

We will consider a class of general reflection laws that are contractive in the sense that we now pass to explain. Let $f\colon (-\pi/2, \pi/2) \to \mathbb{R}$ be a differentiable function and define

$$\lambda(f) = \sup_{\theta \in (-\pi/2, \pi/2)} |f'(\theta)|.$$

For each $k \geq 1$, define $\mathcal{R}_1^k$ to be the set of all C^k embeddings $f\colon (-\pi/2, \pi/2) \to \mathbb{R}$ with $f(0) = 0$ and $\lambda(f) < 1$. Note that if $f \in \mathcal{R}_1^k$, then f is a strict contraction and $\theta = 0$ is its unique fixed point. Moreover, note that requiring f to be strictly increasing and such that $f \in \mathcal{R}_1^k$ leads to a reflection law satisfying Condition A1b of [21]. The simplest example of such a reflection law is given by the family of linear reflection laws $f(\theta) = \sigma\theta$ with $0 < \sigma < 1$, previously studied in [2, 3, 18, 21].

A invariant set $\Sigma \subset M$ is *hyperbolic* if there exist a non-trivial measurable splitting $T_\Sigma M = E \oplus F$ and two measurable functions $0 < \mu < 1$ and $A > 0$ on Σ such that for all $x \in \Sigma$ and $n \geq 1$, we have

$$\|D\Phi_f^n|_{E(x)}\| \leq A(x)\mu(x)^n$$
$$\|D\Phi_f^{-n}|_{F(\Phi_f^n x)}\| \leq A(x)\mu(x)^n.$$

If the functions μ and A can be replaced by constants and the splitting is continuous, then Σ is called *uniformly hyperbolic*, otherwise it is called *non-uniformly hyperbolic*.

Let D be the set defined by

$$D = \bigcap_{n \geq 0} \Phi_f^n(M \setminus N_\infty^+)$$

and note that every element of D has infinite positive and negative semi-orbits. Indeed, D is the largest invariant subset of M under Φ_f. Following the terminology introduced by Pesin in [23], we call the set

$$A = \overline{\bigcap_{n \geq 0} \Phi_f^n(M \setminus N_\infty^+)},$$

the *attractor* of Φ_f. If D is uniformly hyperbolic, then we say that the attractor A is *hyperbolic*.

We are now ready to provide an overview of the main results in [19]. These concern properties such as hyperbolicity of the invariant sets and the existence of hyperbolic attractors with finitely many ergodic SRB (Sinai-Ruelle-Bowen) measures for the class of billiards described above, i.e. billiards in polygonal tables P with contractive reflection laws $f \in \mathcal{R}_1^k$, for $k \geq 1$.

3.1 Hyperbolicity

In this section we will discuss hyperbolicity of the invariant sets of the billiard map Φ_f. Markarian, Pujals and Sambarino proved in [21] that any invariant set Σ of Φ_f has dominated splitting. Their results hold for a large class of billiards, which includes the polygonal ones. Note that this is in striking contrast with what is observed for conservative polygonal billiards, which exhibit parabolic dynamics.

We start with the following simple characterization for the periodic orbits of the billiard map Φ_f.

Proposition 1 *The periodic points of Φ_f have the following properties:*

1. *every periodic point of period two is parabolic,*
2. *the set $\mathcal{P}$ of all periodic points of period two of Φ_f is attracting,*
3. *every periodic point of period greater than two is hyperbolic.*

From the previous result, we conclude that if a polygon has parallel sides, then its billiard map Φ_f has $\mathcal{P}$ as an attracting set. However, even when $\mathcal{P}$ is non-empty, uniformly hyperbolic invariant subsets Σ may exist.

Proposition 2 *Suppose Σ is an invariant set with the property that there exists $m > 0$ such that no positive semi-orbit of points in Σ contains sequences of consecutive collisions between parallel sides of P consisting of more than m collisions. Then Σ is uniformly hyperbolic.*

We remark that the proposition above holds for every $f \in \mathcal{R}_1^k$, thus improving a previous result by Markarian, Pujals and Sambarino, stating that for any convex polygon without parallel sides, the map Φ_f is hyperbolic as long as $\lambda(f)$ is sufficiently close to zero [21, Corollary 4 and Theorem 23].

As a consequence of the proposition above, we obtain that any invariant subset of the billiard map Φ_f associated with any regular polygon with an odd number of sides is uniformly hyperbolic.

3.2 *Existence of Hyperbolic Attractors with Finitely Many Ergodic SRB Measures*

An invariant measure μ for a diffeomorphism f on a manifold is SRB if μ is hyperbolic and the conditional measures of μ on unstable manifolds are absolutely continuous with respect to the Riemannian measures on these leaves. The relevance of SRB measures is due to the fact that ergodic SRB measures are physical.

In what follows, we will provide conditions for the existence of a hyperbolic attractor for Φ_f with finitely many ergodic SRB measures. This implies that the hyperbolic periodic points are dense. Our approach is based on the use of a general theorem, due to Pesin [23], providing conditions for the existence of countably many ergodic SRB measures for hyperbolic attractors of piecewise smooth maps. This result was later improved to a finite number of ergodic SRB measures by Sataev in [24]. It is known that the conditions of such theorem are satisfied for the Lorenz, generalized Lozi and Belykh families of maps.

Recall the definition of the set A given in the beginning of this section. The following result gives us a criterium for A to be a hyperbolic attractor for Φ_f.

Proposition 3 *A is a hyperbolic attractor if and only if the polygon does not have parallel sides facing each other.*

Define the sequence of sets S_n^+ through the recursion

$$S_{n+1}^+ = S_n^+ \cup \Phi_f^{-1}(S_n^+), \qquad n \geq 1.$$

Let $p(S_n^+)$ be the maximum number of smooth components of S_n^+ intersecting at one point. This quantity is well defined, since for any reflection law $f \in \mathcal{R}_1^k$, S_n^+ is a union of finitely many C^k curves. Moreover, for the particular case of a convex polygonal billiard, we have $p(S_1^+) = 2$, while for a non-convex polygonal billiard one must have $p(S_1^+) \geq 2$. Let $\alpha(\Phi_f^n)$ denote the smallest expansion rate of the nth iterate of the billiard map along the unstable direction. Finally, denote by $\mathcal{B}$ the subset of contractive C^2 reflection laws f in $\mathcal{R}_1^2$ such that f and its inverse f^{-1} have bounded second derivatives.

The following theorem provides us with a condition for a hyperbolic attractor of Φ_f to have finitely many ergodic SRB measures.

Theorem 1 *Consider a polygon without parallel sides facing each other and suppose that* $f \in \mathcal{B}$*. If there is* $m \geq 1$ *such that*

$$p(S_m^+) < \alpha(\Phi_f^m),$$

then A has finitely many ergodic SRB measures.

The existence of a hyperbolic attractor with finitely many ergodic SRB measures is indeed a robust property under small changes in the reflection law.

Theorem 2 *Consider a polygon without parallel sides facing each other,* $f_0 \in \mathcal{R}_1^1$ *or* $f_0 = 0$*, and*

$$n > \frac{\log p(S_1^+)}{\log \alpha(\Phi_{f_0})}$$

such that

$$\Phi_{f_0}^k(S_1^+) \cap S_1^+ = \emptyset \quad for \quad 1 \leq k \leq n-1.$$

Then there is $\delta > 0$ *such that for* $f \in \mathcal{B}$ *satisfying* $\lambda(f - f_0) < \delta$*, the billiard map* Φ_f *has a hyperbolic attractor with finitely many ergodic SRB measures.*

The existence of a hyperbolic attractor with finitely many ergodic SRB measures is a generic property among the space of n-gons with strongly contractive reflection laws.

Theorem 3 *There exists an open, dense and full measure subset of the space of* n*-gons such that for every polygon P in such set, there is* $\lambda_0 > 0$ *such that if* $\lambda(f) < \lambda_0$ *and* $f \in \mathcal{B}$*, then* Φ_f *has a hyperbolic attractor with finitely many ergodic SRB measures.*

3.3 Some Special Polygons

We will now specialize the results above concerning hyperbolicity and the existence of hyperbolic attractors with finitely many ergodic SRB measures to billiard tables with particular shapes.

We start by considering the case of regular polygons, for which we have obtained the following result.

Theorem 4 *Consider a regular polygon with an odd number of sides. There is* $\lambda_0 > 0$ *such that if* $\lambda(f) < \lambda_0$ *and* $f \in \mathcal{B}$*, then the billiard map* Φ_f *has a hyperbolic attractor with finitely many ergodic SRB measures.*

The techniques used in the proof of the theorem above do not apply to the case of regular polygons with an even number of sides though. However, in this last case we are still able to provide a set of conditions on the reflection law f under which any invariant set not intersecting the attracting set $\mathcal{P}$ is uniformly hyperbolic.

Proposition 4 *Consider a regular polygon with an even number $N \geq 6$ of sides and let $f \in \mathcal{R}_1^k$ be an odd function such that $f' > 0$, $\lambda(f) \leq 1/2$ and $f(\delta \frac{2\pi}{N}) \leq \delta f(\frac{2\pi}{N})$ for every $0 \leq \delta \leq 1$. Then there exists a positive constant C such that every orbit of Φ_f having more than C collisions between parallel sides belongs to the basin of attraction of $\mathcal{P}$. In particular, every invariant set Σ of Φ_f not intersecting the basin of attraction of $\mathcal{P}$ is uniformly hyperbolic.*

We will now consider the case of acute triangles. Recall that a triangle is called acute if all its internal angles ϕ_i are less than $\pi/2$. The following theorem generalizes a result in [3] for equilateral triangles ($\phi_i = \pi/3$) and reflection laws of the form $f(\theta) = \sigma\theta$ with $\sigma < 1/3$.

Theorem 5 *For any acute triangle and $f \in \mathcal{B}$ satisfying*

$$\lambda(f) < \frac{2}{\pi} \min_i \left(\frac{\pi}{2} - \phi_i \right),$$

the billiard map Φ_f has a hyperbolic attractor with finitely many ergodic SRB measures.

Finally, we consider the case of rectangular billiard tables P. Note that in this case the attracting set $\mathcal{P}$ is never empty. Despite this fact, we are still able to prove that every invariant set Σ not intersecting the basin of attraction of $\mathcal{P}$ is hyperbolic.

Recall the definition of the set D given in the beginning of this section and note that $D \setminus B(\mathcal{P})$ is invariant. Indeed, $D \setminus B(\mathcal{P})$ is always hyperbolic, being uniformly hyperbolic if the hypothesis of Proposition 2 holds.

Proposition 5 *The set $D \setminus B(\mathcal{P})$ is hyperbolic.*

We note that the previous proposition holds also for polygonal tables that are finite unions of rectangles with sides parallel to two fixed orthogonal axes such as, for instance, L-shaped polygons.

It is possible to check that $B(\mathcal{P}) = M \setminus N_\infty^+$ for every $f \in \mathcal{R}_1^k$ with $f' < 0$. Therefore, we restrict ourselves to the case $f' > 0$ from now on. By rescaling P, we may assume without loss of generality that the sides of P have length 1 and $0 < h \leq 1$. Define

$$f_1(\theta) = -f(\theta) \quad \text{for } \theta \in [-\pi/2, \pi/2],$$

and

$$f_2(\theta) = f\left(\operatorname{sgn}(\theta)\frac{\pi}{2} - \theta\right) \quad \text{for } \theta \in [-\pi/2, 0) \cup (0, \pi/2].$$

The geometrical meaning f_1 and f_2 is the following. If (s_0, θ_0) and (s_1, θ_1) are two consecutive collisions, then $\theta_1 = f_1(\theta_0)$ when the two collisions occur at parallel sides of P, and $\theta_1 = f_2(\theta_0)$ when the two collisions occur at adjacent sides of P.

From the properties of f, one can easily deduce that i) f_2 is strictly decreasing, ii) the restriction of f_2 to each interval $(-\pi/2, 0)$ and $(0, \pi/2)$ is a strict contraction with Lipschitz constant less than $\lambda(f)$, and iii) f_2 has two fixed points $\theta^- < 0 < \theta^+$.

Let $\theta_*^- < 0 < \theta_*^+$ be the two solutions of

$$\mathcal{F}(\theta) := \sum_{n=0}^{\infty} \tan\left(\left|f_1^n(\theta)\right|\right) = \frac{1}{h}, \quad \theta \in (-\pi/2, \pi/2)$$

and let $\tilde{\theta}$ be the quantity given by

$$\tilde{\theta} := \min\left\{f_2\left(f\left(\frac{\pi}{2}\right)\right), -f_2\left(f\left(-\frac{\pi}{2}\right)\right)\right\}.$$

Let T be the subset of $M \setminus N_\infty^+$ consisting of elements whose positive semi-trajectory bounces between two parallel sides of the rectangle P. Note that T is a forward invariant set containing $\mathcal{P}$.

Finally, let

$$\hat{M} = M \setminus (T \cup N_\infty^+).$$

The set $\hat{M}$ consists of elements with infinite positive semi-trajectory having at least a pair of consecutive collisions at adjacent sides of P.

We state below a sufficient condition for the existence of uniform hyperbolic invariant sets.

Proposition 6 *Suppose that* $f' > 0$ *and*

$$\tilde{\theta} > \max\{-\theta_*^-, \theta_*^+\}.$$

We have that

1. $\hat{M}$ *is a forward invariant set,*
2. *every invariant subset of* $\hat{M}$ *is uniformly hyperbolic,*
3. $B(\mathcal{P}) = T$.

Finally, we give sufficient conditions for the basis of attraction of $\mathcal{P}$ to be the whole set of points in M for which every iterate of Φ_f is well defined.

Proposition 7 *Suppose that* $f' > 0$ *and*

$$h \leq \tan\tilde{\theta}$$
$$f_1(\theta_*^+) < \theta^- \quad \text{and} \quad \theta^+ < f_1(\theta_*^-).$$

Then $B(\mathcal{P}) = M \setminus N_\infty^+$.

Acknowledgments The authors were supported by Fundação para a Ciência e a Tecnologia through the Program POCI 2010 and the Project "Randomness in Deterministic Dynamical Systems and Applications" (PTDC-MAT-105448-2008).

References

1. Altmann, E.G., Del Magno, G., Hentschel, M.: Non-Hamiltonian dynamics in optical microcavities resulting from wave-inspired corrections to geometric optics. Europhys. Lett. EPL **84**, 10008–10013 (2008)
2. Arroyo, A., Markarian, R., Sanders, D.P.: Bifurcations of periodic and chaotic attractors in pinball billiards with focusing boundaries. Nonlinearity **22**, 1499–1522 (2009)
3. Arroyo, A., Markarian, R., Sanders, D.P.: Structure and evolution of strange attractors in non-elastic triangular billiards. Chaos **22**, 026107 (2012)
4. Birkhoff, G.: Dynamical Systems, vol. 9. American Mathematical Society Colloquium Publication. American Mathematical Society, Providence (1927)
5. Boshernitzan, M., Galperin, G., Krüger, T., Troubetzkoy, S.: Periodic billiard orbits are dense in rational polygons. Trans. Am. Math. Soc. **350**, 3523–3535 (1998)
6. Bunimovich., L.: On the ergodic properties of certain billiards. Anal. Appl. **8**, 254–255 (1974)
7. Bunimovich., L. Billiards and other hyperbolic systems. In: Encyclopedia of Mathematical Sciences, vol. 100, pp. 192–233. Springer, New York (2000)
8. Chernov, N.I., Markarian, R.: Chaotic billiards, vol. 127. Mathematical Surveys and Monographs. American Mathematical Society, Providence (2006)
9. Donnay, V.: Using integrability to produce chaos: billiards with positive entropy. Commun. Math. Phys. **141**, 225–257 (1991)
10. Galperin, G.A., Stepin, A.M.: Vorobets, Ya.B.: Periodic billiard trajectories in polygons: generating mechanisms. Russ. Math. Surv. **47**, 5–80 (1992)
11. Gutkin, E.: Billiards in polygons. Physica D **19**, 311–333 (1986)
12. Gutkin, E.: Billiards in polygons: survey of recent results. J. Stat. Phys. **83**, 7–26 (1996)
13. Gutkin, E.: Billiard dynamics: an updated survey with the emphasis on open problems. Chaos **22**, 026116 (2012)
14. Galperin, G., Krüger, T., Troubetzkoy, S.: Local instability of orbits in polygonal and polyhedral billiards. Commun. Math. Phys. **169**, 463–473 (1995)
15. Katok, A., Strelcyn, J.-M-.: Invariant Manifolds, Entropy and Billiards; Smooth Maps with Singularities, vol. 1222, Lecture Notes in Mathematics. Springer, Heidelberg (1986)
16. Katok, A., Zemlyakov, A.: Topological transitivity of billiards in polygons. Math. Notes **18**, 760–764 (1975)
17. Kerckhoff, S., Masur, H., Smillie, J.: Ergodicity of billiard flows and quadratic differentials. Ann. Math. **124**, 293–311 (1986)
18. Del Magno, G.: Lopes Dias, J., Duarte, P., Gaivão, J. P., Pinheiro, D.: Chaos in the square billiard with a modified reflection law. Chaos **22**, 026106 (2012)
19. Del Magno, G.: Lopes Dias, J., Duarte, P., Gaivão, J.P., Pinheiro, D.: SRB Measures for Polygonal Billiards with Contracting Reflection Laws. Commun. Math. Phys. **329**, 687–723 (2014)
20. Markarian, R.: Billiards with Pesin region of measure one. Commun. Math. Phys. **118**, 87–97 (1988)
21. Markarian, R., Pujals, E.J., Sambarino, M.: Pinball billiards with dominated splitting. Ergod. Theory Dyn. Syst. **30**, 1757–1786 (2010)
22. Masur, H.: Closed trajectories for quadratic differentials with an application to billiards. Duke Math. J. **53**, 307–313 (1986)
23. Pesin, Ya.B.: Dynamical systems with generalized hyperbolic attractors: hyperbolic, ergodic and topological properties. Ergod. Theory Dyn. Syst. **12**, 123–151 (1992)
24. Sataev, E.A.: Invariant measures for hyperbolic mappings with singularities. Russ. Math. Surv. **47**, 191–251 (1992)
25. Sinai, Y.: On the foundations of the ergodic hypothesis for a dynamical system of statistical mechanics. Sov. Math. Dokl. **4**, 1818–1822 (1963)
26. Sinai, Y.: Dynamical systems with elastic reflections: ergodic properties of dispersing billiards. Usp. Mat. Nauk. **25**, 141–192 (1970)

27. Smillie, J.: The dynamics of billiard flows in rational polygons. In: Encyclopedia of Mathematical Sciences, vol. 100, pp. 360–382. Springer, New York (2000)
28. Szasz, S. (ed.): Hard Ball Systems and the Lorentz Gas. Encyclopedia of Mathematical, vol. 101. Springer, Berlin (2000)
29. Tabachnikov, S.: Geometry and Billiards American Mathematical Society. Providence, RI (2005)
30. Wojtkowski, M.: Principles for the design of billiards with nonvanishing Lyapunov exponents. Commun. Math. Phys. **105**, 391–414 (1986)

Asymptotic Representation of Solutions of Linear Autonomous Difference Equations

Hideaki Matsunaga

Abstract We establish an explicit asymptotic representation formula for solutions of linear autonomous difference equations with infinite delay. As an application, we investigate the limit of solutions of a certain delay difference equation in the critical case where the equation loses its asymptotic stability.

Keywords Linear difference equations · Asymptotic periodic solutions · Delay

1 Introduction

Let $\mathbf{N}$, $\mathbf{Z}^+$, $\mathbf{Z}^-$, $\mathbf{Z}$ and $\mathbf{C}$ be the set of natural numbers, nonnegative integers, nonpositive integers, integers and complex numbers, respectively. For $d \in \mathbf{N}$, we denote by $\mathbf{C}^d$ the space of all d-column vectors whose components are complex numbers, with the Euclidean norm $|\cdot|$.

In this paper, following Ref. [3], we summarize several results on the asymptotic behavior of solutions of the Volterra difference equation with infinite delay

$$x(n+1) = \sum_{j=-\infty}^{n} A(n-j)x(j), \quad n \in \mathbf{Z}^+, \tag{1}$$

where $A(j)$, $j \in \mathbf{Z}^+$, are $d \times d$ complex matrices satisfying $\sum_{j=0}^{\infty} |A(j)|e^{\gamma j} < \infty$ for some (fixed) constant $\gamma > 0$. As an application, we investigate the limit of solutions of the delay difference equation

$$x(n+1) - x(n) + Bx(n-k) = 0, \quad n \in \mathbf{Z}^+,$$

H. Matsunaga (✉)
Department of Mathematical Sciences, Osaka Prefecture University, Sakai 599-8531, Japan
e mail: hideaki@ms.osakafu-u.ac.jp

L. Alsedà i Soler et al. (eds.), *Difference Equations, Discrete Dynamical Systems and Applications*, Springer Proceedings in Mathematics & Statistics 180, DOI 10.1007/978-3-662-52927-0_15

in the critical case where the above equation loses its asymptotic stability; here, B is a 2×2 real constant matrix and k is a positive integer.

2 An Explicit Asymptotic Formula for Solutions of (1)

Let us consider the function space $\mathscr{B}_\gamma$ defined by

$$\mathscr{B}_\gamma = \{\phi : \mathbf{Z}^- \to \mathbf{C}^d \mid \sup_{\theta \in \mathbf{Z}^-} |\phi(\theta)| e^{\gamma\theta} < \infty\}, \quad \|\phi\| = \sup_{\theta \in \mathbf{Z}^-} |\phi(\theta)| e^{\gamma\theta}.$$

Then Eq. (1) can be formulated as a linear functional difference equation of the form

$$x(n+1) = L(x_n), \quad n \in \mathbf{Z}^+, \tag{2}$$

on $\mathscr{B}_\gamma$, where $L : \mathscr{B}_\gamma \to \mathbf{C}^d$ is a bounded linear functional defined by $L(\phi) = \sum_{j=0}^{\infty} A(j)\phi(-j)$ for $\phi \in \mathscr{B}_\gamma$ and x_n is an element in $\mathscr{B}_\gamma$ defined as $x_n(\theta) = x(n+\theta)$ for $\theta \in \mathbf{Z}^-$. For any $\phi \in \mathscr{B}_\gamma$, there exists a unique function $x : \mathbf{Z} \to \mathbf{C}^d$ such that $x(\theta) = \phi(\theta)$ for all $\theta \in \mathbf{Z}^-$ and x satisfies (2) (equivalently, (1)) on $\mathbf{Z}^+$. We call this x the solution of (2) (or (1)) through $(0, \phi)$, and denote it by $x(\cdot, \phi)$.

For any $n \in \mathbf{Z}^+$, we define the solution operator $T(n)$ of (2) (or (1)) by

$$T(n)\phi = x_n(\cdot, \phi), \quad \phi \in \mathscr{B}_\gamma. \tag{3}$$

Then $T(n)$ is a bounded linear operator on $\mathscr{B}_\gamma$ and satisfies the semigroup property $T(n+m) = T(n)T(m)$ for $n, m \in \mathbf{Z}^+$, which implies that $T(n) = T^n$ for $n \in \mathbf{Z}^+$, where $T := T(1)$. Rewriting (3) with $n = 1$, we get the following expression of $T\phi$:

$$[T\phi](\theta) = \begin{cases} \sum_{j=0}^{\infty} A(j)\phi(-j) & (\theta = 0), \\ \phi(\theta + 1) & (\theta \le -1), \end{cases} \quad \phi \in \mathscr{B}_\gamma.$$

Let $\mathbf{C}^{d*}$ be the space of all d-dimensional row vectors with complex components with the norm $|\cdot|$ satisfying $|x^* x| \le |x^*||x|$ for any $x \in \mathbf{C}^d$ and $x^* \in \mathbf{C}^{d*}$. For any (fixed) constant $\tilde{\gamma}$ with $0 < \tilde{\gamma} < \gamma$, we introduce the function space $\mathscr{B}^\sharp_{\tilde{\gamma}}$ defined by

$$\mathscr{B}^\sharp_{\tilde{\gamma}} = \{\psi : \mathbf{Z}^+ \to \mathbf{C}^{d*} \mid \sup_{s \in \mathbf{Z}^+} |\psi(s)| e^{-\tilde{\gamma}s} < \infty\}, \quad \|\psi\| = \sup_{s \in \mathbf{Z}^+} |\psi(s)| e^{-\tilde{\gamma}s}.$$

Let us consider the bilinear form $\langle \cdot, \cdot \rangle$ on $\mathscr{B}^\sharp_{\tilde{\gamma}} \times \mathscr{B}_\gamma$ defined by

$$\langle \psi, \phi \rangle = \psi(0)\phi(0) + \sum_{j=1}^{\infty} \sum_{s=0}^{j-1} \psi(s+1) A(j) \phi(s-j), \quad \phi \in \mathscr{B}_\gamma, \ \psi \in \mathscr{B}^\sharp_{\tilde{\gamma}}.$$

Then the formal adjoint operator $T^{\sharp}$ on $\mathscr{B}^{\sharp}_{\bar{\gamma}}$ which satisfies $\langle \psi, T\phi \rangle = \langle T^{\sharp}\psi, \phi \rangle$ for any $\phi \in \mathscr{B}_{\gamma}$ and $\psi \in \mathscr{B}^{\sharp}_{\bar{\gamma}}$ is given by

$$[T^{\sharp}\psi](s) = \begin{cases} \sum_{j=0}^{\infty} \psi(j)A(j) & (s = 0), \\ \psi(s-1) & (s \geq 1), \end{cases} \qquad \psi \in \mathscr{B}^{\sharp}_{\bar{\gamma}}.$$

Denote by $\sigma(T)$ and $P_{\sigma}(T)$ the spectrum and the point spectrum of T, respectively. Let $\mathbf{C}_{-\gamma} = \{\lambda \in \mathbf{C} \mid |\lambda| > e^{-\gamma}\}$, and introduce the characteristic equation defined by

$$\det \Delta(\lambda) = 0, \quad \Delta(\lambda) = \lambda E_d - \sum_{j=0}^{\infty} \lambda^{-j} A(j),$$

where E_d is the $d \times d$ unit matrix. We call $\lambda \in \mathbf{C}_{-\gamma}$ such that $\det \Delta(\lambda) = 0$ a characteristic root of (1). The sets $\sigma(T)$ and $\sigma(T^{\sharp})$ are characterized as follows.

Proposition 1 ([3, Propositions 2.1 and 3.1]) *The following relations hold.*

$$\sigma(T) \cap \mathbf{C}_{-\gamma} = P_{\sigma}(T) \cap \mathbf{C}_{-\gamma} = \{\lambda \in \mathbf{C}_{-\gamma} \mid \det \Delta(\lambda) = 0\},$$
$$\sigma(T^{\sharp}) \cap \mathbf{C}_{-\bar{\gamma}} = P_{\sigma}(T^{\sharp}) \cap \mathbf{C}_{-\bar{\gamma}} = \{\lambda \in \mathbf{C}_{-\bar{\gamma}} \mid \det \Delta(\lambda) = 0\}.$$

Let $\Sigma^{cu} = \{\lambda \in \sigma(T) \mid |\lambda| \geq 1\}$. Since $\det \Delta(\lambda)$ is an analytic function on $\mathbf{C}_{-\gamma}$, we see that Σ^{cu} is a finite set, and thus $\mathscr{B}_{\gamma}$ is decomposed as a direct sum

$$\mathscr{B}_{\gamma} = \mathscr{B}^{s}_{\gamma} \oplus \mathscr{B}^{cu}_{\gamma},$$

where $\mathscr{B}^{cu}_{\gamma}$ and $\mathscr{B}^{s}_{\gamma}$ are both closed subspaces which are invariant under T, and the restrictions $T|_{\mathscr{B}^{cu}_{\gamma}}$ and $T|_{\mathscr{B}^{s}_{\gamma}}$ satisfy $\sigma(T|_{\mathscr{B}^{cu}_{\gamma}}) = \Sigma^{cu}$ and $\sigma(T|_{\mathscr{B}^{s}_{\gamma}}) = \sigma(T) \backslash \Sigma^{cu} = \{\lambda \in \sigma(T) \mid |\lambda| < 1\}$. Denote by Π^{cu} the projection from $\mathscr{B}_{\gamma}$ onto $\mathscr{B}^{cu}_{\gamma}$. Then we get the following result on the asymptotic behavior of solutions of (1).

Lemma 1 ([3, Lemma 2.1]) *If $\Sigma^{cu} = \emptyset$, then the zero solution of (1) is exponentially asymptotically stable. On the other hand, if $\Sigma^{cu} \neq \emptyset$, then the solution $x(n, \phi)$ of (1) tends to $x_*(n, \phi) = x(n, \Pi^{cu}\phi)$ exponentially as $n \to \infty$. More precisely, there exist constants $K_1 \geq 1$ and μ_1 with $0 < \mu_1 < 1$ such that*

$$|x(n, \phi) - x_*(n, \phi)| \leq K_1 \mu_1^n \|\phi\|, \quad n \in \mathbf{Z}^+.$$

Let λ_0 be a characteristic root of (1), and denote by $\mathscr{M}_{\lambda_0}(T)$ the generalized eigenspace of T for λ_0. If there exists a positive integer p such that $\mathscr{N}((T - \lambda_0 I)^p) = \mathscr{N}((T - \lambda_0 I)^{p+1})$ (and hence $\mathscr{M}_{\lambda_0}(T) = \mathscr{N}((T - \lambda_0 I)^p)$), then the ascent of λ_0 for T is said to be finite; here $\mathscr{N}((T - \lambda_0 I)^p)$ is the null space of $(T - \lambda_0 I)^p$. In this case, the smallest integer p such that $\mathscr{N}((T - \lambda_0 I)^p) = \mathscr{N}((T - \lambda_0 I)^{p+1})$ is called the ascent of λ_0. In fact, the following result holds.

Proposition 2 ([3, Proposition 2.3]) *Let $\lambda_0 \in \mathbf{C}_{-\gamma}$ and* $\det \Delta(\lambda_0) = 0$*, and let q be the order of λ_0 as a pole of the function $\Delta(\lambda)^{-1}$. Then λ_0 is a pole of the operator-valued function $(\lambda I - T)^{-1}$ and there exists a positive integer p such that $p \leq q$ and $\mathcal{M}_{\lambda_0}(T) = \mathcal{N}((T - \lambda_0 I)^p)$; in other words, the ascent of λ_0 does not exceed q. Also, for the above cited λ_0 and p, we have*

$$\mathcal{B}_\gamma = \mathcal{R}((T - \lambda_0 I)^p) \oplus \mathcal{N}((T - \lambda_0 I)^p),$$

where $\mathcal{R}((T - \lambda_0 I)^p)$ is the range space of $(T - \lambda_0 I)^p$.

For any $\tau \in \mathbf{Z}$ and $k \in \mathbf{N}$, we set

$$\binom{\tau}{0} = 1, \quad \binom{\tau}{k} = \frac{\tau(\tau - 1)\cdots(\tau - k + 1)}{k!}.$$

Also, for any $k \in \mathbf{N}$ and $\lambda \in \mathbf{C}$, we introduce functions $w_k(\lambda)$, $w_k^\sharp(\lambda)$, a $d \times d$ matrix $P_k(\lambda)$ and a $(kd) \times (kd)$ matrix $D_k(\lambda)$ given by

$$\begin{aligned}
[w_k(\lambda)](\theta) &= \binom{\theta}{k-1}\lambda^{\theta-k+1}, \quad \theta \in \mathbf{Z}^-, \\
[w_k^\sharp(\lambda)](s) &= [w_k(\lambda)](-s), \quad s \in \mathbf{Z}^+, \\
P_k(\lambda) &= \frac{1}{(k-1)!}\frac{d^{k-1}}{d\lambda^{k-1}}\Delta(\lambda), \\
D_k(\lambda) &= \begin{pmatrix} P_1(\lambda) & P_2(\lambda) & \cdots & P_k(\lambda) \\ O & P_1(\lambda) & \cdots & P_{k-1}(\lambda) \\ \vdots & \ddots & \ddots & \vdots \\ O & \cdots & O & P_1(\lambda) \end{pmatrix},
\end{aligned}$$

respectively. Notice that $w_k(\lambda)\xi \in \mathcal{B}_\gamma$ for $\xi \in \mathbf{C}^d$ and $\eta w_k^\sharp(\lambda) \in \mathcal{B}_{\tilde{\gamma}}^\sharp$ for $\eta \in \mathbf{C}^{d*}$. The null spaces $\mathcal{N}((T - \lambda I)^m)$ and $\mathcal{N}((T^\sharp - \lambda I)^m)$ are characterized as follows.

Proposition 3 ([3, Propositions 2.4 and 3.2]) *The following statements hold:*

(i) Let $\lambda \in \sigma(T) \cap \mathbf{C}_{-\gamma}$ and $m \in \mathbf{N}$. Then ϕ belongs to $\mathcal{N}((T - \lambda I)^m)$ if and only if it is written as

$$\phi = \sum_{k=1}^{m} w_k(\lambda)\xi_k,$$

where ξ_k, $k = 1, \ldots, m$, belong to $\mathbf{C}^d$ satisfying $D_m(\lambda)\,\mathrm{col}\,(\xi_1, \ldots, \xi_m) = 0$.

(ii) Let $\lambda \in \sigma(T^\sharp) \cap \mathbf{C}_{-\tilde{\gamma}}$ and $m \in \mathbf{N}$. Then α belongs to $\mathcal{N}((T^\sharp - \lambda I)^m)$ if and only if it is written as

$$\alpha = \sum_{k=1}^{m} \eta_{m+1-k} w_k^\sharp(\lambda),$$

where η_k, $k = 1, \ldots, m$, *belong to* $\mathbf{C}^{d*}$ *satisfying* $(\eta_1, \ldots, \eta_m)D_m(\lambda) = 0$.

Let $\Sigma^{cu} = \{\lambda_1, \ldots, \lambda_r\}$. Then $\mathscr{B}_\gamma^{cu} = \mathscr{M}_{\lambda_1}(T) \oplus \cdots \oplus \mathscr{M}_{\lambda_r}(T)$, and hence

$$\mathscr{B}_\gamma = \mathscr{B}_\gamma^s \oplus \mathscr{M}_{\lambda_1}(T) \oplus \cdots \oplus \mathscr{M}_{\lambda_r}(T).$$

The ascent of each λ_i is finite by Proposition 2. Let p_i be the ascent of $\lambda_i, i = 1, \ldots, r$. Then $\mathscr{M}_{\lambda_i}(T) = \mathscr{N}((T - \lambda_i I)^{p_i})$; consequently $\mathscr{M}_{\lambda_i}(T)$ has the same dimension, say s_i, as the null space $\mathscr{N}(D_{p_i}(\lambda_i))$ by Proposition 3. Thus the dimension of $\mathscr{B}_\gamma^{cu}$ is equal to $s := s_1 + \cdots + s_r$. Notice that the dimension of the subspace

$$\mathscr{N}^\sharp = \mathscr{N}((T^\sharp - \lambda_1 I)^{p_1}) \oplus \cdots \oplus \mathscr{N}((T^\sharp - \lambda_r I)^{p_r})$$

of $\mathscr{B}_\gamma^\sharp$ is exactly the same as s, since $\dim \mathscr{N}((T^\sharp - \lambda_i I)^{p_i}) = \dim \mathscr{N}((D_{p_i}(\lambda_i))^*) = s_i$ by Proposition 3. Let $\{\phi_1, \ldots, \phi_s\}$ be a basis for $\mathscr{B}_\gamma^{cu}$, and set $\Phi = (\phi_1, \ldots, \phi_s)$. Similarly, let $\{\psi_1, \ldots, \psi_s\}$ be a basis for $\mathscr{N}^\sharp$, and set $\Psi = \mathrm{col}\,(\psi_1, \ldots, \psi_s)$. We call Φ and Ψ a basis vector for $\mathscr{B}_\gamma^{cu}$ and $\mathscr{N}^\sharp$, respectively. We use the notation $\langle \Psi, \Phi \rangle$ to denote the $s \times s$ matrix $(\langle \psi_i, \phi_j \rangle), i, j = 1, \ldots, s$. Also, we denote the column vector $\mathrm{col}\,(\langle \psi_1, \phi \rangle, \ldots, \langle \psi_s, \phi \rangle)$ by $\langle \Psi, \phi \rangle$ for any $\phi \in \mathscr{B}_\gamma$. The following results yield a characterization of the component $\mathscr{B}_\gamma^s$ in the decomposition $\mathscr{B}_\gamma = \mathscr{B}_\gamma^s \oplus \mathscr{B}_\gamma^{cu}$ and the representation form of the projection Π^{cu}.

Proposition 4 ([3, Corollary 3.1]) ϕ *belongs to* $\mathscr{B}_\gamma^s$ *if and only if* $\langle \alpha, \phi \rangle = 0$ *for all* $\alpha \in \mathscr{N}^\sharp$.

Theorem 1 ([3, Theorem 3.1]) *Let* Φ *and* Ψ *be a basis vector for* $\mathscr{B}_\gamma^{cu}$ *and* $\mathscr{N}^\sharp$, *respectively. Then* $\langle \Psi, \Phi \rangle$ *is nonsingular, and the projection* Π^{cu} *is given by*

$$\Pi^{cu}\phi = \Phi \langle \Psi, \Phi \rangle^{-1} \langle \Psi, \phi \rangle, \quad \phi \in \mathscr{B}_\gamma.$$

For any $k \in \mathbf{N}$ and $\lambda \in \mathbf{C}$, we consider a function $\tilde{w}_k(\lambda)$ defined by

$$[\tilde{w}_k(\lambda)](\tau) = \binom{\tau}{k-1} \lambda^{\tau - k + 1}, \quad \tau \in \mathbf{Z}.$$

Obviously, $[w_k(\lambda)](\theta) = [\tilde{w}_k(\lambda)](\theta)$ for $\theta \in \mathbf{Z}^-$ and we call $\tilde{w}_k(\lambda)$ the canonical prolongation of $w_k(\lambda)$. For the basis vector Φ for $\mathscr{B}_\gamma^{cu}$ cited above, we set

$$\tilde{\Phi} = (\tilde{\phi}_1, \ldots, \tilde{\phi}_s), \quad \tilde{\phi}_j = \sum_{i=1}^{r} \sum_{k=1}^{p_i} \tilde{w}_k(\lambda_i) \xi_{k,i,j},$$

where $\xi_{k,i,j} \in \mathbf{C}^d$ satisfies $D_{p_i}(\lambda_i)\,\mathrm{col}\,(\xi_{1,i,j}, \ldots, \xi_{p_i,i,j}) = 0$. Clearly, $\tilde{\Phi}(\theta) = \Phi(\theta)$ for $\theta \in \mathbf{Z}^-$ and we call $\tilde{\Phi}$ the canonical prolongation of Φ. We now establish the following result which is an explicit asymptotic formula for solutions of (1).

Theorem 2 ([3, Theorem 4.1]) *Let Φ and Ψ be a basis vector for $\mathscr{B}_\gamma^{cu}$ and $\mathscr{N}^\sharp$, respectively, and let $\phi \in \mathscr{B}_\gamma$. Then*

$$x_*(n, \phi) = \tilde{\Phi}(n)\langle \Psi, \Phi\rangle^{-1}\langle \Psi, \phi\rangle, \quad n \in \mathbf{Z}^+,$$

where $x_*(\cdot, \phi)$ is the solution of (1) through $(0, \Pi^{cu}\phi)$, and $\tilde{\Phi}$ is the canonical prolongation of Φ. In particular, the solution $x(\cdot, \phi)$ of (1) satisfies

$$\lim_{n\to\infty} |x(n, \phi) - \tilde{\Phi}(n)\langle \Psi, \Phi\rangle^{-1}\langle \Psi, \phi\rangle| = 0 \quad (exponentially).$$

Remark 1 If $\langle \Psi, \Phi\rangle$ is the unit matrix, then the solution $x_*(\cdot, \phi)$ of (1) is given by

$$x_*(n, \phi) = \tilde{\Phi}(n)\langle \Psi, \phi\rangle, \quad n \in \mathbf{Z}^+.$$

Notice that by considering $\langle \Psi, \Phi\rangle^{-1}\Psi$ instead of Ψ as a basis vector for $\mathscr{N}^\sharp$ if necessary, we may assume that $\langle \Psi, \Phi\rangle$ is the unit matrix.

3 Application

Let us consider the delay difference equation

$$x(n+1) - x(n) + Bx(n-k) = 0, \quad n \in \mathbf{Z}^+, \tag{4}$$

where $x(n) \in \mathbf{R}^2$, B is a 2×2 real constant matrix and k is a positive integer. The author and Hara [2] established the following explicit necessary and sufficient condition for the asymptotic stability of (4) which is an extension of the well-known criterion due to Levin and May [1] for the scalar case. In this paper we only consider the case where the matrix B has complex eigenvalues. Without loss of generality, we may assume that the matrix B is given by

$$B = b\begin{pmatrix} \cos\beta & -\sin\beta \\ \sin\beta & \cos\beta \end{pmatrix}, \tag{5}$$

where b and β are real numbers with $0 < |\beta| \le \pi/2$.

Proposition 5 ([2, Theorem1]) *Suppose that the matrix B is given by (5). Then the zero solution of (4) is asymptotically stable if and only if*

$$0 < b < 2\cos\frac{k\pi + |\beta|}{2k+1}.$$

In the critical case where $b = 2\cos((k\pi + |\beta|)/(2k+1))$, Eq. (4) loses its asymptotic stability, and thus we can expect that the solution of (4) tends to some (periodic) orbit as $n \to \infty$. In fact, applying Theorem 2 to (4), we have the following result.

Theorem 3 *Suppose that the matrix B is given by (5). Let $x(\cdot, \varphi)$ be a solution of (4) with the initial condition $x(\theta) = \varphi(\theta)$ for $-k \le \theta \le 0$, where $\varphi(\theta) \in \mathbf{R}^2$. If $b = 2\cos((k\pi + |\beta|)/(2k+1))$, then*

$$\lim_{n\to\infty} |x(n,\varphi) - \tilde{\Phi}(n)c| = 0 \quad (\textit{exponentially}),$$

where

$$\tilde{\Phi}(n) = \begin{pmatrix} \cos\omega n & -\sin\omega n \\ \sin\omega n & \cos\omega n \end{pmatrix}, \quad \omega = \frac{2|\beta| - \pi}{2k+1},$$

$$c = \left(E_2 - k\tilde{\Phi}(-k-1)B\right)^{-1}\left(\varphi(0) - \sum_{j=1}^{k} \tilde{\Phi}(j-k-1)B\varphi(-j)\right).$$

Remark 2 In Theorem 3, if ω/π is rational, then $\tilde{\Phi}(n)c$ is periodic. On the other hand, if ω/π is irrational, then $\tilde{\Phi}(n)c$ densely fills out a circle with center 0 and radius $|c|$. The explicit asymptotic representation of some (periodic) solutions of (4) in the critical case was first obtained by Nagabuchi [4, Theorem 3] via complicated and tedious calculations based on fundamental results in linear algebra.

To prove Theorem 3, we regard Eq. (4) as Eq. (1) with

$$A(j) = \begin{cases} E_2 & (j=0) \\ O & (j = 1, 2, \ldots, k-1, k+1, k+2, \ldots) \\ -B & (j=k). \end{cases}$$

Since $A(j) = O$ for $j = k+1, k+2, \ldots$, we can treat the phase space $\mathscr{B}_\gamma$ (and $\mathscr{B}^\sharp_{\tilde{\gamma}}$) for any $\gamma > 0$ (and $\tilde{\gamma}$ with $0 < \tilde{\gamma} < \gamma$). The bilinear form $\langle\cdot,\cdot\rangle$ on $\mathscr{B}^\sharp_{\tilde{\gamma}} \times \mathscr{B}_\gamma$ becomes

$$\langle\psi,\phi\rangle = \psi(0)\phi(0) + \sum_{s=0}^{k-1} \psi(s+1)(-B)\phi(s-k), \quad \phi \in \mathscr{B}_\gamma, \ \psi \in \mathscr{B}^\sharp_{\tilde{\gamma}}.$$

Lemma 2 ([2]) *Let $\Delta(\lambda) = (\lambda - 1)E_2 + \lambda^{-k}B$ and let B be given by (5). If $b = 2\cos((k\pi + |\beta|)/(2k+1))$, then* $\det \Delta(\lambda) = 0$ *has two simple roots $e^{\pm i\omega}$, and the remaining roots are inside the unit disk, where $\omega = (2|\beta| - \pi)/(2k+1)$.*

Proof of Theorem 3 From Lemma 2, it follows that $\Sigma^{cu} = \{e^{i\omega}, e^{-i\omega}\}$, and each element in Σ^{cu} is a simple root of $\det \Delta(\lambda) = 0$. Hence the ascent for each element in Σ^{cu} is equal to 1, which implies that $\mathscr{B}^{cu}_\gamma = \mathscr{N}(T - e^{i\omega}I) \oplus \mathscr{N}(T - e^{-i\omega}I)$. We will compute a basis vector of $\mathscr{B}^{cu}_\gamma$. It follows that

$$\begin{aligned} P_1(e^{i\omega}) = \Delta(e^{i\omega}) &= (e^{i\omega} - 1)E_2 + be^{-ik\omega} \begin{pmatrix} \cos\beta & -\sin\beta \\ \sin\beta & \cos\beta \end{pmatrix} \\ &= (e^{i\omega} - 1)E_2 - (e^{i\omega} - 1)e^{-i|\beta|} \begin{pmatrix} \cos\beta & -\sin\beta \\ \sin\beta & \cos\beta \end{pmatrix} \\ &= (e^{i\omega} - 1)e^{-i|\beta|} i \sin|\beta| \begin{pmatrix} 1 & -\mathrm{sgn}(\beta)i \\ \mathrm{sgn}(\beta)i & 1 \end{pmatrix}. \end{aligned}$$

We consider the case $0 < \beta \le \pi/2$. (In case $-\pi/2 \le \beta < 0$, the proof is similar.) Notice that $\xi = \mathrm{col}\,(1, -i)$ satisfies $P_1(e^{i\omega})\xi = 0$. Applying Proposition 3 (i), we have that a basis of $\mathcal{N}(T - e^{i\omega} I)$ is

$$\hat{\phi}_1(\theta) = [w_1(e^{i\omega})](\theta) \begin{pmatrix} 1 \\ -i \end{pmatrix} = e^{i\omega\theta} \begin{pmatrix} 1 \\ -i \end{pmatrix}, \quad \theta \in \mathbf{Z}^-.$$

Also, since $P_1(e^{-i\omega}) = \Delta(e^{-i\omega}) = \overline{\Delta(e^{i\omega})}$, $\xi = \mathrm{col}\,(1, i)$ satisfies $P_1(e^{-i\omega})\xi = 0$. This yields that a basis of $\mathcal{N}(T - e^{-i\omega} I)$ is

$$\hat{\phi}_2(\theta) = [w_1(e^{-i\omega})](\theta) \begin{pmatrix} 1 \\ -i \end{pmatrix} = e^{-i\omega\theta} \begin{pmatrix} 1 \\ i \end{pmatrix}, \quad \theta \in \mathbf{Z}^-,$$

and consequently, $\mathcal{B}_\gamma^{cu} = \mathrm{span}\{\hat{\phi}_1, \hat{\phi}_2\}$. We define a new basis for $\mathcal{B}_\gamma^{cu}$ by

$$\phi_1 = \frac{1}{2}(\hat{\phi}_1 + \hat{\phi}_2), \quad \phi_2 = \frac{1}{2i}(\hat{\phi}_2 - \hat{\phi}_1).$$

Then a basis vector Φ for $\mathcal{B}_\gamma^{cu}$ is expressed as

$$\Phi = (\phi_1, \phi_2), \quad \phi_1(\theta) = \begin{pmatrix} \cos\omega\theta \\ \sin\omega\theta \end{pmatrix}, \quad \phi_2(\theta) = \begin{pmatrix} -\sin\omega\theta \\ \cos\omega\theta \end{pmatrix}, \quad \theta \in \mathbf{Z}^-.$$

In a similar fashion, applying Proposition 3 (ii), we see that a basis vector Ψ for $\mathcal{N}^\sharp = \mathcal{N}(T^\sharp - e^{i\omega} I) \oplus \mathcal{N}(T^\sharp - e^{-i\omega} I)$ is expressed as

$$\Psi = \begin{pmatrix} \psi_1 \\ \psi_2 \end{pmatrix}, \quad \psi_1(s) = (\cos\omega s, \sin\omega s), \quad \psi_2(s) = (-\sin\omega s, \cos\omega s), \quad s \in \mathbf{Z}^+.$$

Let $\tilde{\Phi}$ be the canonical prolongation of Φ. Then it follows that $\Psi(s) = \tilde{\Phi}(-s)$ for $s \in \mathbf{Z}^+$, which, together with the commutativity of $\tilde{\Phi}$ and B, implies that

$$\langle \Psi, \Phi \rangle = \Psi(0)\Phi(0) + \sum_{s=0}^{k-1} \Psi(s+1)(-B)\Phi(s-k)$$

$$= E_2 - \sum_{s=0}^{k-1} \tilde{\Phi}(-s-1)B\tilde{\Phi}(s-k)$$
$$= E_2 - k\tilde{\Phi}(-k-1)B$$

and

$$\langle \Psi, \varphi \rangle = \Psi(0)\varphi(0) + \sum_{s=0}^{k-1} \Psi(s+1)(-B)\varphi(s-k)$$
$$= \varphi(0) - \sum_{s=0}^{k-1} \tilde{\Phi}(-s-1)B\varphi(s-k)$$
$$= \varphi(0) - \sum_{j=1}^{k} \tilde{\Phi}(j-k-1)B\varphi(-j).$$

Applying Theorem 2, we conclude that $x(n, \varphi)$ tends to $\tilde{\Phi}(n)c$ exponentially as $n \to \infty$, where $c = \langle \Psi, \Phi \rangle^{-1} \langle \Psi, \varphi \rangle$. This completes the proof of Theorem 3. □

References

1. Levin, S.A., May, R.M.: A note on difference-delay equations. Theor. Popul. Biol. **9**, 178–187 (1976)
2. Matsunaga, H., Hara, T.: The asymptotic stability of a two-dimensional linear delay difference equation. Dyn. Contin. Discret. Impuls. Syst. **6**, 465–473 (1999)
3. Matsunaga, H., Murakami, S., Nagabuchi, Y., Nakano, Y.: Formal adjoint equations and asymptotic formula for solutions of Volterra difference equations with infinite delay. J. Differ. Equ. Appl. **18**, 57–88 (2012)
4. Nagabuchi, Y.: Asymptotic periodic solutions for a two-dimensional linear delay difference system. Dyn. Contin. Discret. Impuls. Syst. Ser. A Math. Anal. **9**, 187–200 (2002)

Translation Arcs and Stability in Two Dimensions

Rafael Ortega

Abstract Given a closed orbit γ of a system of differential equations in the plane

$$\dot{x} = X(x), \quad x \in \mathbb{R}^2,$$

the index of the vector field X around γ is one. This classical result has a counterpart in the theory of discrete systems in the plane. Consider the equation

$$x_{n+1} = h(x_n), \quad x_n \in \mathbb{R}^2,$$

where h is an orientation-preserving embedding and assume that there is a recurrent orbit that is not a fixed point. Then there exists a Jordan curve γ such that the fixed point index of h around this curve is one. The proof is based on the theory of translation arcs, initiated by Brouwer. These notes are dedicated to discuss some consequences of the above result, specially in stability theory. We will compute the indexes associated to a stable invariant object and show that Lyapunov stability implies persistence (in two dimensions). The invariant sets under consideration will be fixed points, periodic orbits and Cantor sets.

Keywords Stability · Translation arc · Degree · Cantor set

1 Introduction

We are going to discuss the dynamics of a system of the type

$$x_{n+1} = h(x_n), \quad x_n \in \mathbb{R}^2 \tag{1}$$

R. Ortega (✉)
Departamento de Matemática Aplicada, Universidad de Granada,
18071 Granada, Spain
e-mail: rortega@ugr.es

L. Alsedà i Soler et al. (eds.), *Difference Equations, Discrete Dynamical Systems and Applications*, Springer Proceedings in Mathematics & Statistics 180, DOI 10.1007/978-3-662-52927-0_16

where $h : \mathbb{R}^2 \to \mathbb{R}^2$ is a one-to-one continuous map. Notice that h is not necessarily onto and so the image $h(\mathbb{R}^2)$ can be a proper subset of $\mathbb{R}^2$. In general orbits are defined in the future, $x_n = h^n(x_0)$, $n \geq 0$, but not always in the past. We will also assume that the map h is orientation preserving. This class of maps will be denoted by $\mathscr{E}_+$, indicating that h is an orientation preserving topological embedding. We refer to the appendix for more details on the topological aspects.

As a counterpoint to the discrete model (1) we can consider the system of differential equations

$$\dot{x} = X(x), \quad x \in \mathbb{R}^2 \tag{2}$$

where $X : \mathbb{R}^2 \to \mathbb{R}^2$ is a continuous vector field such that there is uniqueness for the initial value problem associated to (2). We all know that the behavior of the continuous system is much simpler. In particular Poincaré–Bendixson theory implies that the dynamics for (2) cannot be very intricate. In general this type of results cannot be extended to the discrete situation. In these notes we will be concerned with a result for continuous systems that is somehow exceptional because it can be extended to discrete systems. This result, already obtained by Poincaré, deals with the computation of the index around a closed orbit of (2). The discrete version is the so-called arc translation lemma, originally due to Brouwer. We will discuss these classical results and later derive some new consequences in stability theory.

2 The Index Around a Closed Orbit

Let Γ be a closed orbit of the system (2) and let $R_i(\Gamma)$ denote the bounded component of $\mathbb{R}^2 \setminus \Gamma$. It is well known that the vector field X has an equilibrium on $R_i(\Gamma)$. Indeed a stronger conclusion can be obtained, namely

$$d(X, R_i(\Gamma), 0) = 1,$$

where d is the Brouwer degree on the plane. A proof of this result can be found in the book [18] and an alternative proof will be presented later. The book [15] contains a very complete presentation of degree theory. By now we just recall how to compute geometrically the degree $d(X, R_i(\Gamma), 0)$. The curve Γ can be parameterized by a continuous function $\beta : [0, 1] \to \mathbb{R}^2$ with $\Gamma = \beta([0, 1])$, β is one-to-one in $[0, 1[$, $\beta(0) = \beta(1)$. It is also assumed that β has positive orientation. The vector field X does not vanish on Γ and so we can find a continuous argument $\theta : [0, 1] \to \mathbb{R}$ satisfying

$$X(\beta(t)) = ||X(\beta(t))||(\cos\theta(t), \sin\theta(t)).$$

Then

$$d(X, R_i(\Gamma), 0) = \frac{1}{2\pi}(\theta(1) - \theta(0)).$$

3 Translation Arcs

We go back to discrete systems. Our first task will be to construct an object that somehow plays the role of the orbits in continuous dynamics. An arc is a set $\alpha \subset \mathbb{R}^2$ that is homeomorphic to the compact interval $[0, 1]$. The end points of the arc will be denoted by $p \neq q$ and sometimes we will use the notation $\alpha = \widehat{pq}$ for the oriented arc. Given $h \in \mathscr{E}_+$ we say that $\alpha = \widehat{pq}$ is a translation arc for h if $h(p) = q$ and

$$h(\alpha \setminus \{q\}) \cap (\alpha \setminus \{q\}) = \emptyset.$$

We illustrate this definition with a typical situation (Fig. 1).

Notice that the translation arc α cannot contain fixed points and the open arcs $\dot{\alpha} = \alpha \setminus \{p, q\}$ and $h(\dot{\alpha})$ cannot intersect. In some special cases the two arcs can have the same end points. This case can only occur if $\{p, q\}$ is a two cycle (Fig. 2).
Next we discuss some simple examples:

1. Assume first that h is a translation, say

$$h(x, y) = (x + 1, y).$$

The horizontal segment joining $p = (0, 0)$ and $q = (1, 0)$ is a translation arc. The broken line passing by $p = (0, 0)$, $(0, 1)$, $(1, 1)$ and $q = (1, 0)$ is not a translation arc, in this case $\alpha \cap h(\alpha)$ is the segment joining $(1, 1)$ and q (Fig. 3).

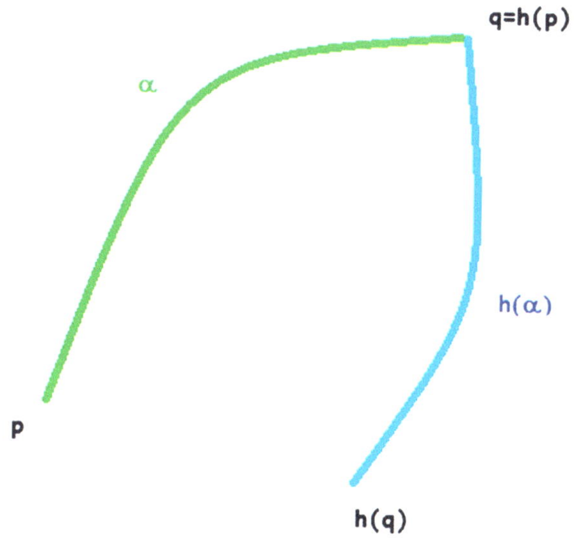

Fig. 1 A typical translation arc

2. Assume that h is the rotation of 180 degrees in the counterclockwise sense. The set $\alpha = \{(\cos\theta, \sin\theta) : \theta \in [0, \pi]\}$ is a translation arc. In this case the end points $p = (1, 0)$ and $q = (-1, 0)$ become two cycles.

3. Assume that the continuous system (2) defines a global flow $\{\phi_t\}_{t\in\mathbb{R}}$ in the plane and take $h = \phi_\tau$ for some fixed $\tau > 0$. We notice that ϕ_τ is a homeomorphism of the plane that is isotopic to the identity. In particular, $h \in \mathscr{E}_+$. Assume that p is a point in the plane that is not an equilibrium of X and let α be the piece of orbit running from p to $q = \phi_\tau(p)$, that is

$$\alpha = \{\phi_t(p) : t \in [0, \tau]\}.$$

This is an arc if the orbit is not closed or if it is closed and $\tau < T$, where $T > 0$ is the minimal period. This segment of orbit is a translation arc whenever the orbit is not closed or $\tau \leq \frac{1}{2}T$.

After these examples we go back to the general setting and ask ourselves about the existence of translation arcs. Next result shows that they exist for any $h \neq id$. By a topological disk we understand a subset D of the plane that is homeomorphic to the disk $\Delta = \{x \in \mathbb{R}^2 : ||x|| \leq 1\}$. Notice that ∂D is a Jordan curve and $\text{int}(D) = R_i(\partial D)$.

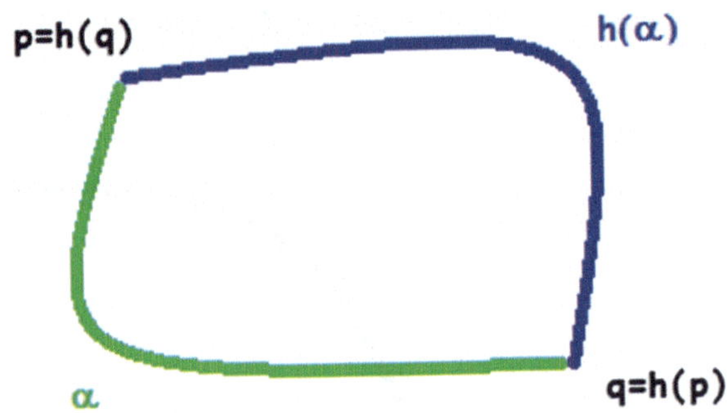

Fig. 2 A translation arc with 2-periodic end points

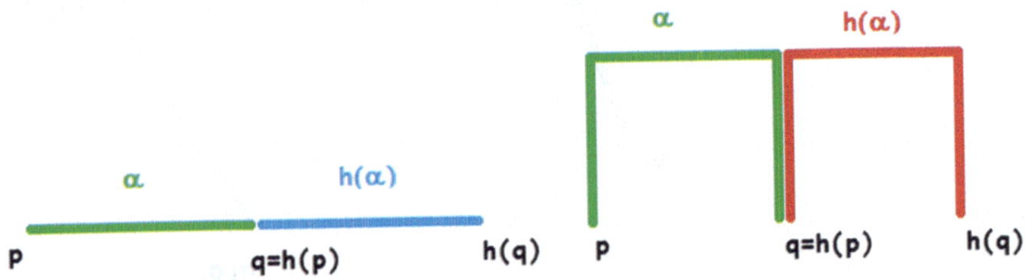

Fig. 3 In the *left* side $\alpha \cap h(\alpha) = \{q\}$ but in the *right* side this intersection is a segment

Fig. 4 The arc α passes through the points $x_1, \ldots, x_k$, all lying in the disk D

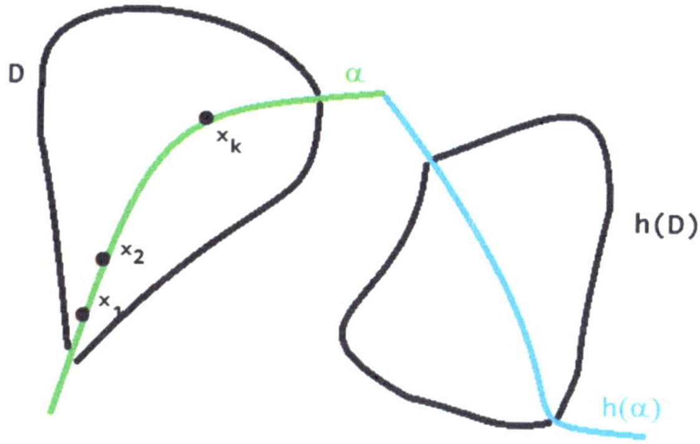

Lemma 1 *Assume that $h \in \mathscr{E}_+$ and D is a topological disk with $D \cap h(D) = \emptyset$. In addition assume that the points $x_1, \ldots, x_k$ belong to the interior of D. Then there exists a translation arc α with $x_1, \ldots, x_k \in \dot{\alpha}$. (Fig. 4).*

This result is a variation of a lemma in [8]. A proof can be found in Chap. 3 of [23].

Given $h \in \mathscr{E}_+$ and a translation arc α, the successive iterations of α can lead to two different situations: either $\alpha \cap h^n(\alpha) = \emptyset$ for each $n \geq 2$ or $\alpha \cap h^n(\alpha) \neq \emptyset$ for some $n \geq 2$. The prototype for the first situation is example 1 (translations), while the second appears in example 2 (rotations). The second situation is somehow analogous to the closed orbit in a continuous system. To obtain a result on the index for the discrete case we need to replace the vector field X by some map related to h. The simplest one is $f := id - h$, where id denotes the identity in the plane. The zeros of f are precisely the fixed points of h, in analogy with the equilibria of the continuous system which are the zeros of the vector field.

4 Lemma on Translation Arcs (Brouwer)

Given $h \in \mathscr{E}_+$, assume that for some $n \geq 2$ there exists a translation arc α with

$$\alpha \cap h^n(\alpha) \neq \emptyset.$$

Then there exists a Jordan curve Γ, contained in $\alpha \cup h(\alpha) \cup \cdots \cup h^n(\alpha)$, and such that

$$d(id - h, R_i(\Gamma), 0) = 1.$$

Notice that this degree is well defined because h cannot have fixed points on $\bigcup_{k \geq 0} h^k(\alpha)$. (Fig. 5). The proof of this lemma is delicate. Classical proofs (see for instance [25]) employ some intuitive method for computing the degree. This type of argument, very common in planar dynamics, is replaced by more rigorous arguments in [7, 13, 14]. In particular the proof by M. Brown is very elegant and, although it

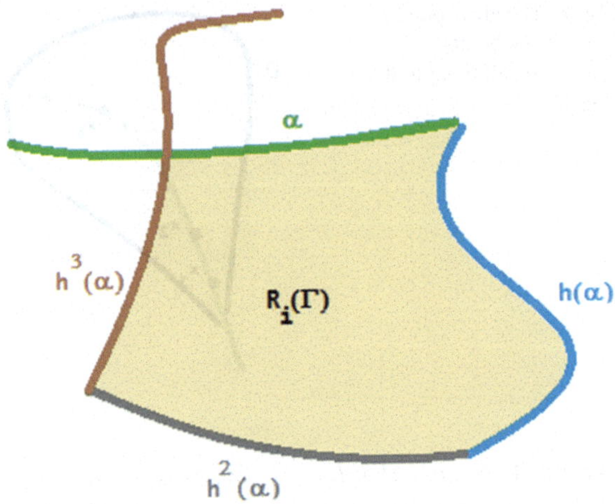

Fig. 5 Brouwer's lemma with $n = 3$. The Jordan curve Γ is composed by $h(\alpha)$, $h^2(\alpha)$ and sub-arcs taken from α and $h^3(\alpha)$

only covers the case of homeomorphisms h, with $h(\mathbb{R}^2) = \mathbb{R}^2$, it can be adapted to embeddings. See [20] and [23].

We are always assuming that h is orientation-preserving and this condition is essential for the previous lemma. This is shown by the following example. Consider the map

$$h : \begin{cases} x_1 = \lambda(y)x, \\ y_1 = -y, \end{cases}$$

where $\lambda : \mathbb{R} \to \mathbb{R}$ is a continuous function satisfying

$$\lambda(y) > 1 \quad \text{if } y \in]-2, 2[, \quad \lambda(y) = 1 \quad \text{if } |y| \geq 2.$$

This is an orientation-reversing homeomorphism and the dynamics is easy to describe. The origin is a fixed point and the orbits lying on $x = 0$ or $|y| \geq 2$ are two cycles, the remaining orbits are unbounded. The set

$$\alpha = \left\{(2\cos t, 2\sin t) : t \in \left[-\frac{\pi}{2}, \frac{\pi}{2}\right]\right\}$$

is a translation arc with $\alpha \cap h^2(\alpha) \neq \emptyset$. Notice that this intersection is composed by the end points of α. Since the origin is the only fixed point of h and lies in the unbounded component of $\mathbb{R}^2 \setminus (\alpha \cup h(\alpha) \cup h^2(\alpha))$, the degree of $id - h$ vanishes in any topological disk whose boundary is contained in $\alpha \cup h(\alpha) \cup h^2(\alpha)$. This shows that the conclusion of Brouwer's lemma cannot be valid if h is orientation-reversing.

It is interesting to recover the result stated in Sect. 2 from Brouwer's Lemma. To this end we first recall two properties of degree. Given a bounded set Ω and a continuous function $f : \overline{\Omega} \to \mathbb{R}^2$ with $f(x) \neq 0$ if $x \in \partial\Omega$, the *continuity property of degree* says that there exists $\eta > 0$ such that if $g : \overline{\Omega} \to \mathbb{R}^2$ is any continuous function with

$$||f(x) - g(x)|| \leq \eta \ \text{ for each } x \in \partial\Omega,$$

then $d(f, \Omega, 0) = d(g, \Omega, 0)$. Notice that η must be small enough to guarantee that g does not vanish on $\partial\Omega$.

The second property refers to the composition with linear maps. Given a 2×2 matrix L,

$$d(Lf, \Omega, 0) = \text{sign}(\det L) d(f, \Omega, 0).$$

We are ready for the proof of the result in Sect. 2. For simplicity we will assume that the vector field X defines a global flow $\{\phi_t\}_{t\in\mathbb{R}}$ on the plane. First we prove that the degree of this vector field coincides with the degree of $id - \phi_t$ for t positive and small. To prove this we integrate the differential equation and obtain the formula

$$\frac{1}{t}(\phi_t(\xi) - \xi) = \frac{1}{t}\int_0^t X(\phi_s(\xi))ds.$$

Then $\frac{1}{t}(\phi_t - id)$ converges to X uniformly on compact sets. In consequence, if Ω is an open and bounded subset of the plane and X does not vanish on the boundary $\partial\Omega$, for small t the map ϕ_t will not have fixed points on $\partial\Omega$ and

$$d(X, \Omega, 0) = d(\frac{1}{t}(\phi_t - id), \Omega, 0). \tag{3}$$

This last degree is the same as $d(id - \phi_t, \Omega, 0)$. Assume now that Γ is a closed orbit. We fix $\xi \in \Gamma$ and consider the arc $\alpha = \{\phi_s(\xi) : 0 \leq s \leq t\}$. We know that α is a translation arc for $h = \phi_t$ if t is small. Moreover, since Γ is closed, some iterate of this arc will have a non-empty intersection with the initial arc α. We are now in the conditions of Brouwer's lemma and there exists a Jordan curve γ such that

$$d(id - h, R_i(\gamma), 0) = 1. \tag{4}$$

But the curve γ is contained in Γ and so they have to coincide, $\gamma = \Gamma$. Putting together the two identities (3), (4) with $\Omega = R_i(\Gamma)$, we conclude that the degree of X is 1.

5 Stability and Index of Fixed Points

Let us discuss some consequences of the previous result in stability theory. Assume that $h : \mathcal{U} \subset \mathbb{R}^2 \to \mathbb{R}^2$ is a map defined on an open set $\mathcal{U}$ and let $p \in \mathcal{U}$ be a fixed point of h, $p = h(p)$. We say that p is *stable* if given any neighborhood $\mathcal{V}$ of p, there exists another neighborhood $\mathcal{W} \subset \mathcal{U}$ such that

$$h^n(\mathcal{W}) \subset \mathcal{V} \ \text{ for each } n \geq 0.$$

We say that the fixed point p is *isolated* if there exists a neighborhood $\mathcal{N}$ such that p is the only fixed point of h in $\mathcal{N}$; that is,

$$\text{Fix}\,(h) \cap \mathcal{N} = \{p\},$$

where $\text{Fix}\,(h) := \{x \in \mathcal{U} : h(x) = x\}$.

To illustrate these notions we consider two simple examples. Assume first that h is a rotation around p, the disks centered at p are invariant under h and this implies that p is stable. Also, p is isolated because it is the only fixed point of h. In the second example h is a symmetry and p is a point in the axis of symmetry. For the same reasons the point p is stable but now it is not isolated because all points lying on the axis of symmetry are fixed.

In the next result we prove that stable and isolated fixed points have index one.

Theorem 1 *Assume that $h : \mathcal{U} \subset \mathbb{R}^2 \to \mathbb{R}^2$ is a continuous and one-to-one map that is orientation-preserving and let p be an isolated and stable fixed point of h, then*

$$d(id - h, \Omega, 0) = 1,$$

where Ω is any open and bounded set with $p \in \Omega$, $\overline{\Omega} \subset \mathcal{U}$ and $\text{Fix}\,(h) \cap \overline{\Omega} = \{p\}$.

It is interesting to notice that this result can also be seen as a discrete counterpart of a classical result in the theory of continuous systems: given an isolated equilibrium of a vector field X, if this equilibrium is stable then the index of X around the equilibrium is one. A proof of Theorem 1 using Brouwer's lemma was obtained by Dancer and myself in [12], previously Krasnoselskii stated the result (without proof) in [16]. We will present a simplified proof but first we discuss some consequences of the theorem. We also notice that the above theorem was extended to the orientation-reversing case by Ruiz del Portal [26]. In higher dimensions there are stable fixed points with arbitrary index, see [4] and the references therein.

6 Instability Criteria

Traditionally the instability of fixed points is obtained using linearization or Lyapunov functions. In two dimensions the previous theorem allows to prove the instability of a fixed point via degree computations. As an example consider a smooth map that can be expressed (in complex notation) as

$$h(z, \overline{z}) = z + z^3 + R(z, \overline{z}),$$

where $R(z, \overline{z}) = o(|z|^3)$ as $z \to 0$. Then $z = 0$ is an unstable fixed point of

$$z_{n+1} = z_n + z_n^3 + R(z_n, \overline{z}_n).$$

To prove this we first apply the inverse function theorem and notice that h is in the conditions of Theorem 1 when it is restricted to a small neighborhood of the origin. The expansion

$$z - h(z, \bar{z}) = -z^3 + o(|z|^3) \text{ as } |z| \to 0$$

implies that $z = 0$ is an isolated fixed point. Let us consider the parameterized curve $\gamma_\rho(t) = \rho e^{2\pi i t}$, $t \in [0, 1]$, with $\rho > 0$ very small. Then

$$(id - h)(\gamma_\rho(t)) = \rho^3(e^{6\pi i t + \pi i} + o(1)) \text{ as } \rho \to 0.$$

The continuous argument of $(id - h) \circ \gamma_\rho$ satisfies

$$\theta_\rho(t) = 6\pi t + \pi + o(1) \text{ as } \rho \to 0.$$

From here it is easy to deduce that the degree on the disk $D_\rho = \{z \in \mathbb{C} : |z| < \rho\}$ is

$$d(id - h, D_\rho, 0) = 3$$

for small ρ. This implies the instability of $z = 0$.

Many other results on unstable fixed points can be obtained using the known algorithms to compute the degree of a planar map.

7 Persistence and Stability

In this section we work with the class of homeomorphisms of the plane, denoted by $\mathcal{H}$. The sub-class of orientation-preserving homeomorphisms will be denoted by $\mathcal{H}_+$. Notice that $\mathcal{H}_+ = \mathcal{H} \cap \mathcal{E}_+$. Given $h \in \mathcal{H}$ and a non-empty compact invariant set $\Lambda \subset \mathbb{R}^2$, $\Lambda = h(\Lambda)$, we say that Λ is *persistent* if given $\varepsilon > 0$ there exists $\delta > 0$ such that for any $\tilde{h} \in \mathcal{H}$ with

$$||h - \tilde{h}||_\infty < \delta$$

there exists a compact set $\tilde{\Lambda} \subset \mathbb{R}^2$ that is invariant under $\tilde{h}$, $\tilde{\Lambda} = \tilde{h}(\tilde{\Lambda})$ and satisfies

$$D_H(\Lambda, \tilde{\Lambda}) \leq \varepsilon.$$

The notation D_H refers to the Hausdorff distance in the space of non-empty compact subsets of the plane and $|| \cdot ||_\infty$ is the uniform norm, that is

$$||h - \tilde{h}||_\infty = \sup_{x \in \mathbb{R}^2} ||h(x) - \tilde{h}(x)||.$$

The theorem of the previous section implies the persistence of stable and isolated fixed points.

Corollary 1 *Given $h \in \mathscr{H}_+$ and $p = h(p)$ a fixed point that is stable and isolated, the invariant set $\Lambda = \{p\}$ is persistent.*

Proof Since p is an isolated fixed point, for small $\varepsilon > 0$ we know that

$$\text{Fix }(h) \cap \overline{\Omega}_\varepsilon = \{p\},$$

where $\Omega_\varepsilon = \{x \in \mathbb{R}^2 : \ ||x|| < \varepsilon\}$. From Theorem 1,

$$d(f, \Omega_\varepsilon, 0) = 1,$$

where $f = id - h$. Let us take $g = id - \tilde{h}$ where $\tilde{h} \in \mathscr{H}$. Since $||f - g||_\infty = ||h - \tilde{h}||_\infty$, we can apply the continuity property of the degree that was stated in Sect. 4 and select $\delta = \eta$ in order to find a fixed point $\tilde{p}$ of $\tilde{h}$ with $\tilde{p} \in \Omega_\varepsilon$. The proof is complete with $\tilde{\Lambda} = \{\tilde{p}\}$.

In the previous result it is essential that the fixed point is isolated. An example of a stable fixed point that is not persistent as invariant set is presented in [24].

8 Proof of the Theorem

To prove Theorem 1 we distinguish two cases:

(i) *p is asymptotically stable.*
In this case the theorem will be a consequence of degree theory and it is even valid in arbitrary (finite) dimension.

(ii) *p is not asymptotically stable.*
From our perspective this will be the more interesting case. We will prove the theorem using Brouwer's lemma. The proof and the result cannot be extended to higher dimensions.

8.1 Asymptotically Stable Fixed Points

Although the contents of this subsection are valid in arbitrary dimension, we remain in the plane for notational convenience. We will consider a continuous map $h : \mathscr{U} \subset \mathbb{R}^2 \to \mathbb{R}^2$ defined on an open set. Notice that we do not need to assume that h is one-to-one.

A fixed point $p = h(p)$ is *asymptotically stable* if it is stable and there exists a neighborhood $\mathscr{R}$ of p such that

$$h^n(x) \to p \quad \text{as } n \to +\infty$$

for each $x \in \mathscr{R}$.

From this definition we notice that Fix $(h) \cap \mathscr{R} = \{p\}$. This implies that asymptotically stable fixed points are isolated. Also, it is not hard to prove that the attraction is uniform on compact subsets of $\mathscr{R}$. Next result is taken from Krasnoselskii's book [16].

Proposition 1 *In the previous assumptions for h, assume that $p = h(p)$ is an asymptotically stable fixed point. Then*

$$d(id - h, \Omega, 0) = 1$$

for each open and bounded set Ω satisfying $p \in \Omega$, $\overline{\Omega} \cap$ Fix $(h) = \{p\}$.

Proof Browder's asymptotic fixed point theorem (see [5]) implies that if D and Δ are closed disks in the plane with $\Delta \subset D$ and $h : D \to \mathbb{R}^2$ is a continuous map such that all the iterates $h^n(\Delta)$ are well defined for $n \geq 0$ and, for some $N > 0$,

$$h^n(\Delta) \subset \Delta \quad \text{if} \quad n \geq N, \tag{5}$$

then h has a fixed point in Δ. A refinement of this result says that if the condition (5) is replaced by

$$h^n(\Delta) \subset \text{int}\,(\Delta) \quad \text{if} \quad n \geq N, \tag{6}$$

then

$$d(id - h, D, 0) = 1.$$

A complete proof of this result can be found in the book [17]. We are ready for the proof of the proposition.

Let us fix a small closed disk D centered at p and such that $D \subset \mathscr{U}$. Since p is stable we can find a smaller disk Δ centered at p and such that all forward iterates $h^n(\Delta)$ are well defined and contained in the interior of D; that is,

$$h^n(\Delta) \subset \text{int}\,(D) \quad \text{if } n \geq 0.$$

After restricting the size of Δ we can also assume that it is contained in $\mathscr{R}$. Then the compact disk Δ is uniformly attracted by p. This means that

$$h^n(x) \to p \quad \text{as } n \to +\infty,$$

uniformly in $x \in \Delta$. In consequence the condition (6) will hold if N is large enough. We conclude that $d(id - h, \text{int}(\Delta), 0) = 1$. The identity $d(id - h, \text{int}(\Delta), 0)) = d(id - h, \Omega, 0)$ follows by excision. Notice that $\Delta \cap \text{Fix}(h) \subset \mathcal{R} \cap \text{Fix}(h) = \{p\}$.

8.2 Stability Without Attraction

We are now assuming that h is continuous, one-to-one and orientation-preserving. However, since the map is not defined in the whole plane, we cannot say that it belongs to $\mathcal{E}_+$. Our first task will be to replace h by a map H in $\mathcal{E}_+$. To do this we first recall that a stable fixed point in the plane always possesses small neighborhoods that are positively invariant, open and simply connected. We refer to the appendix for more details. Since p is an isolated fixed point we find a neighborhood $\mathcal{N}$ with

$$\mathcal{N} \cap \text{Fix}(h) = \{p\}.$$

Let $\mathcal{W}$ be an open neighborhood of p satisfying

$$\mathcal{W} \subset \mathcal{U} \cap \mathcal{N}, \quad \mathcal{W} \text{ is simply connected }, h(\mathcal{W}) \subset \mathcal{W}.$$

As a consequence of Riemann theorem on conformal maps we know that $\mathcal{W}$ is homeomorphic to the open disk or, equivalently, to the whole plane. It is possible to assume that the homeomorphism $\sigma : \mathbb{R}^2 \cong \mathcal{W}$ satisfies $\sigma(p) = p$ and we define $H = \sigma^{-1} \circ h_{\mathcal{W}} \circ \sigma$ where $h_{\mathcal{W}} : \mathcal{W} \to \mathcal{W}$ is the restriction of h,

$$H : \ \mathbb{R}^2 \cong \mathcal{W} \longrightarrow \mathcal{W} \cong \mathbb{R}^2.$$

By definition H is continuous and one-to-one. Moreover, it is conjugate to the restriction of h to $\mathcal{W}$. This implies that H is orientation-preserving and the stability properties of p as a fixed point are the same for the two maps. We summarize the relevant properties of H,

- $H \in \mathcal{E}_+$
- p is a stable fixed point of H
- p is not asymptotically stable
- $\text{Fix}(H) = \{p\}$
- $d(id - h, \Omega, 0) = d(id - H, G, 0)$, where G is any open and bounded subset of the plane with $p \in G$.

Notice that the last property is a consequence of the excision property and the invariance of the fixed point index under conjugation. More precisely, given a homeomorphism $\psi : G_1 \cong G_2$ between two bounded and open subsets of $\mathbb{R}^2$,

$$d(id - \psi^{-1} \circ f \circ \psi, G_1, 0) = d(id - f, G_2, 0)$$

for any continuous function $f : \overline{G}_2 \to \overline{G}_2$ with $\text{Fix}(f) \cap \partial G_2 = \emptyset$.

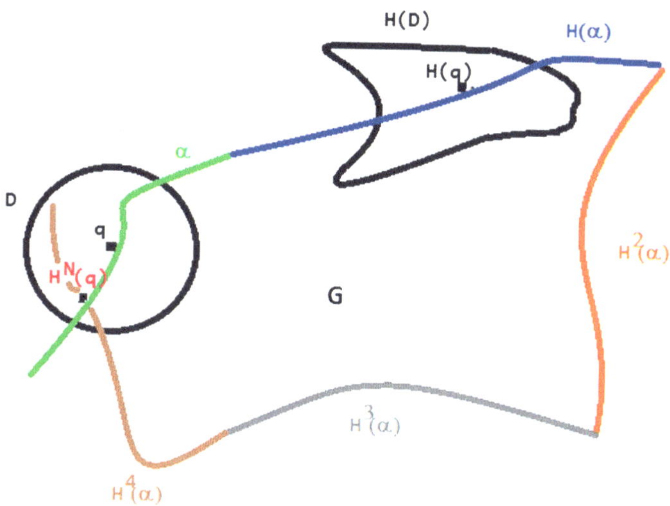

Fig. 6 In this example $N = 4$ and the degree is one in the region G

From now on we work with the map H. Since p is stable but not asymptotically stable, we can find a point $q \in \mathbb{R}^2 \setminus \{p\}$ that is recurrent. This means that $q \in L_\omega(q, H)$, where $L_\omega(q, H)$ is the ω-limit set. We recall that this limit set is composed by all accumulation points of the sequence $\{H^n(p)\}_{n \geq 0}$. More details can be found in the appendix. The point q is not fixed under H and so we can find a small disk D around q satisfying

$$D \cap H(D) = \emptyset.$$

Since q is recurrent there exists an integer $N \geq 2$ such that $H^N(q) \in \text{int}\,(D)$. If we apply Lemma 1, we find a translation arc α with $q, H^N(q) \in \dot{\alpha}$. After iterating N times the arc α under H, we observe that the point $H^N(q)$ belongs simultaneously to $\dot{\alpha}$ and $H^N(\dot{\alpha})$ and we are in the conditions of Brouwer's lemma (Fig. 6).

9 Stability of Invariant Sets

From now on we assume that the map h is a homeomorphism of the plane and we indicate it by $h \in \mathscr{H}$. In particular it is assumed that h is onto, $h(\mathbb{R}^2) = \mathbb{R}^2$. Let Λ be a compact subset of $\mathbb{R}^2$ that is invariant under h,

$$h(\Lambda) = \Lambda.$$

The notion of stability can be adapted to this general setting. We say that Λ is *stable* if given any neighborhood $\mathscr{U}$ of Λ, there exists another neighborhood $\mathscr{V}$ such that

$$h^n(\mathcal{V}) \subset \mathcal{U} \quad \text{for each } n \geq 0.$$

If Λ is a singleton, $\Lambda = \{p\}$, p is a fixed point and the two notions of stability coincide. When Λ is a finite set then it is composed by a finite number of periodic orbits, say

$$\Lambda = \Lambda_1 \cup \cdots \cup \Lambda_r$$

where

$$\Lambda_i = \{p_i, h(p_i), \ldots, h^{m_i-1}(p_i)\}$$

and $h^{m_i}(p_i) = p_i$, $i = 1, \ldots, r$. Each point p_i is fixed under the iterate h^{m_i} and Λ_i is stable if and only if p_i is a stable fixed point of h^{m_i}. In particular, if p_i is isolated as a fixed point of h^{2m_i} and Λ is stable we can apply Theorem 1 to the orientation-preserving homeomorphism h^{2m_i} to deduce that

$$d(id - h^{2m_i}, G, 0) = 1$$

if G is a bounded and open subset of the plane with $p_i \in G$ and Fix $(h^{2m_i}) \cap \overline{G} = \{p_i\}$. This is a more or less trivial consequence of Theorem 1 but we will obtain similar results when Λ is a Cantor set.

10 Stable Cantor Sets

Assume that $\Lambda \subset \mathbb{R}^2$ is a Cantor set that is invariant under h,

$$h(\Lambda) = \Lambda.$$

We recall that a Cantor set is a compact, perfect and totally disconnected metric space. We will also assume that Λ is *transitive*. This means that, for some $p \in \Lambda$,

$$L_\omega(p, h) = \Lambda.$$

There are many classical examples where transitive Cantor sets appear. In Smale's horseshoe there appear a Cantor set with the dynamics of the shift, see [1]. In the theory of twist maps, Aubry-Mather sets are invariant Cantor sets with Denjoy dynamics, see [19]. However, in these examples the set Λ is not stable. To present an example of stable Cantor set we consider the following construction, already described by Cartwright in [10] (Fig. 7). Assume that D is a disk and inside D there are two disjoint disks D_0 and D_1. We repeat this process recursively so that D_0 contains two disks D_{00} and D_{01}, D_1 contains D_{10} and D_{11}, and so on. The intersection of the sets $D, D_0 \cup D_1, D_{00} \cup D_{01} \cup D_{10} \cup D_{11} \ldots$ is a Cantor set Λ. We construct a homeomorphism h satisfying

$$h(D) = D$$

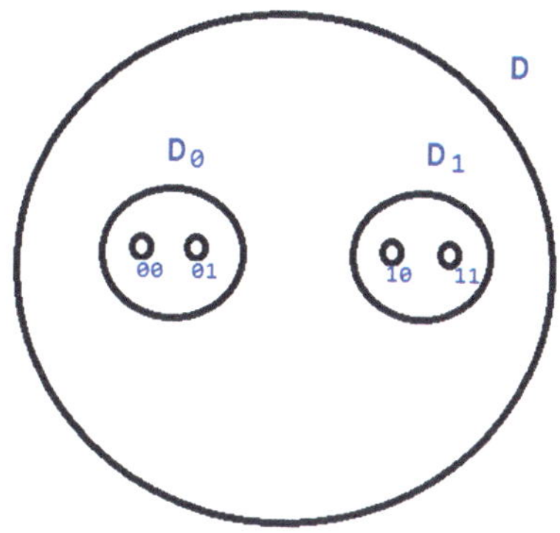

Fig. 7 Construction of a Cantor set as a limit of families of 2^n disks. The disks are labelled by the words of length n with the symbols 0 and 1. For $n = 2$, we have 00, 01, 10, 11

$$h(D_0) = D_1, \quad h(D_1) = D_0$$

$$h(D_{00}) = D_{10}, \; h(D_{10}) = D_{01}, \; h(D_{01}) = D_{11}, \; h(D_{11}) = D_{00}, \ldots$$

Notice that, at each level of the construction, the iterates of h map every disk in all the others before returning. In particular $h(D) = D$, $h^2(D_0) = D_0$, $h^4(D_{00}) = D_{00}$,... Since the sets D, $D_0 \cup D_1$, $D_{00} \cup D_{01} \cup D_{10} \cup D_{11}$... are invariant under h, the same will happen to the intersection Λ. Moreover Λ is stable because it has a basis of invariant neighborhoods. Finally we notice that Λ is *minimal*. This means that, for each $p \in \Lambda$,

$$L_\omega(p, h) = \Lambda.$$

This is clear because the iterates of a disk visit all the disks at the same level. There can be some questioning as to whether this construction is feasible. This is easily solved after interpreting D as a section of the flow on a solid torus described in the next figure (Fig. 8).

Buescu and Stewart proved that the dynamics on any stable and transitive Cantor set is almost periodic, the so-called adding machine (see [9]). In particular Λ has to be minimal. The Bernoulli shift on a Cantor set is transitive but not minimal and so it is always unstable in the plane. Notice that this is the case in Smale's horseshoe. Cantor sets with Denjoy dynamics are minimal but not almost periodic (see [21]). For this reason Aubry-Mather sets cannot be stable.

In [3] Bell and Meyer proved that any stable and transitive Cantor set in the plane can be approximated by periodic points. In the same paper they constructed an example showing that this is false in three dimensions. Next we state the interesting result by Bell and Meyer.

Theorem 2 *Assume that $h \in \mathscr{H}$ and Λ is a Cantor set in $\mathbb{R}^2$ that is invariant under h. In addition it is assumed that Λ is stable and transitive. Then, for each $p \in \Lambda$ there exists a sequence $\{x_n\}$, $x_n \in \mathbb{R}^2$, converging to p and such that $h^{\sigma_n}(x_n) = x_n$ for some integer $\sigma_n \geq 1$.*

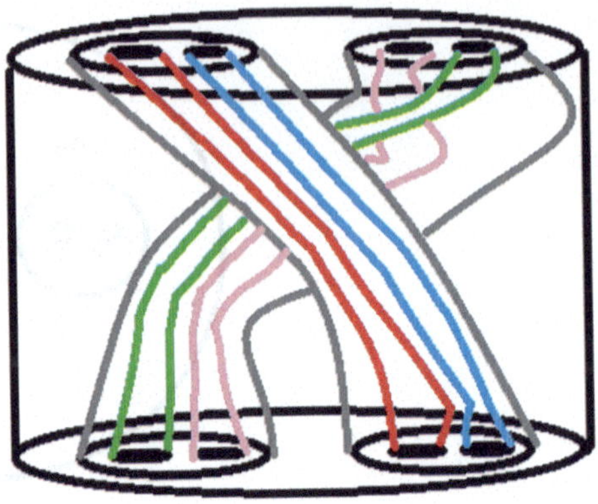

Fig. 8 The *top* and *bottom* of the cylinder are identified as a common disk. In this way we obtain a *solid torus* that is invariant under the flow. Inside this torus another invariant torus is obtained by gluing the two *grey* cylinders connecting the smaller disks. This torus winds twice before closing. The process is repeated recursively, in the next step there appear an invariant torus after gluing the four colored thinner cylinders

We know that Λ is minimal and so it cannot contain periodic points. For this reason the points x_n do not lie in Λ and $\sigma_n \to +\infty$. In particular Λ cannot be an isolated invariant set of h. The proof by Bell and Meyer combines the result by Buescu and Stewart mentioned above with the following classical fixed point theorem. We recall that a continuum is a non-empty compact and connected metric space.

Theorem 3 *Assume that $h \in \mathscr{H}_+$ and K is a continuum of the plane with $h(K) = K$. In addition assume that $\mathbb{R}^2 \setminus K$ is connected. Then h has a fixed point in K.*

This result was proved by Cartwright and Littlewood in the beautiful paper [11]. Later Bell extended it to orientation-reversing homeomorphisms in [2]. In the recent paper [24], written in collaboration with Ruiz-Herrera, it is shown that additional information on stable Cantor sets can be obtained if one replaces the use of Theorem 3 by Brouwer's lemma. In retrospect this seems natural because Cartwright-Littlewood theorem can be obtained as a consequence of Brouwer's lemma (see [6]). To state the next result we employ the notation $B(p, \delta)$ for the open ball of center p and radius δ.

Theorem 4 *Assume that the same conditions of Theorem 2 hold. Then, for each $p \in \Lambda$ and each $\delta > 0$, there exists a topological disk D and an integer $N = N(p, \delta) \geq 1$ such that $D \subset B(p, \delta)$ and*

$$d(id - h^N, \operatorname{int}(D), 0) = 1.$$

As a consequence of this result we will deduce that stable and transitive Cantor sets have some remarkable properties: they are always persistent and non-isolated as invariant sets. We already mentioned that the non-isolated character follows from the theorem by Bell and Meyer. We can now see Bell-Meyer's result as a corollary of Theorem 4, because h^N has a fixed point in D. Next we present a second corollary that seems to be new.

Corollary 2 *In the conditions of Theorem 4, the set Λ is persistent as an invariant set.*

This is again a consequence of the continuity property of the degree that was discussed in Sect. 4. Given $\varepsilon_1 > 0$ we find a finite set of points in Λ, $p_1, \ldots, p_r$, such that for every $p \in \Lambda$,

$$\min_{i=1,\ldots,r} ||p - p_i|| < \varepsilon_1.$$

Then we apply Theorem 4 at each $p_i \in \Lambda$ and find topological disks D_i and integers $N_i \geq 1$ with $p_i \in \text{int}(D_i)$, $D_i \subset B(p_i, \varepsilon_1)$ and

$$d(id - h^{N_i}, \text{int}(D_i), 0) = 1.$$

Given $\tilde{h} \in \mathscr{H}$ with $||h - \tilde{h}||_\infty$ small, it is possible to prove that

$$\max_{x \in \partial D_i} ||h^{N_i}(x) - \tilde{h}^{N_i}(x)||$$

is also small for each $i = 1, \ldots, r$. This implies that

$$d(id - \tilde{h}^{N_i}, \text{int}(D_i), 0) = d(id - h^{N_i}, \text{int}(D_i), 0) = 1.$$

In consequence $\tilde{h}$ has a periodic point $\tilde{p}_i$ lying in D_i and the set

$$\tilde{\Lambda} = \{\tilde{h}^k(\tilde{p}_i) : 0 \leq k < N_i, \ i = 1, \ldots, r\}$$

is invariant under $\tilde{h}$. By some continuity arguments the number ε_1 can be adjusted so that the Hausdorff distance between Λ and $\tilde{\Lambda}$ satisfies $D_H(\Lambda, \tilde{\Lambda}) \leq \varepsilon$. The details can be found in [24]. It is convenient to remark that the persistence of Λ is understood in the class of invariant sets, the perturbed set $\tilde{\Lambda}$ is not necessarily a Cantor set. Notice that in the above proof the perturbed set was finite.

We finish with a sketch of the proof of Theorem 4. The complete details can be found in [24].

Proof (Theorem 4) **First step**: Given $p \in \Lambda$ and $\delta > 0$ there exist an open and simply connected set $\omega \subset \mathbb{R}^2$ and an integer $N \geq 1$ such that

$$p \in \omega \subset \overline{\omega} \subset B(p, \delta), \quad h^N(\omega) \subset \omega.$$

The proof of this step combines the ideas of the proof of Lemma 2 in the appendix with the minimality of Λ, that follows from Buescu-Stewart result.

Second step: There exists a point $q \in \omega \cap \Lambda$ that is recurrent under h^{2N}; that is, $q \in L_\omega(q, h^{2N})$.

To prove this step we consider the set

$$R_{2N} = \{q \in \Lambda : \ q \in L_\omega(q, h^{2N})\}$$

and prove that it is non-empty. Since Λ is invariant under h^{2N}, it must contain a minimal set M (with respect to h^{2N}). Then $\emptyset \neq M \subset R_{2N} \subset \Lambda$. Once we know that R_{2N} is non-empty we observe that it is invariant under h. From $h(R_{2N}) = R_{2N}$ we deduce that also the closure is invariant, $h(\overline{R}_{2N}) = \overline{R}_{2N}$. Since Λ is minimal with respect to h, $\overline{R}_{2N} = \Lambda$ and so the recurrent points (with respect to h^{2N}) are dense in Λ.

Third step: Conclusion via Brouwer's lemma.
Let $\sigma : \mathbb{R}^2 \cong \omega$ be a homeomorphism with $\sigma(Q) = q$. Define $H = \sigma^{-1} \circ h_\omega^{2N} \circ \sigma$, where $h_\omega^{2N} : \omega \to \omega$ is the restriction of h^{2N}. Then $H \in \mathscr{E}_+$ and Q is recurrent under H, $Q \in L_\omega(Q, H)$. Since q belongs to Λ, it is not a fixed point of h^{2N}. Then Q is not a fixed point of H and a disk D around Q can be found such that $D \cap H(D) = \emptyset$. Let $M \geq 2$ be an integer such that $H^M(Q) \in D$. We draw a translation arc α passing through Q and $H^M(Q)$. Then $\alpha \cap H^M(\alpha)$ is non-empty and there exists a Jordan curve $\Gamma \subset \alpha \cup \cdots \cup H^M(\alpha)$ with

$$d(id - H, R_i(\Gamma), 0) = 1.$$

We transport this curve to ω, $\gamma = \sigma(\Gamma)$ and notice that $\overline{R_i(\gamma)} \subset \omega$ because ω is simply connected. The invariance of the fixed point index under conjugation implies that

$$d(id - h^{2N}, R_i(\gamma), 0) = d(id - H, R_i(\Gamma), 0) = 1$$

and the proof is complete with $D = \overline{R_i(\gamma)}$.

Acknowledgments Supported by the research project MTM2011-23652, Spain

Appendix

Remarks on Embeddings

Given a continuous and one-to-one map $h : \mathbb{R}^2 \to \mathbb{R}^2$, the theorem of invariance of the domain implies that h is open. In particular $h(\mathbb{R}^2)$ is an open subset of $\mathbb{R}^2$.

The map h is orientation-preserving ($h \in \mathscr{E}_+$) if

$$d(h, G, h(p)) = 1,$$

where the degree is computed on any open and bounded set $G \subset \mathbb{R}^2$ with $p \in G$. It can be proved that the value of this degree is independent of the choice of G and p. Moreover it can only take the values $+1$ or -1. In the second case h is orientation-reversing. Embeddings that are isotopic to the identity lie in $\mathscr{E}_+$. Those isotopic to the symmetry $S(x_1, x_2) = (x_1, -x_2)$ are orientation-reversing. We recall that two embeddings h_0 and h_1 are isotopic if there exists a continuous map

$$H : \mathbb{R}^2 \times [0, 1] \to \mathbb{R}^2, \quad (x, \lambda) \mapsto H_\lambda(x)$$

such that H_λ is one-to-one for each $\lambda \in [0, 1]$, $H_0 = h_0$ and $H_1 = h_1$.

Lemmas for the Proof of Theorem 1

Lemma 2 *Assume that $\mathscr{U}$ is an open subset of the plane and $h : \mathscr{U} \subset \mathbb{R}^2 \to \mathbb{R}^2$ is a continuous and one-to-one map with a stable fixed point $p = h(p)$. Then, given any neighborhood $\mathscr{W}$ of p, there exists an open and simply connected set $\omega \subset \mathscr{U} \cap \mathscr{W}$ such that $p \in \omega$ and $h(\omega) \subset \omega$.*

Proof It is extracted from [27], page 185. Let us fix a closed disk Δ centered at p and contained in $\mathscr{U} \cap \mathscr{W}$. The stability of the fixed point allows us to find an open and connected set G with $p \in G$ and $h^n(G) \subset \Delta$ for each $n \geq 0$. The open and connected set

$$\Omega := \bigcup_{n \geq 0} h^n(G)$$

is positively invariant under h and contained in Δ. At first sight Ω could be the searched set, because it is contained in $\mathscr{U} \cap \mathscr{W}$ and $h(\Omega) \subset \Omega$. However this set can have holes. Let $\widehat{\Omega}$ be the smallest simply connected domain containing Ω. Intuitively this set is constructed by filling the holes in Ω, a more formal construction can be found in Lemma 2.6 of [22]. Since h is an embedding, $h(\widehat{\Omega}) = \widehat{h(\Omega)}$ and so we can take $\omega = \widehat{\Omega}$.

Lemma 3 *Assume that $h : \mathbb{R}^2 \to \mathbb{R}^2$ is a continuous and one-to-one map with a fixed point $p = h(p)$ that is stable but not asymptotically stable. Then, for each neighborhood $\mathscr{W}$ of p there exists a point $q \in \mathscr{W} \setminus \{p\}$ that is recurrent.*

Proof By a contradiction argument assume that there exists a neighborhood $\mathscr{W}$ of p such that $\mathscr{W} \setminus \{p\}$ does not contain recurrent points. Since p is stable we can find two closed balls β and B centered at p and such that $\beta \subset B \subset \mathscr{W}$ and

$$h^n(\beta) \subset B \quad \text{for each } n \geq 0.$$

We claim that $p \in L_\omega(x, h)$ for each $x \in \beta$. Indeed, the positive orbit $\{h^n(x)\}_{n \geq 0}$ is contained in B and so it is bounded. In consequence the limit set $L_\omega(x, h)$ is non-empty, compact and invariant under h. This limit set has to contain a minimal set M.

All points y in M must satisfy $L_\omega(y, h) = M$ and so they are recurrent. Summing up, we can say that $L_\omega(x, h)$ is contained in B and has at least one recurrent point. Since p is the only recurrent point in this ball, the claim $p \in L_\omega(x, h)$ has been proved. Now we observe that, due to the stability of p,

$$L_\omega(x, h) = \{p\}.$$

Then all points in β are attracted by p and this point should be asymptotically stable.

References

1. Arrowsmith, D.K., Place, C.M.: An Introduction to Dynamical Systems. Cambridge University Press, Cambridge (1990)
2. Bell, H.: A fixed point theorem for planar homeomorphisms. Bull. Am. Math. Soc. **82**, 778–780 (1976)
3. Bell, H., Meyer, K.R.: Limit periodic functions, adding machines and solenoids. J. Dyn. Diff. Equ. **7**, 409–422 (1995)
4. Bonatti, C., Villadelprat, J.: The index of stable critical points. Topol. Appl. **126**, 263–271 (2002)
5. Browder, F.E.: On a generalization of the Schauder fixed point theorem. Duke Math. J. **26**, 291–303 (1959)
6. Brown, M.: A short short proof of the Cartwright-Littlewood theorem. Proc. Am. Math. Soc. **65**, 372 (1977)
7. Brown, M.: A new proof of Brouwer's lemma on translation arcs. Houston Math. J. **10**, 35–41 (1984)
8. Brown, M.: Homeomorphisms of two-dimensional manifolds. Houston Math. J. **11**, 455–469 (1985)
9. Buescu, J., Kulczycki, M., Stewart, I.: Liapunov stability and adding machines revisited. Dyn. Syst. **21**, 379–384 (2006)
10. Cartwright, M.L.: Almost-periodic flows and solutions of differential equations. Proc. Lond. Math. Soc. **17**, 355–380 (1967). (Corrigenda: p. 768)
11. Cartwright, M.L., Littlewood, J.E.: Some fixed point theorems. Ann. Math. **54**, 1–37 (1951)
12. Dancer, E.N., Ortega, R.: The index of Lyapunov stable fixed points in two dimensions. J. Dyn. Diff. Equ. **6**, 631–637 (1994)
13. Fathi, A.: An orbit closing proof of Brouwers lemma on translation arcs. L'enseignement Mathématique **33**, 315–322 (1987)
14. Franks, J.: A new proof of the Brouwer plane translation theorem. Ergod. Theory Dyn. Syst. **12**, 217–226 (1992)
15. Granas, A., Dugundji, J.: Fixed Point Theory. Springer, New York (2003)
16. Krasnoselskii, M.A.: Translation Along Trajectories of Differential Equations. American Mathematical Society, Providence, RI (1968)
17. Krasnoselskii, M.A., Zabreiko, P.P.: Geometrical Methods of Nonlinear Analysis. Springer, Berlin (1984)
18. Lefschetz, S.: Differential Equations: Geometric Theory. Dover Publications, New York (1977)
19. Mather, J.N.: Existence of quasi-periodic orbits for twist homeomorphisms on the annulus. Topology **21**, 457–467 (1982)
20. Murthy, P.: Periodic solutions of two-dimensional forced systems: the Massera theorem and its extensions. J. Dyn. Diff. Equ. **10**, 275–302 (1998)
21. Nemytskii, V.V., Stepanov, V.V.: Qualitative Theory of Differential Equations. Princeton University Press, Princeton (1960)

22. Ortega, R.: The number of stable periodic solutions of time-dependent Hamiltonian systems with one degree of freedom. Ergod. Theory Dyn. Syst. **18**, 1007–1018 (1998)
23. Ortega, R.: Topology of the Plane and Periodic Differential Equations. www.ugr.es/local/ecuadif/fuentenueva.htm
24. Ortega, R., Ruiz-Herrera, A.: Index and Persistence of Stable Cantor Sets. Rend. Istit. Mat. Univ. Trieste, **44**, 33–44 (2012)
25. Pliss, V.A.: Nonlocal Problems in the Theory of Oscillations. Academic Press, New York (1966)
26. Ruiz del Portal, F.R.: Planar isolated and stable fixed points have index =1. J. Diff. Equ. **199**, 179–188 (2004)
27. Siegel, C.L., Moser, J.K.: Lectures on Celestial Mechanics. Springer, Berlin (1971)

Integrated Exponential Functions, the Weyl Fractional Calculus and the Laplace Transform

Pedro J. Miana

Abstract In this paper we give new relations between the Weyl fractional calculus, the usual convolution product and the Laplace transform. To express them, we consider and study in detail integrated exponential functions, a particular class of Kummer functions. We also extend a equality due to S. Goldstein in the last section.

Keywords Laplace transform · Fractional calculus · Kummer functions · Convolution products

1 Introduction

The monography [1] is an excellent introduction to the vector-valued Laplace transform. However, in this paper, we come back to the scalar-valued Laplace transform with some ideas which were learnt from the vector-valued case. We prove new results between fractional calculus, convolution product and the Laplace transform.

In fact, there is not an unique fractional calculus. There exist several definitions which are applied to several class of functions. In this paper, we work with one of them, the Weyl fractional calculus, W_+^α with $\alpha \in \mathbb{R}$. In [4], we have given the expression of α-times derivative (in the sense of Weyl) of a convolution product, i.e., $W_+^\alpha(f * g)$ (see here the Theorem 2.3). We prove here the α-times integral of a convolution product, $W_+^{-\alpha}(f * g)$ (Theorem 2.2). A direct consequence of these results is the equality

$$W_+^\alpha(f * g) + W_+^{-\alpha}(W_+^\alpha f * W_+^\alpha g) = (W_+^\alpha f * g) + (f * W_+^\alpha g),$$

see Corollary 2.4 for more details.

P.J. Miana has been partially supported by Project MTM2013-42105-P and MTM2016-77710-P, DGI-FEDER, of the MCYTS and Project E-64-FEDER, D.G. Aragón.

P.J. Miana (✉)
Department of Mathematics, Universidad de Zaragoza, 50009 Zaragoza, Spain
e-mail: pjmiana@unizar.es

L. Alsedà i Soler et al. (eds.), *Difference Equations, Discrete Dynamical Systems and Applications*, Springer Proceedings in Mathematics & Statistics 180, DOI 10.1007/978-3-662-52927-0_17

Integrated exponential functions $(S_\alpha(t,z))_{t\geq 0, z\in \mathbf{C}}$ are defined in a straightforward way in the Sect. 2. They are known functions [5], extended in different ways [8] and verifies nice properties (Proposition 3.1). Here we prove more results which looks like to be new, for example Lemma 3.3 or Theorem 3.5. The Theorem 3.4 shows that integrated exponential functions are scalar-versions of integrated semigroups, see definition in [1] or [4].

Integrated exponential functions are closer to the convolution product and the Laplace transform, $\mathscr{L}$. In fact, integrated exponential functions may be written like a convolution product and they are also the Laplace transforms of Bochner–Riesz functions. This gives to $S_\alpha(t,z)$ a double behavior in each one of both variables. However there is a useful symmetry between both variables (Lemma 3.2).

In the third section, we check the Laplace transforms of fractional derivatives and integrals which also involves convolution product. Integrated exponential functions appear in some of them (Theorems 4.3 and 4.4).

In the last section, we show a dual formula between the Weyl fractional derivation and the Laplace transform: if $f, g \in \mathscr{S}_+$ and $\alpha \geq 0$, then

$$\int_0^\infty W_+^\alpha g(s) s^\alpha \mathscr{L}(f)(s)ds = \int_0^\infty \mathscr{L}(g)(s)s^\alpha W_+^\alpha f(s)ds,$$

(Theorem 5.2). This formula extends the case $\alpha = 0$ given in [2].

We present a self-contained paper and we give some direct proofs to avoid the lack of completeness.

Notation. $\mathbf{R}$ and $\mathbf{C}$ are the set of real and complex numbers; $\Re z$ is the real part of a complex number z; $\mathbf{C}^+ := \{z \in \mathbf{C} \mid \Re z > 0\}$.

Γ is the Euler function, and χ_E is the characteristic function on the subset E. $L^1(\mathbf{R}^+)$ is the set of Lebesgue functions f such that

$$\int_0^\infty |f(t)|dt < \infty.$$

2 The Weyl Fractional Calculus

We present basic results and notations about the Weyl fractional calculus and convolution product on $[0,\infty)$. We also prove new results, see the Theorem 2.2 and Corollary 2.4. We will use them in later sections.

We denote by $\mathscr{D}_+$ the set of test functions of compact support in $[0,\infty)$, $\mathscr{D}_+ \equiv C_c^\infty([0,\infty))$ and by $\mathscr{S}_+$ the Schwartz class on $[0,\infty)$, i.e., infinitely differentiable functions which verifies

$$\sup_{t\geq 0}\left| t^m \frac{d^n}{dt^n} f(t)\right| < \infty,$$

for any $m, n \in \mathbf{N}\cup\{0\}$.

Given $f \in \mathscr{S}_+$, the *Weyl fractional integral* of f of order $\alpha > 0$ is defined by

$$W_+^{-\alpha} f(u) := \frac{1}{\Gamma(\alpha)} \int_u^\infty (t-u)^{\alpha-1} f(t)dt, \quad u \geq 0,$$

with $\alpha > 0$. This operator $W_+^{-\alpha} : \mathscr{S}_+ \to \mathscr{S}_+$ is one to one, and its inverse, W_+^{α}, the *Weyl fractional derivative* of order α, holds

$$W_+^{\alpha} f(t) = \frac{(-1)^n}{\Gamma(n-\alpha)} \frac{d^n}{dt^n} \int_t^\infty (s-t)^{n-\alpha-1} f(s)ds, \quad t \geq 0,$$

with $n = [\alpha] + 1$, see for example [5, 7].

It is easy to check that if $\alpha \in \mathbf{N}$ then $W_+^{\alpha} f = (-1)^{\alpha} f^{(\alpha)}$ and $W_+^{\alpha+\beta} f = W_+^{\alpha}(W_+^{\beta} f)$ with $\alpha, \beta \in \mathbf{R}$, with $W_+^0 = Id$ and $f \in \mathscr{S}_+$. In particular, let $\lambda \in \mathbf{C}^+$ and $e_\lambda(s) := e^{-\lambda s}$ with $s \geq 0$. It is clear that $e_\lambda \in \mathscr{S}_+$ and

$$W_+^{-\alpha}(e_\lambda)(s) = \lambda^{-\alpha} e^{-\lambda s}, \quad s \geq 0.$$

Then $W_+^{\alpha}(e_\lambda) = \lambda^{\alpha} e_\lambda$ for $\alpha \in \mathbf{R}$.

Proposition 2.1 *Take $\alpha \in \mathbf{R}$ and $f \in \mathscr{S}_+$. Then we have*

$$W_+^{\alpha}(sf(s))(t) = tW_+^{\alpha}(f)(t) - \alpha W_+^{\alpha-1}(f)(t), \qquad t > 0.$$

Proof If $\alpha < 0$, it is shown in [5, p. 246]; if $\alpha > 0$ we have that

$$W_+^{-\alpha}(sW_+^{\alpha} f(s) - \alpha W_+^{\alpha-1} f(s))(t) = tf(t) + \alpha W_+^{-1} f(t) - \alpha W_+^{-1} f(t) = tf(t),$$

with $t > 0$.

Consider on $\mathscr{D}_+$ the usual convolution product $*$, given by

$$(f * g)(t) = \int_0^t f(t-s)g(s)ds, \qquad f, g \in \mathscr{D}_+,$$

for $t \geq 0$. Now we are interested in obtain $W_+^{\alpha}(f * g)$ ($\alpha \in \mathbf{R}$). These formuli for convolution products are analogous to the Leibniz formula for pointwise product.

Theorem 2.2 *Let $f, g \in \mathscr{D}_+$ be with $\alpha > 0$. Then we have that*

$$\begin{aligned} W_+^{-\alpha}(f * g)(t) &= (W_+^{-\alpha} f * g)(t) + (f * W_+^{-\alpha} g)(t) \\ &+ \int_t^\infty g(u) \int_t^\infty (x+u-t)^{\alpha-1} f(x)dxdu \\ &- \int_0^t g(u) \int_{t-u}^t (x+u-t)^{\alpha-1} f(x)dxdu \end{aligned}$$

with $t \geq 0$.

Proof Using the Fubini theorem, we get that

$$\begin{aligned}
W_+^{-\alpha}(f*g)(t) &= \frac{1}{\Gamma(\alpha)}\int_t^\infty (s-t)^{\alpha-1}\int_0^s f(x)g(s-x)dxds \\
&= \frac{1}{\Gamma(\alpha)}\int_0^t f(x)\int_t^\infty (s-t)^{\alpha-1}g(s-x)dsdx \\
&+ \frac{1}{\Gamma(\alpha)}\int_t^\infty f(x)\int_x^\infty (s-t)^{\alpha-1}g(s-x)dsdx \\
&= f*W_+^{\alpha}g(t) + \frac{1}{\Gamma(\alpha)}\int_t^\infty (s-t)^{\alpha-1}\int_t^s f(x)g(s-x)dxds
\end{aligned}$$

where, again, we apply the Fubini theorem in the second summand.

We change the variable $u = s - x$ in the second integral, and get that

$$\begin{aligned}
&\int_t^\infty (s-t)^{\alpha-1}\int_t^s f(x)g(s-x)dxds = \int_t^\infty (s-t)^{\alpha-1}\int_0^{s-t} f(s-u)g(u)duds \\
&= \int_0^\infty g(u)\int_{u+t}^\infty (s-t)^{\alpha-1}f(s-u)dsdu = \int_0^\infty g(u)\int_t^\infty (x+u-t)^{\alpha-1}f(x)dxdu \\
&= \int_0^t g(u)\int_t^\infty (x+u-t)^{\alpha-1}f(x)dxdu + \int_t^\infty g(u)\int_t^\infty (x+u-t)^{\alpha-1}f(x)dxdu.
\end{aligned}$$

Now, we obtain in the first summand

$$\begin{aligned}
&\int_0^t g(u)\int_t^\infty (x+u-t)^{\alpha-1}f(x)dxdu \\
&= \int_0^t g(u)\int_{t-u}^\infty (x+u-t)^{\alpha-1}f(x)dxdu - \int_0^t g(u)\int_{t-u}^t (x+u-t)^{\alpha-1}f(x)dxdu \\
&= \Gamma(\alpha)(W^{-\alpha}f*g)(t) - \int_0^t g(u)\int_{t-u}^t (x+u-t)^{\alpha-1}f(x)dxdu,
\end{aligned}$$

and we conclude the proof.

Next formula gives the Weyl derivation of a convolution product. This formula contains the known case $\alpha \in \mathbf{N}$, i.e.,

$$(f*g)^{(n)}(t) = f^{(n)}*g(t) + \sum_{j=0}^{n-1} f^{(n-1-j)}(0)g^{(j)}(t), \quad t \geq 0.$$

Theorem 2.3 ([4, Proposition 1.1]) *Let $f, g \in \mathscr{D}_+$ be and $\alpha > 0$. Then we have that*

$$\begin{aligned}
W_+^{\alpha}(f*g)(t) &= \frac{1}{\Gamma(\alpha)}\int_0^t W_+^{\alpha}g(u)\int_{t-u}^t (x+u-t)^{\alpha-1}W_+^{\alpha}f(x)dxdu \\
&- \frac{1}{\Gamma(\alpha)}\int_t^\infty W_+^{\alpha}g(u)\int_t^\infty (x+u-t)^{\alpha-1}W_+^{\alpha}f(x)dxdu
\end{aligned}$$

with $t \geq 0$.

A consequence of Theorems 2.2 and 2.3 is the following corollary.

Corollary 2.4 *Let $f, g \in \mathscr{D}_+$ be and $\alpha > 0$. Then we have that*

$$W_+^\alpha(f * g) + W_+^{-\alpha}(W_+^\alpha f * W_+^\alpha g) = (W_+^\alpha f * g) + (f * W_+^\alpha g).$$

In fact, the Theorem 2.3 may be applied for functions which do not belong to $\mathscr{D}_+$. For example, let $(R_t^\alpha)_{t\geq 0}$ be Bochner–Riesz functions with $\alpha > 0$ defined by $R_0^\alpha = 0$, and

$$R_t^\alpha(s) := \begin{cases} \dfrac{(t-s)^{\alpha-1}}{\Gamma(\alpha)}, & 0 < s < t, \\ 0, & t \leq s, \end{cases}$$

for $t > 0$. Formally, $W_+^\alpha(R_t^\alpha) \equiv \delta_t$ holds, where δ_t is Dirac measure in the point t. Following the same ideas than in the proof of [4, Proposition 1.1], we obtain that

$$\begin{aligned} W_+^\alpha(R_t^{\alpha-1} * f)(s) &= \frac{\chi_{(t,\infty)}(s)}{\Gamma(\alpha)} \int_{s-t}^{s} (r+t-s)^{\alpha-1} W_+^\alpha f(r)dr \\ &\quad - \frac{\chi_{(0,t)}(s)}{\Gamma(\alpha)} \int_{s}^{\infty} (r+t-s)^{\alpha-1} W_+^\alpha f(r)dr, \quad s > 0, \qquad (1) \end{aligned}$$

for $f \in \mathscr{D}_+, \alpha > 0$ and $t \geq 0$.

3 Integrated Exponential Functions

In this section, we consider the exponential integrated function, S_α defined by

$$S_\alpha(t,z) := \frac{1}{\Gamma(\alpha)} \int_0^t (t-s)^{\alpha-1} e^{-zs} ds, \quad z \in \mathbf{C}, \ t \geq 0,$$

for $\alpha > 0$. In [5, p. 48, p. 314], there are shown a great number of properties to the function $E_t(\nu, a)$ with

$$E_t(\nu, a) = S_\nu(t, -a),$$

and they show some connection with some special functions, one of this is the incomplete gamma function. In the following proposition, we collect three of these properties.

Proposition 3.1 ([5]) *Let $\alpha, t > 0$, be and $z, w \in \mathbf{C}$. Then we have that*

(i) $S_\alpha(t,z) = \dfrac{e^{-zt}}{\Gamma(\alpha)}\displaystyle\int_0^t u^{\alpha-1}e^{zu}du.$

(ii) $\displaystyle\int_0^t S_\alpha(s,z)S_\beta(t-s,w)ds = \frac{1}{z-w}\left(S_{\alpha+\beta}(t,z) - S_{\alpha+\beta}(t,w)\right)$, *if* $z \neq w$.

(iii) $\displaystyle\int_0^t S_\alpha(s,z)S_\beta(t-s,z)ds = tS_{\alpha+\beta}(t,z) - (\alpha+\beta)S_{\alpha+\beta+1}(t,z).$

In fact, we may obtain integrated exponential functions from a convolution product. If $\alpha > 0$, we consider the functions $j^\alpha(s) := \dfrac{s^{\alpha-1}}{\Gamma(\alpha)}$. Then

$$S_\alpha(t,z) = \frac{1}{\Gamma(\alpha)}\int_0^t (t-s)^{\alpha-1}e^{-zs}ds = (j^\alpha * e_z)(t), \quad t \geq 0, \tag{1}$$

with $e_z(t) := e^{-zt}$ for $z \in \mathbf{C}$ and $t \geq 0$. From this equality, the Proposition 3.1 (ii), (iii) may be proved directly and we may also show that

$$S_\beta(t) = \frac{1}{\Gamma(\beta-\alpha)}\int_0^t (t-s)^{\beta-\alpha-1}S_\alpha(s)ds, \tag{2}$$

for $\beta > \alpha$ and $t \geq 0$. In the next lemma, we give a nice symmetry between both parameters in the integrated exponential function.

Lemma 3.2 *If α, t and $s > 0$, then $s^\alpha S_\alpha(t,s) = t^\alpha S_\alpha(s,t)$ holds.*

Proof We change the variable $y = \dfrac{s}{t}x$, we have that

$$s^\alpha S_\alpha(t,s) = \frac{s^\alpha}{\Gamma(\alpha)}\int_0^t (t-x)^{\alpha-1}e^{-sx}dx = \frac{t^\alpha}{\Gamma(\alpha)}\int_0^s (s-y)^{\alpha-1}e^{-sy}dy = t^\alpha S_\alpha(s,t),$$

and we obtain the equality.

In fact, the function $S_\alpha(t,z)$ is a particular case of a Kummer function or a confluent hypergeometric functions ${}_1F_1(a,c;z)$ where

$${}_1F_1(a,c;z) := \sum_{n=0}^{\infty}\frac{(a)_n}{(c)_n}\frac{z^n}{n!}, \quad c \neq 0, -1, -2, \ldots,$$

and $(a)_n = \alpha(\alpha+1)(\alpha+n-1) = \dfrac{\Gamma(\alpha+n)}{\Gamma(n)}$. We have that

$$S_\alpha(t,z) = \frac{t^\alpha}{\Gamma(\alpha+1)}\,{}_1F_1(1, 1+\alpha; -zt), \quad t > 0, z \in \mathbf{C},$$

for $\alpha > 0$, see [3, p. 268].

Lemma 3.3 *Let $\alpha, t, r > 0$ be and $z \in \mathbf{C}$. Then we have*

(i) $S_\alpha(t+r,z) - e^{-zt}S_\alpha(r,z) = \dfrac{1}{\Gamma(\alpha)}\displaystyle\int_0^t (t+r-s)^{\alpha-1}e^{-zs}ds.$

(ii) $e^{-zr}S_\alpha(t,z) + e^{-zt}S_\alpha(r,z) - S_\alpha(t+r,z)$
$$= \frac{1}{\Gamma(\alpha)}\left(\int_r^{t+r} - \int_0^t (t+r-s)^{\alpha-1}e^{-zs}ds\right).$$

Proof To prove the part (i), we change the variable, $u = t+r-s$, to obtain

$$\begin{aligned}
&\frac{1}{\Gamma(\alpha)}\int_0^t (t+r-s)^{\alpha-1}e^{-zs}ds = \frac{e^{-z(t+r)}}{\Gamma(\alpha)}\int_r^{t+r} u^{\alpha-1}e^{-zu}du\\
&= e^{-z(t+r)}e^{z(t+r)}S_\alpha(t+r,z) - e^{-z(t+r)}e^{zr}S_\alpha(r,z)\\
&= S_\alpha(t+r,z) - e^{-zt}S_\alpha(r,z),
\end{aligned}$$

and we apply the Proposition 3.1 (i). Using the part (i), we have that

$$\begin{aligned}
&\frac{1}{\Gamma(\alpha)}\int_r^{t+r}(t+r-s)^{\alpha-1}e^{-zs}ds - \frac{1}{\Gamma(\alpha)}\int_0^t (t+r-s)^{\alpha-1}e^{-zs}ds\\
&= S_\alpha(t+r,z) - \left(S_\alpha(t+r,z) - e^{-zr}S_\alpha(z,t)\right) - \left(S_\alpha(t+r,z) - e^{-zt}S_\alpha(r,z)\right)\\
&= e^{-zr}S_\alpha(t,z) + e^{-zt}S_\alpha(r,z) - S_\alpha(t+r,z),
\end{aligned}$$

and we get the equality (ii).

Theorem 3.4 *Let $\alpha > 0$ be and $z \in \mathbf{C}$. Then*

$$S_\alpha(t,z)S_\alpha(s,z) = \frac{1}{\Gamma(\alpha)}\left(\int_s^{t+s} - \int_0^t (t+s-r)^{\alpha-1}S_\alpha(r,z)dr\right),$$

holds with $t, s \geq 0$.

Proof We change the variable $u = v + x$, to obtain

$$\begin{aligned}
\Gamma(\alpha)^2 S_\alpha(t,z)S_\alpha(s,z) &= \int_0^t\int_0^s (t-v)^{\alpha-1}(s-x)^{\alpha-1}e^{-z(v+x)}dxdv\\
&= \int_0^t (t-v)^{\alpha-1}\int_v^{s+v}(s+v-u)^{\alpha-1}e^{-zu}dudv\\
&= \int_0^t (t-v)^{\alpha-1}\int_0^{s+v}(s+v-u)^{\alpha-1}e^{-zu}dudv\\
&\quad - \int_0^t (t-v)^{\alpha-1}\int_0^v (s+v-u)^{\alpha-1}e^{-zu}dudv.
\end{aligned}$$

In the first summand, we get that

$$\int_0^t (t-v)^{\alpha-1}\int_0^{s+v}(s+v-u)^{\alpha-1}e^{-zu}dudv$$
$$=\Gamma(\alpha)\int_0^t (t-v)^{\alpha-1}S_\alpha(s+v,z)dv=\Gamma(\alpha)\int_s^{t+s}(t+s-r)^{\alpha-1}S_\alpha(r,z)dr,$$

where we change the variable $r=v+s$.

In the second summand, we apply the Fubini theorem, we change the variable $t-v=r-u$ in the second integral, and apply the Fubini theorem again to get

$$\begin{aligned}
&\int_0^t (t-v)^{\alpha-1}\int_0^v (s+v-u)^{\alpha-1}e^{-zu}dudv\\
&=\int_0^t e^{-zu}\int_u^t (t-v)^{\alpha-1}(s+v-u)^{\alpha-1}dvdu\\
&=\int_0^t e^{-zu}\int_u^t (r-u)^{\alpha-1}(s+t-r)^{\alpha-1}drdu\\
&=\int_0^t (s+t-r)^{\alpha-1}\int_0^r (r-u)^{\alpha-1}e^{-zu}drdu\\
&=\Gamma(\alpha)\int_0^t (s+t-r)^{\alpha-1}S_\alpha(r,z)dr,
\end{aligned}$$

and we conclude the proof.

To end this section, we prove the following result.

Theorem 3.5 *Let* $\alpha>0$, $t,s\geq 0$, *be and* $z\in\mathbf{C}$. *Then we have that*

$$\frac{1}{\Gamma(\alpha+1)}\left(\int_s^{t+s}-\int_0^t (t+s-r)^\alpha S_\alpha(r,z)dr\right)=\left(S_\alpha(s,z)-\frac{s^\alpha}{\Gamma(\alpha+1)}\right)S_{\alpha+1}(t,z).$$

Proof By the equality (2) and the Theorem 3.4, we obtain

$$\begin{aligned}
S_\alpha(s,z)S_{\alpha+1}(t,z)&=\int_0^t S_\alpha(s,z)S_\alpha(r,z)dr\\
&=\frac{1}{\Gamma(\alpha)}\int_0^t\left(\int_s^{r+s}-\int_0^r (r+s-u)^{\alpha-1}S_\alpha(u,z)du\right)dr.
\end{aligned}$$

We apply the Fubini theorem in both integrals and we obtain that

$$
\begin{aligned}
S_\alpha(s,z)S_{\alpha+1}(t,z) &= \frac{1}{\Gamma(\alpha)}\int_s^{t+s} S_\alpha(u,z)\int_{u-s}^{t}(r+s-u)^{\alpha-1}drdu \\
&\quad - \frac{1}{\Gamma(\alpha)}\int_0^{t} S_\alpha(u,z)\int_{u}^{t}(r+s-u)^{\alpha-1}drdu \\
&= \frac{1}{\Gamma(\alpha+1)}\left(\int_s^{t+s} - \int_0^{t} S_\alpha(u,z)(t+s-u)^{\alpha}du\right) + \frac{s^\alpha}{\Gamma(\alpha+1)}S_{\alpha+1}(t,z).
\end{aligned}
$$

We conclude the proof.

An immediate consequence of this theorem is the next corollary.

Corollary 3.6 *Let $\alpha > 0$, $t, s \geq 0$, be and $z \in \mathbf{C}$. Then we have that*

$$
\left(S_\alpha(t,z) - \frac{t^\alpha}{\Gamma(\alpha+1)}\right) S_{\alpha+1}(s,z) = \left(S_\alpha(s,z) - \frac{s^\alpha}{\Gamma(\alpha+1)}\right) S_{\alpha+1}(t,z).
$$

4 The Laplace Transform

Given $f \in L^1(\mathbf{R}^+)$, the Laplace transform of f, $\mathscr{L}(f)$, is defined by

$$
\mathscr{L}(f)(z) := \int_0^\infty f(t)e^{-zt}dt, \quad \Re z \geq 0,
$$

see, for example [1]. If f is a function in two variables $f = f(t,s)$, $\mathscr{L}(f;t)$ and $\mathscr{L}(f;s)$ are Laplace transforms, if there exist, in each parameter.

Laplace transform is an injective algebra homomorphism from $L^1(\mathbf{R}^+)$ (with the usual convolution product $*$) in the set of holomorphic functions in the half plane $\mathbf{C}^+$, continuous and bounded functions in $\overline{\mathbf{C}^+}$, $\mathscr{H}_0(\mathbf{C}^+)$, (with the pointwise product)

$$
\mathscr{L} : L^1(\mathbf{R}^+) \to \mathscr{H}_0(\mathbf{C}^+).
$$

We collect two equalities about the Weyl fractional calculus and the Laplace transform in the next proposition, see for example [5, p. 67] and [6].

Proposition 4.1 *Take $\alpha > 0$, $z \in \mathbf{C}^+$ and $f \in \mathscr{S}_+$. Then*

(i) $W_+^{-\alpha}(\mathscr{L}(f))(z) = \mathscr{L}(t^{-\alpha}f)(z)$ *if* $t^{-\alpha}f(t) \in L^1(\mathbf{R}^+)$.
(ii) $W_+^{\alpha}(\mathscr{L}(f))(z) = \mathscr{L}(t^{\alpha}f)(z)$.

Integral exponential functions $S_\alpha(t,z)$ are closer to the Laplace transform: we may take Bochner–Riesz functions, $(R_t^\alpha)_{t\geq 0}$, with $\alpha > 0$ defined in the first section and

$$
S_\alpha(t,z) = \frac{1}{\Gamma(\alpha)}\int_0^t (t-s)^{\alpha-1}e^{-zs}ds = \mathscr{L}(R_t^\alpha)(z), \quad z \in \mathbf{C}, \tag{1}
$$

for $t > 0$. Moreover, we get that

$$\mathscr{L}(f)(z) = \int_0^\infty f(t)e^{-zt}dt = \int_0^\infty W_+^\alpha f(t)S_\alpha(t,z)dt, \quad \Re z \geq 0, \tag{2}$$

where $f \in \mathscr{S}_+$ and $\alpha > 0$.

Proposition 4.2 *Let $f, g \in \mathscr{S}_+$ be and $\alpha > 0$. Then we have that*

(i) $\mathscr{L}(W_+^{-\alpha} f)(z) = \int_0^\infty f(t)S_\alpha(t,z)dt$ *for* $\Re z \geq 0$.

(ii) $\mathscr{L}(W_+^{-\alpha}(f * g))(z) = \int_0^\infty \int_0^\infty f(t)g(s)S_\alpha(t+s,z)dsdt$, *for* $\Re z \geq 0$.

(iii) $\mathscr{L}(S_\alpha(t,z);t)(w) = \dfrac{1}{w^\alpha(w+z)}$ *for* $\Re z \geq 0$, $\Re w > 0$.

Proof The three parts (i), (ii) and (iii) are proved directly using the definition of $S_\alpha(t,z)$, basic properties of the Laplace transform and the Fubini theorem.

To express $\mathscr{L}(W_+^\alpha f)$ in terms of $\mathscr{L}(f)$, we should use special functions, in particular Kummer functions. However, we may give the following results about the Laplace transform of certain Weyl fractional derivatives.

Theorem 4.3 *Take $f, g \in \mathscr{S}_+$ and $\alpha > 0$. Then we have that*

$$\mathscr{L}(W_+^\alpha f)\mathscr{L}(g) + \mathscr{L}(f)\mathscr{L}(W_+^\alpha g) = \mathscr{L}(W_+^\alpha(f * g)) + \mathscr{L}(W_+^{-\alpha}(W_+^\alpha f * W_+^\alpha g)).$$

Moreover the equality

$$\mathscr{L}(W_+^\alpha(f * g))(z) = \mathscr{L}(W_+^\alpha f)(z)\mathscr{L}(g)(z) + \mathscr{L}(f)(z)\mathscr{L}(W_+^\alpha g)(z) - \int_0^\infty \int_0^\infty W_+^\alpha f(t)W_+^\alpha g(s)S_\alpha(t+s,z)dsdt,$$

holds for $\Re z \geq 0$.

Proof The first equality is obtained from the Corollary 2.4 where we apply the Laplace transform. The second equality is the first one where we use the Proposition 4.2(ii).

In the case of Bochner–Riesz functions $(R_t^\alpha)_{t\geq 0}$ we prove the next result.

Theorem 4.4 *Given $f \in \mathscr{S}_+$ and $\alpha, t > 0$ the equality*

$$\int_0^\infty e^{-zs}W_+^\alpha(R_t^{\alpha-1} * f)(s)ds = e^{-zt}\mathscr{L}(f)(z) + S_\alpha(t,z)\mathscr{L}(W_+^\alpha f)(z) - \int_0^\infty W_+^\alpha f(r)S_\alpha(t+r,z)dr,$$

holds for $\Re z \geq 0$.

Proof We apply the equality (1) to have that

$$\int_0^\infty e^{-zs} W_+^\alpha (R_t^{\alpha-1} * f)(s) ds = \frac{1}{\Gamma(\alpha)} \int_t^\infty e^{-zs} \int_{s-t}^s (r+t-s)^{\alpha-1} W_+^\alpha f(r) dr ds$$
$$- \int_0^t e^{-zs} \int_s^\infty (r+t-s)^{\alpha-1} W_+^\alpha f(r) dr ds.$$

By the Fubini theorem, we obtain

$$\int_0^\infty e^{-zs} W_+^\alpha (R_t^{\alpha-1} * f)(s) ds = \int_0^t W_+^\alpha f(r) \int_t^{t+r} \frac{(r+t-s)^{\alpha-1}}{\Gamma(\alpha)} e^{-zs} ds dr$$
$$+ \int_t^\infty W_+^\alpha f(r) \int_r^{t+r} \frac{(r+t-s)^{\alpha-1}}{\Gamma(\alpha)} e^{-zs} ds dr$$
$$- \int_0^t W_+^\alpha f(r) \int_0^r \frac{(r+t-s)^{\alpha-1}}{\Gamma(\alpha)} e^{-zs} ds dr$$
$$- \int_t^\infty W_+^\alpha f(r) \int_0^t \frac{(r+t-s)^{\alpha-1}}{\Gamma(\alpha)} e^{-zs} ds dr.$$

Then we show that

$$\mathscr{L}(W_+^\alpha (R_t^{\alpha-1} * f))(z) = \int_t^\infty W_+^\alpha f(r) \left(\int_r^{t+r} - \int_0^t \frac{(r+t-s)^{\alpha-1}}{\Gamma(\alpha)} e^{-zs} ds \right) dr$$
$$+ \int_0^t W_+^\alpha f(r) \left(\int_t^{t+r} - \int_0^r \frac{(r+t-s)^{\alpha-1}}{\Gamma(\alpha)} e^{-zs} ds \right) dr.$$

Now, we use the Lemma 3.3 (ii) and the equality (2) to have that

$$\mathscr{L}(W_+^\alpha (R_t^{\alpha-1} * f))(z)$$
$$= \int_0^\infty W_+^\alpha f(r) \left(e^{-zr} S_\alpha(t,z) + e^{-zt} S_\alpha(r,z) - S_\alpha(t+r,z) \right) dr$$
$$= e^{-zt} \mathscr{L}(f)(z) + S_\alpha(t,z) \mathscr{L}(W_+^\alpha f)(z) - \int_0^\infty W_+^\alpha f(r) S_\alpha(t+r,z) dr.$$

We conclude the proof.

5 Dual Formula Between the Laplace Transform and the Weyl Fractional Derivative

In [2] it is proven the well-known formula

$$\int_0^\infty f(s)\mathscr{L}(g)(s)ds = \int_0^\infty \mathscr{L}(f)(s)g(s)ds, \tag{1}$$

for $f, g \in \mathscr{S}_+$. In this section, we consider the Weyl derivative to prove the dual formula

$$\int_0^\infty W_+^\alpha g(s)s^\alpha \mathscr{L}(f)(s)ds = \int_0^\infty \mathscr{L}(g)(s)s^\alpha W_+^\alpha f(s)ds, \tag{2}$$

for $f, g \in \mathscr{S}_+$ and $\alpha \geq 0$ (Theorem 5.2). We need first the next proposition.

Proposition 5.1 *Given $f \in \mathscr{S}_+$ and $\alpha > 0$, the equality*

$$s^\alpha \mathscr{L}(f)(s) = \frac{1}{\Gamma(\alpha)}\int_0^s (s-x)^{\alpha-1}\mathscr{L}(W_+^\alpha f(t)t^\alpha)(x)dx,$$

holds for $s \geq 0$.

Proof We use equality (2) and Lemma 3.2 to get

$$\begin{aligned} s^\alpha \mathscr{L}(f)(s) &= \int_0^\infty W_+^\alpha f(t)s^\alpha S_\alpha(t,s)dt = \int_0^\infty W_+^\alpha f(t)t^\alpha S_\alpha(s,t)dt \\ &= \int_0^\infty W_+^\alpha f(t)t^\alpha \int_0^s (s-x)^{\alpha-1}e^{-tx}dxdt. \end{aligned}$$

By the Fubini theorem, we have that

$$\begin{aligned} s^\alpha \mathscr{L}(f)(s) &= \int_0^s (s-x)^{\alpha-1}\int_0^\infty W_+^\alpha f(t)t^\alpha e^{-tx}dtdx \\ &= \frac{1}{\Gamma(\alpha)}\int_0^s (s-x)^{\alpha-1}\mathscr{L}(W_+^\alpha f(t)t^\alpha)(x)dx \end{aligned}$$

for $s > 0$ and we conclude the proof.

Theorem 5.2 *Given $f, g \in \mathscr{S}_+$ and $\alpha \geq 0$. Then we have that*

$$\int_0^\infty g(s)s^\alpha \mathscr{L}(f)(s)ds = \int_0^\infty \mathscr{L}(W_+^{-\alpha}g)(t)t^\alpha W_+^\alpha f(t)dt.$$

Moreover

$$\int_0^\infty W_+^\alpha g(s)s^\alpha \mathscr{L}(f)(s)ds = \int_0^\infty \mathscr{L}(g)(s)s^\alpha W_+^\alpha f(s)ds.$$

Proof By the Proposition 5.1, we have that

$$\begin{aligned}\int_0^\infty g(s)s^\alpha \mathscr{L}(f)(s)ds &= \int_0^\infty g(s)\frac{1}{\Gamma(\alpha)}\int_0^s (s-y)^{\alpha-1}\mathscr{L}(W_+^\alpha f(x)x^\alpha)(y)dyds \\ &= \int_0^\infty \mathscr{L}(W_+^\alpha f(x)x^\alpha)(y)\frac{1}{\Gamma(\alpha)}\int_y^\infty (s-y)^{\alpha-1}g(s)dsdy \\ &= \int_0^\infty \mathscr{L}(W_+^\alpha f(x)x^\alpha)(y)W_+^{-\alpha}g(y)dy = \int_0^\infty W_+^\alpha f(x)x^\alpha \mathscr{L}(W_+^{-\alpha}g)(x)dx\end{aligned}$$

where we apply the Fubini theorem and the equality (1). The first equality is proven.
Using the first equality, we obtain that

$$\begin{aligned}\int_0^\infty W_+^\alpha g(s)s^\alpha \mathscr{L}(f)(s)ds &= \int_0^\infty \mathscr{L}(W_+^{-\alpha}(W_+^\alpha g))(x)x^\alpha W_+^\alpha f(x)dx \\ &= \int_0^\infty \mathscr{L}(g)(x)x^\alpha W_+^\alpha f(x)dx.\end{aligned}$$

This conclude the proof.

In the particular case $g(t) = e^{-\lambda t}$ with $\lambda \in \mathbf{C}^+$, $t > 0$, we use the formula (2) to obtain the unknown equality

$$\int_0^\infty e^{-\lambda s}s^\alpha \mathscr{L}(f)(s)ds = \int_0^\infty \frac{s^\alpha}{\lambda^\alpha(s+\lambda)}W_+^\alpha f(s)ds,$$

which is known in the case $\alpha = 0$, see [9].

References

1. Arendt, W., Batty, C.J.K., Hieber, M., Neubrander, F.: Vector-Valued Laplace Transforms and Cauchy Problems. Monographs in Mathematics, vol. 96. Birkhäuser Verlag, Basel (2001)
2. Golstein, S.: Operational representation of Whitakker's confluyent hypergeometric functions and Weber's parabolic cylinder function. Proc. Lond. Math. Soc. **34**(2), 103–125 (1932)
3. Magnus, W., Oberhettinger, F., Soni, R.P.: Formulas and Theorems for the Special Functions of Mathematical Physics. Springer, New York (1966)
4. Miana, P.J.: α-times integrated semigroups and fractional derivation. Forum Math. **14**, 23–46 (2002)
5. Miller, K.S., Ross, B.: An Introduction to the Fractional Calculus and Fractional Differential Equations. Wiley, New York (1993)

6. Raina, R.K., Koul, C.L.: On Weyl fractional calculus. Proc. Am. Math. Soc. **73**(2), 188–192 (1979)
7. Samko, S.G., Kilbas, A.A., Marichev, O.I.: Fractional Integrals and Derivatives. Theory and Applications. Gordon-Breach, New York (1993)
8. Srivastava, H.M., Saigo, M., Raina, R.K.: Some existence and connection theorems associated with the Laplace transform and a certain class of integral operators. J. Math. Anal. Appl. **172**, 1–10 (1993)
9. Yürekli, O.: A theorem on the generalized Stieltjes transform and its applications. J. Math. Anal. Appl. **168**, 63–71 (1992)

A QRT-System of Two Order One Homographic Difference Equations: Conjugation to Rotations, Periods of Periodic Solutions, Sensitiveness to Initial Conditions

Guy Bastien and Marc Rogalski

Abstract We study the "homographic" system of order one difference equations in $\mathbb{R}_*^{+^2}$

$$u_{n+1}u_n = c + \frac{d}{v_n}, \quad v_{n+1}v_n = c + \frac{d}{u_{n+1}},$$

for $c, d > 0$. We prove that the orbit $(M_n)_n = \big((u_n, v_n)\big)_n$ of a point $M_0 = (u_0, v_0)$ is contained in an invariant cubic curve, and that the restriction to the positive part of this cubic of the associated dynamical system is conjugated to a rotation on the circle. For a dense invariant set of initial points the solutions are periodic, and if $c = 1$ (this is always possible) every integer $n \geq N(d)$ is the minimal period of some periodic solution. Moreover, every $n \geq 11$ is the minimal period of some solution for some $d > 0$, and we find exactly the set of such minimal periods between 2 and 10. We study the associated dynamical system, and prove that there is a chaotic behavior on every compact set of $\mathbb{R}_*^{+^2}$ not containing the equilibrium.

Keywords Difference equations · Periodic solutions · Sensitiveness to initial conditions

G. Bastien
CNRS, Institut Mathématique de Jussieu, University Pierre Et Marie Curie, Paris, France
e-mail: guy.bastien@imj-prg.fr

M. Rogalski (✉)
Laboratoire Paul Painlevé, CNRS, Institut Mathématique de Jussieu, University of Lille 1, Villeneuve-d'Ascq, France
e-mail: marc.rogalski@imj-prg.fr

L. Alsedà i Soler et al. (eds.), *Difference Equations, Discrete Dynamical Systems and Applications*, Springer Proceedings in Mathematics & Statistics 180, DOI 10.1007/978-3-662-52927-0_18

1 A Geometric Definition for an Homographic System of Difference Equations

First we remark that in the system of the abstract we can suppose $c = 1$ (put $u_n = u'_n\sqrt{c}$ and $v_n = v'_n\sqrt{c}$). From now on we take $c = 1$.

1.1 From a Family of Cubic Curves to a System of Difference Equations

Let be the family of cubic curves $\mathscr{C}_K$ in the plane, equations of which are

$$xy(x+y) + (x+y) + d - Kxy = 0, \quad \text{with} \quad d > 0,\ K \in \mathbb{R}. \tag{1}$$

We define a map $F : \mathbb{R}_*^{+^2} \to \mathbb{R}_*^{+^2}$ by the following geometric construction: if $M = (x, y) \in \mathbb{R}_*^{+^2}$, we consider the curve $\mathscr{C}_K$ which contains M; the horizontal line passing through M cuts $\mathscr{C}_K$ in a second point M'; now the vertical line passing through M' cuts again $\mathscr{C}_K$, and this intersection is $F(M)$ (remark that the infinite points in horizontal and vertical directions are on the curve). It is easy to see that $F(x, y) := (X, Y)$ is defined by

$$\begin{cases} Xx = 1 + \dfrac{d}{y}, \\ Yy = 1 + \dfrac{d}{X} \end{cases} \tag{2}$$

or

$$(X, Y) = \left(\frac{y+d}{xy}, \frac{dxy + y + d}{y(y+d)} \right). \tag{3}$$

The map F is defined on $\mathbb{R}_*^{+^2}$, with values in $\mathbb{R}_*^{+^2}$, and it is easy to see that F is an homeomorphism of $\mathbb{R}_*^{+^2}$ onto itself, satisfying

$$F^{-1} = S \circ F \circ S, \tag{4}$$

where S is the symmetry with respect to the diagonal. By definition the cubic curves $\mathscr{C}_K$ are *invariant* under the action of F, and the quantity

$$G(x, y) := x + y + \frac{1}{x} + \frac{1}{y} + \frac{d}{xy} \tag{5}$$

is *invariant* under the action of F: the curve $\mathscr{C}_K$ is the K-level set of G.

If $M_0 := (u_0, v_0) \in \mathbb{R}_*^{+2}$, then its iterated points $M_n := (u_n, v_n) = F^n(M_0)$ are the solutions of the system of two order one difference equations in $\mathbb{R}_*^{+2}$

$$\begin{cases} u_{n+1}\, u_n = 1 + \dfrac{d}{v_n}, \\ v_{n+1}\, v_n = 1 + \dfrac{d}{u_{n+1}}, \end{cases} \tag{6}$$

or

$$\begin{cases} u_{n+1} = \dfrac{v_n + d}{u_n v_n} \\ v_{n+1} = \dfrac{d u_n v_n + v_n + d}{v_n (v_n + d)}. \end{cases} \tag{7}$$

Thus the orbit of M_0 is included into the cubic $\mathscr{C}_K$ passing through M_0, and the function G is an invariant for the system (6): the quantity $u_n + v_n + \dfrac{1}{u_n} + \dfrac{1}{v_n} + \dfrac{d}{u_n v_n}$ is independant of the integer n.

In fact, the map F is a particular case of the so called QRT-maps, introduced in [8] and particularly studied in [6]. But our goal is to study the behavior of the solutions of system (6), and in particular to find the possible periods of periodic points, and the chaotic behavior of the map F. For this, we prefer to use methods analogous to these used in [2] or [12] instead of to use the general theory of QRT-maps.

We start with the search of the fixed points of F, and of the critical points of the function G.

1.2 Existence, Unicity and Equality of the Critical Point of G and the Fixed Point of F

The equations of the critical points of G in $\mathbb{R}_*^{+2}$ are

$$x^2 y - y - d = 0, \quad y^2 x - x - d = 0.$$

By difference one has $(x - y)(xy + d) = 0$, and so $x = y := t$ satisfies the equation $t^3 - t - d = 0$ which has exactly one solution $\ell > 0$. We denote $L := (\ell, \ell)$ this critical point of G.

A fixed point (x, y) of F satisfies the same equations as these of the critical point of G, so it has the form (s, s) where s satisfies the same equation $s^3 - s - d = 0$. So we have already proved a part of the following result.

Lemma 1 *The map F has exactly one fixed point $L = (\ell, \ell)$ where $\ell \in \left]\max(1, \sqrt[3]{d}), 1 + \dfrac{d}{2}\right[$ is the positive solution of the equation*

$$t^3 - t - d = 0. \tag{8}$$

The function G tends to $+\infty$ *at the infinite point of* $\mathbb{R}_*^{+^2}$, *and has a unique critical point which is the equilibrium L, where G attains is strict minimum*

$$K_m = \frac{4\ell + 3d}{\ell^2} = 3\ell + \frac{1}{\ell} > 4. \tag{9}$$

Proof The set in $\mathbb{R}_*^{+^2}$ defined by $\{G \leq M\}$ is compact, for $M > 0$, because on it we have $x + y \leq M$ and $xy \geq \frac{d}{M}$. This proves that G tends to $+\infty$ at the infinite point of $\mathbb{R}_*^{+^2}$.

If $P(t) = t^3 - t - d$, then $P' = 3t^2 - 1$ vanishes at $\sqrt{\frac{1}{3}}$, and one has $P(1) = -d$ and $P(\sqrt[3]{d}) = -\sqrt[3]{d}$. Since the curve $y = P(t)$ is, on $\left[\sqrt{\frac{1}{3}}, +\infty\right[$, above its tangent at the point $(1, -d)$, it is easy to see that $\ell \in \left]max(1, \sqrt[3]{d}), 1 + \frac{d}{2}\right[$. Formulas (9) are obvious, and the other points are previously proved. □

Now Lemma 1 has an important consequence, which is a direct application of a result of [3] generalized in [5].

Proposition 2 *The solutions of system (6) are permanent; if* $(u_0, v_0) \neq L$, *then the solution diverges. The equilibrium L is localy stable. Moreover, for* $K > K_m$ *the positive component* $\mathscr{C}_K^+$ *of the cubic* $\mathscr{C}_K$ *is diffeomorphic to the circle* $\mathbb{T}$ *and surrounds the point L.*

2 The Group Law on the Cubic $\mathscr{C}_K$ and the Dynamical System (2)

We will interpret the restriction of the map F to the positive part $\mathscr{C}_K^+$ of $\mathscr{C}_K$ with the chord-tangent law on the cubic.

2.1 *Study of the Cubic Curve* $\mathscr{C}_K$

The following lemma gives the essential facts about the cubic curve $\mathscr{C}_K$.

Lemma 3 *For* $K > K_m$, *we have the following properties:*
(1) *the cubic* $\mathscr{C}_K$ *is non-singular;*
(2) *the cubic* $\mathscr{C}_K$ *has three asymptotes: the two axis* $x = 0$ *and* $y = 0$, *and the line* $x + y = K$, *which is an inflection tangent at the infinite point* $D := (1, -1, 0)$ *(in projective coordinates);*

(3) *the cubic $\mathscr{C}_K$ cuts the axes at the points $A := (-d, 0)$ and $B := (0, -d)$;*
(4) *the positive component $\mathscr{C}_K^+$ of the cubic $\mathscr{C}_K$ is located in the triangular domain $x > 0,\ y > 0,\ x + y < K$; the part $\mathscr{C}_K \backslash \{\mathscr{C}_K^+ \cup \{A\} \cup \{B\}\}$ of the cubic is contained in five triangular domains: $x < 0,\ y < 0,\ x + y > -d$; $x > 0,\ x + y < -d$; $y > 0,\ x + y < -d$; $x < 0,\ x + y > K$; $y < 0,\ x + y > K$. The part $\overline{\mathscr{C}_K} \backslash \mathscr{C}_K^+$ is connected in $\mathbb{P}^2(\mathbb{R})$ ($\overline{\mathscr{C}_K}$ is the extension of $\mathscr{C}_K$ in $\mathbb{P}^2(\mathbb{R})$).*

Proof Points (2) and (3) are easy. Point (4) becomes from Lemma 1 and from an other form for the equation of $\mathscr{C}_K$: $xy(x + y - K) = -(x + y + d)$, so one has only to compare the signs of xy, $x + y - K$ and $x + y + d$. See the form of the cubic curve in Fig. 1, which is proved in Lemma 17 below, in Sect. 4.3. The set $\overline{\mathscr{C}_K} \backslash \mathscr{C}_K^+$ is connected by its infinite points.

For point (1), the equations of a singular point (x, y, t) are $f_x' = 0$, $f_y' = 0$, $f_t' = 0$, where $f(x, y, t) = xy(x + y) + (x + y)t^2 + dt^3 - Kxyt$. We obtain

$$\begin{cases} 2xy + y^2 + t^2 - Kyt = 0, \\ 2yx + x^2 + t^2 - Kxt = 0, \\ 3dt^2 + 2(x + y)t - Kxy = 0. \end{cases} \tag{10}$$

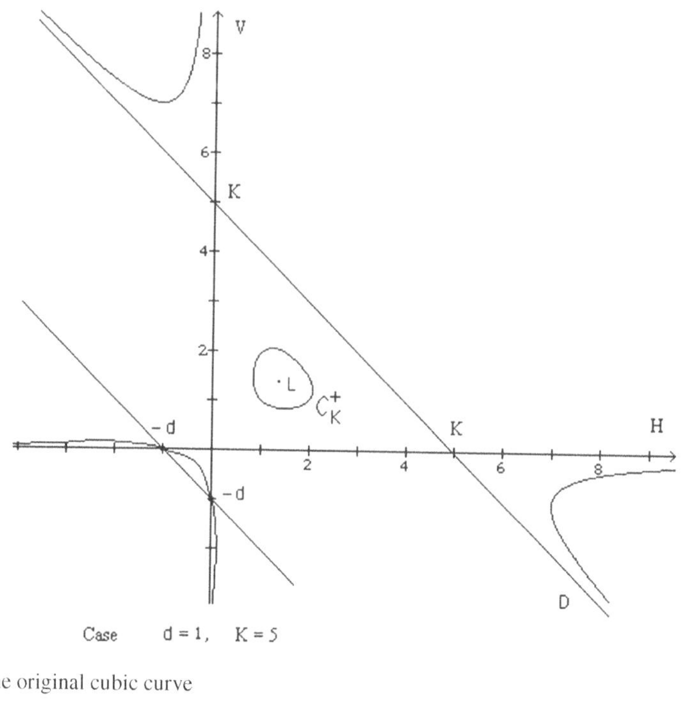

Fig. 1 The original cubic curve

Obviously $t = 0$ is not possible. The difference of the two first equations gives the relation $(y - x)(x + y - Kt) = 0$.

First suppose that $x \neq y$. Then we have two different (and symmetric) real singular points on the inflection asymptote, and so the curve splits in this line and some symmetric hyperbola with equation $xy - \alpha t^2 = 0$. We write the equation of $\mathscr{C}_K$ under the form

$$(xy - \alpha t^2)(x + y - Kt) + (\alpha + 1)(x + y)t^2 + (d - \alpha K)t^3 = 0.$$

If we take a finite point on the line $x + y - Kt = 0$ we must have $(\alpha + 1)K + d - \alpha K = 0$, and thus $K = -d$, which is impossible.

Thus $x = y := s$ satisfies the equations $Ks^2 - 4s - 3d = 0$ and $3s^2 - Ks + 1 = 0$. By elimination of K between these two equations, we obtain $s^3 - s - d = 0$. So $s = \ell$, and the first of the previous equations gives $K = \dfrac{4\ell + 3d}{\ell^2} = K_m$. In this case, L is the singular point of $\mathscr{C}_{K_m}$: it is a real isolated point of the curve, and $\mathscr{C}^+_{K_m}$ reduces to $\{L\}$. □

2.2 The Map F and the Group Law on the Cubic $\mathscr{C}_K$

For $K > K_m$ there is on the cubic $\mathscr{C}_K$, more exactly on its extension $\widetilde{\mathscr{C}_K}$ in $\mathbb{P}^2(\mathbb{C})$, a classical chord-tangent group law (see [1] or [7]). Denote H the infinite point in horizontal direction and V the infinite point in vertical direction. If we denote, for $P, Q \in \widetilde{\mathscr{C}_K}$, $P * Q$ the third point (finite or infinite) of $\widetilde{\mathscr{C}_K}$ on the line (PQ) (or on the tangent to $\widetilde{\mathscr{C}_K}$ at P if $P = Q$), the chord-tangent group law, the zero element of which is the point V, is

$$P \underset{V}{+} Q = (P * Q) * V. \tag{11}$$

Note that in this case V is not an inflection point of $\widetilde{\mathscr{C}_K}$; so the relation of alignment of three points $P, Q, R \in \widetilde{\mathscr{C}_K}$ is $P \underset{V}{+} Q \underset{V}{+} R = V * V$. Moreover the real part $\overline{\mathscr{C}_K}$ in $\mathbb{P}^2(\mathbb{R})$ is a subgroup of the complex cubic.

Now from the geometric definition of the map F given in Sect. 1.1 we deduce the following result for the map $\overline{F}$, extension of F to $\mathbb{P}^2(\mathbb{R})$ given by $\overline{F}(x, y, t) = \big(x(y + dt)^2, x(dxy + yt + dt^2), xy(y + dt)\big)$.

Proposition 4 *For $K > K_m$ the restriction of the map $\overline{F}$ to the cubic $\overline{\mathscr{C}_K}$ is nothing but the addition of the point H for the group law $\underset{V}{+}$: one has, for $M \in \overline{\mathscr{C}_K}$*

$$\overline{F}(M) = M \underset{V}{+} H, \quad \text{and} \quad M_n := \overline{F}^n(M_0) = M_0 \underset{V}{+} nH. \tag{12}$$

So a solution of (6) with starting point $M_0 \in \overline{\mathscr{C}_K}$ is periodic with minimal period n iff the infinite point H is exactly of order n in the group law $\underset{V}{+}$ on $\overline{\mathscr{C}_K}$. If a point $M_0 \in \overline{\mathscr{C}_K}$ is n-periodic, it is also the case for all the points of $\overline{\mathscr{C}_K}$. The set $\mathscr{C}_K^+$ is stable under the action of $\overline{F}$, which coincides with F on $\mathscr{C}_K^+$.

Proof Relations (12) are obvious from the geometric definition of F in Sect. 1.1. Then $M \in \overline{\mathscr{C}_K}$ has minimal period n iff

$$nH = V \text{ and } kH \neq V \text{ for } 1 \leq k \leq n-1. \tag{13}$$

But this condition depends only on $\overline{\mathscr{C}_K}$, that is on K, and not on the particular point $M \in \overline{\mathscr{C}_K}$: this proves the last assertion of the proposition. □

Remark 5 In [3], exactly the same cubic curve $\mathscr{C}_K$ was used, as an invariant level set for the order 2 difference equation $x_{n+2}x_n = 1 + \dfrac{a}{x_{n+1}}$, defined by the chord-tangent law $\underset{D}{+}$ on $\overline{\mathscr{C}_K}$, the zero element of which is the infinite point D on the asymptote $x + y = K$: one has in this case $(x_{n+2}, x_{n+1}) = (x_{n+1}, x_n) \underset{D}{+} V$.

2.3 The Group Law on the Cubic $\overline{\mathscr{C}}_K$ and Periodic Solutions of the System (6)

We will see elementary that no solution of (6) has minimal period 2, 3 nor 4.

Lemma 6 *The only solutions of (6) which are 2, 3 or 4 periodic are constant (identical to the point L).*

Proof If a solution is 2-periodic, we have

$$1 + \frac{d}{v_{n+1}} = u_{n+2}u_{n+1} = u_n u_{n+1} = 1 + \frac{d}{v_n},$$

and so v_n is constant, equal to v_0; then $1 + \dfrac{d}{u_{n+1}} = v_0^2$, and thus u_n is constant.

If a solution is 3-periodic, then $u_{n+2}u_{n+1}u_n = a$, a constant, and $v_{n+2}v_{n+1}v_n = b$, b constant. So we have $u_{n+2}\left(1 + \dfrac{d}{v_n}\right) = a$ and $v_{n+2}\left(1 + \dfrac{d}{u_{n+1}}\right) = b$, and thus $\dfrac{d}{v_n} = \dfrac{a}{u_{n+2}} - 1$ and $1 + \dfrac{d}{u_{n+1}} = \dfrac{b}{v_{n+2}} = \dfrac{b}{d}\left(\dfrac{a}{u_{n+4}} - 1\right) = \dfrac{b}{d}\left(\dfrac{a}{u_{n+1}} - 1\right)$; thus $u_{n+1} = \dfrac{ab - d^2}{b + d}$ is constant, and so v_n is also constant.

Now we will search if 4 may be a period of a solution of (6) by studying geometrically the equation $4H = V$.

First it is easy to see the opposite of a point X of $\overline{\mathscr{C}_K}$ for the group law $\underset{V}{+}$:

$$-X = X * B, \quad \text{where} \quad B = V * V = (0, -d, 1). \tag{14}$$

We denote $A = H * H = (-d, 0, 1)$, and $\mathscr{S}$, $\mathscr{S}^+$ and $\mathscr{S}^-$ the three connected real affine components of $\mathscr{C}_K \backslash \mathscr{C}_K^+$ located in the three domains $\{x + y < 0\}$, $\{x + y > K\} \cap \{x < 0\}$ and $\{x + y > K\} \cap \{y < 0\}$ (see Lemmas 3 and 17).

We have easily $2H \in \mathscr{S}^+$. We see that $-2H = (2H) * B$ (from (14)) is on $\mathscr{S}$. So we have $2H \neq -2H$, that is $4H \neq V$: there is no 4-periodic solution of (6) except $\{L\}$. □

In the following, we will transform the cubic curve $\overline{\mathscr{C}_K}$ in a standard cubic with equation $y^2 = 4x^3 - g_2 x - g_3$ and deduce of this that the restriction of the map F to $\mathscr{C}_K^+$ is conjugated to a rotation on the circle (remark that we know already that $\mathscr{C}_K^+$ is diffeomorphic to the circle, from Proposition 2). This result will give an other approach for the question of periodic solutions of (6). For algebraic consistency, we work with the version of the cubic in homogeneous complex coordinates, that is in $\mathbb{P}^2(\mathbb{C})$, and we denote as above $\widetilde{\mathscr{C}_K}$ this extension of the cubic, with equation $xy(x + y) + (x + y)t^2 + dt^3 - Kxyt = 0$.

3 Conjugation of $F_{|\mathscr{C}_K^+}$ to a Rotation on the Circle via Weierstrass' Function $\wp$

We start with the projective transformation $\mathscr{T}_1$ defined by

$$\begin{cases} 2X = x + y, \\ 2Y = y - x, \\ \ \ T = x + y - Kt, \end{cases} \tag{15}$$

or

$$x = X - Y, \ y = X + Y, \ t = \frac{2X - T}{K},$$

in order to transform the diagonal asymptote into the line at infinity and to use the symmetry of the curve. The new cubic has for equation

$$Y^2 T = 8\frac{K + d}{K^3} X^3 + \frac{K^3 - 8K - 12d}{K^3} X^2 T + 2\frac{K + 3d}{K^3} X T^2 - \frac{d}{K^3} T^3.$$

Now we make an affine transformation $\mathscr{T}_2$, where x, y, t are the new coordinates:

$$\begin{cases} X = \lambda x, \\ Y = \lambda^2 y, \qquad \text{where} \qquad \lambda = \dfrac{1}{K^{3/2}}, \ \ \mu = 2\dfrac{K+d}{K^{3/2}}. \\ T = \mu t, \end{cases} \tag{16}$$

We obtain a new cubic with equation

$$y^2 t = 4x^3 + (K^3 - 8K - 12d)x^2 t + 4(K + 3d)(K + d)xt^2 - 4d(K + d)^2 t^3.$$

We put

$$A := K^3 - 8K - 12d, \tag{17}$$

and make an horizontal translation $\mathcal{T}_3$ defined by (with new variables X, Y, T)

$$X = x + \frac{A}{12}t, \quad Y = y, \quad T = t. \tag{18}$$

We obtain a new cubic Γ_K with equation

$$Y^2 T = 4X^3 - g_2 X T^2 - g_3 T^3, \tag{19}$$

where

$$g_2 = \frac{1}{12}(K^6 - 16K^4 - 24dK^3 + 16K^2); \tag{20}$$

the value of g_3 will be unuseful.

We interpret these three changes of variables as transformations between cubics, and so $\phi := \mathcal{T}_3 \circ \mathcal{T}_2 \circ \mathcal{T}_1$ is a projective real transformation of $\overline{\mathscr{C}_K}$ onto Γ_K. So it transforms the two projective real connected parts of $\overline{\mathscr{C}_K}$ in two projective real connected parts of Γ_K, and so the positive compact part $\mathscr{C}_K^+$ of $\mathscr{C}_K$ onto the compact component Γ_K^+ of Γ_K. The three points of $\mathscr{C}_K$ on the diagonal, with coordinates f_1, f_2, f_3, become the three points of Γ_K on the X-axis, with coordinates e_1, e_2, e_3 (see Fig. 2; we will see in Lemma 9 that the "order" of the three points are inverted).

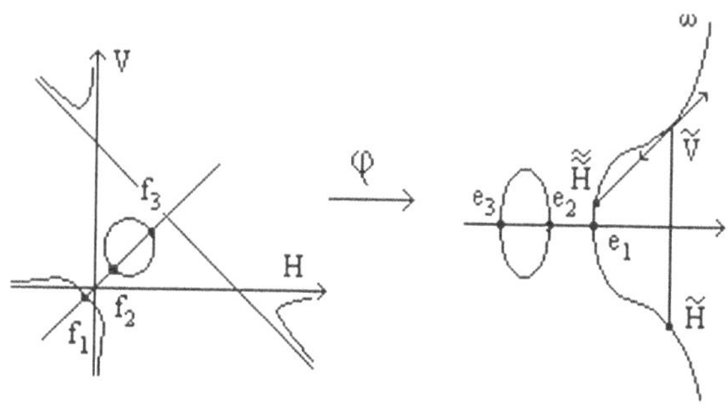

Fig. 2 The projective transform

By the map ϕ, the addition of H on $\overline{\mathscr{C}_K}$ for the chord-tangent law $\underset{V}{+}$ with zero element V is conjugated to the addition of $\widetilde{H}$ on Γ_K for the chord-tangent law $\underset{\widetilde{V}}{+}$ with zero element $\widetilde{V}$, where $\widetilde{H} = \phi(H)$ and $\widetilde{V} = \phi(V)$ (this is a general fact, but easy in our case because ϕ is a linear map in homogeneous coordinates).

It is easy to see that we have

$$\widetilde{V} = \begin{pmatrix} K + d + \frac{A}{12} \\ (K+d)K^{3/2} \\ 1 \end{pmatrix} \quad \text{and} \quad \widetilde{H} = \begin{pmatrix} K + d + \frac{A}{12} \\ -(K+d)K^{3/2} \\ 1 \end{pmatrix} \tag{21}$$

If ω is the infinite point on Γ_K at the vertical direction, we know (see [1, 7]) that the standard group chord-tangent law $\underset{\omega}{+}$ on Γ_K with ω as zero element is isomorphic to the standard group law on $\mathbb{T}^2$ (if we take real and complex points on Γ_K), via the parametrization of Γ_K by the Weierstrass' function $\wp$. But the addition of $\widetilde{H}$ on Γ_K is not for its standard law, but for $\underset{\widetilde{V}}{+}$. So we will make a supplementary isomorphism on Γ_K in order to pass from a law to the other.

Recall that generally the chord-tangent law on a cubic, with zero element Z, is defined by

$$M \underset{Z}{+} P = (M * P) * Z,$$

where $U * V$ denotes the third point of the intersection of the line (UV) with the cubic.

Now we will define a group isomorphism ψ of $(\Gamma_K, \widetilde{V})$ onto (Γ_K, ω) by

$$\psi : \Gamma_K \to \Gamma_K : M \mapsto M \underset{\widetilde{V}}{+} \omega = (M * \omega) * \widetilde{V}. \tag{22}$$

It is obvious that we have $\psi(\widetilde{V}) = \omega$. The fact that ψ transforms the addition $\underset{\widetilde{V}}{+}$ in the addition $\underset{\omega}{+}$ is not an obvious fact. It comes from a general fact about elliptic curves, because ψ is birational (see [10]). But in the particular case of the addition of a point in the Weierstrass' cubic, there is an elementary computer assisted proof, which is given in Appendix (the existence of an elementary proof is asserted in [10, p. 21], without proof).

So the initial addition of H on $(\overline{\mathscr{C}_K}, V)$ is conjugated by $\psi \circ \phi$ to the addition of $\widetilde{\widetilde{H}} = \psi(\widetilde{H})$ on (Γ_K, ω). Let be $X(K)$ the abscissa of the point $\widetilde{\widetilde{H}}$. It is known (see [2], where one uses the parametrization of Γ_K in the complex field by the Weierstrass' function) that the number of rotation of F restricted to $\mathscr{C}_K^+$ is the number $\theta_d(K)$ in $]0, 1/2[$ given by the following integral formula which permits to invert the Weierstrass' function $\wp$

$$2\theta_d(K) = \frac{\int_0^{\sqrt{\frac{e_1-e_3}{\nu}}} \frac{\mathrm{d}u}{\sqrt{(1+u^2)(1+\varepsilon u^2)}}}{\int_0^{+\infty} \frac{\mathrm{d}u}{\sqrt{(1+u^2)(1+\varepsilon u^2)}}}, \tag{23}$$

where one has

$$\nu := X(K) - e_1 > 0 \quad \text{and} \quad \varepsilon := \frac{e_1 - e_2}{e_1 - e_3} > 0 \tag{24}$$

(functions of K). So we have proved the following result.

Theorem 7 *For $d > 0$ and $K \in]K_m, +\infty[$ the restriction of the map F to $\mathscr{C}_K^+$ is conjugated to the rotation on the circle $\mathbb{T}$ with angle $2\pi\theta_d(K) \in]0, \pi[$ given by formula (23). The map $K \mapsto \theta_d(K)$ is analytic on $]K_m, +\infty[$.*

Proof The only thing to prove is the analyticity, and it results easily from the integral formula (23) because all the parameters in the integrals are analytic functions of K. □

4 The Possible Periods of Periodic Solutions of System (6)

We will study the number of rotation $\theta_d(K)$ given by formula (23) for $K > K_m$. Our goal is to find the limit of the function $K \mapsto \theta_d(K)$ when $K \to +\infty$ and when $K \to K_m$.

4.1 The Limit of $\theta_d(K)$ at $+\infty$

It is first necessary to have asymptotic expressions of the numbers e_1, e_2, e_3 which appear in formulas (23) and (24), and which become from f_1, f_2, f_3.

Lemma 8 *For $K > K_m$, the points F_i, $i = 1, 2, 3$ of $\mathscr{C}_K$ on the diagonal exist, and their coordinates f_i satisfy the inequalities*

$$-\frac{d}{2} < f_1 < 0 < f_2 < \ell < f_3 < \frac{K}{2}.$$

When $K \to +\infty$, we have the asymptotic developments

$$f_1 \sim -\sqrt{\frac{d}{K}},\ f_2 \sim \sqrt{\frac{d}{K}},\ f_3 = \frac{K}{2} - \frac{2}{K} + o\left(\frac{1}{K}\right),\ f_2 - f_1 \sim 2\sqrt{\frac{d}{K}},\ 2f_3 - K \sim -\frac{4}{K}. \tag{25}$$

Proof The f_i's are the solutions of the equation $2t^3 - Kt^2 + 2t + d = 0$, which has two positive roots because $K > \min G$, and a negative root. The inequalities of the lemma are obvious on Figs. 1 and 2. Then it is easy to deduce (25) from the three relations

$$f_1 f_2 f_3 = -\frac{d}{2},\ f_1 f_2 + f_2 f_3 + f_3 f_1 = 1,\ f_1 + f_2 + f_3 = \frac{K}{2}. \qquad \square$$

Now we calculate the e_i's.

Lemma 9 *We have the formula*

$$e_i = \frac{\mu}{\lambda} \frac{f_i}{2f_i - K} + \frac{A}{12}, \tag{26}$$

and in particular $e_3 < e_2 < e_1$.

Proof We take the images of the points $(f_i, f_i, 1)$ by $\phi = \mathcal{T}_3 \circ \mathcal{T}_2 \circ \mathcal{T}_1$, and obtain easily (26). Then the decreasing monotony of the function $x \mapsto \dfrac{x}{2x - K}$ on $]-\infty, \dfrac{K}{2}[$ gives the final result. $\square$

Now it is possible to obtain the asymptotic developments when $K \to +\infty$ of the parameters ε and $\sqrt{\dfrac{e_1 - e_3}{\nu}}$ which appear in the integrals of (23), and then to obtain the limit at $+\infty$ of $\theta_d(K)$.

Proposition 10 *One has*

$$\lim_{K \to +\infty} \theta_d(K) = \frac{3}{7}. \tag{27}$$

Proof From formulas (16), (17) and (26), and relations (25), we have easily $\varepsilon \sim \dfrac{16\sqrt{d}}{K^{7/2}}$. We obtain also $e_1 - e_3 \sim \dfrac{K^3}{4}$. Now it is necessary to get the value of $X(K)$. Let α be the first coordinate of $\widetilde{V}$, that is $\alpha = K + d + \dfrac{A}{12}$. The quantity $2\alpha + X(K)$ is the sum of the roots of the equation of degree 3 which express the abscissas of the intersection of Γ_K with its tangent at the point $\widetilde{V}$; let $Y = pX + q$ this tangent, the equation giving the abscissas of the intersections is $(pX + q)^2 = 4X^3 - g_2 X - g_3$, that is $4X^3 - p^2 X^2 + \cdots = 0$. So we have $X(K) + 2\alpha = p^2/4$. But the tangent at $\widetilde{V}$ to Γ_K is the image by ϕ of the tangent at V to $\mathcal{C}_K$, the equation of which is $x = 0$. After transformation we obtain for the final tangent the equation $X - A/12 = \lambda Y$; so $p = 1/\lambda = K^{3/2}$, and then $X(K) + 2\alpha = \dfrac{K^3}{4}$. So we have $X(K) = K^3/4 - 2K - 2d - \dfrac{1}{6}(K^3 - 8K - 12d) = \dfrac{K^3}{12} - \dfrac{2}{3}K$.

Now easy calculations give $\nu = X(K) - e_1 = d - 2\left(1 + \frac{d}{K}\right)\sqrt{\frac{d}{K}}(1 + o(1)) \to d$ when $K \to +\infty$, and so $\frac{e_1 - e_3}{\nu} \sim \frac{K^3}{4d}$.

If we take $\varepsilon \to 0$ as the variable, we have $K \sim \frac{M_1}{\varepsilon^{2/7}}$, and $\sqrt{\frac{e_1 - e_3}{\nu}} \sim \frac{M_2}{\varepsilon^{3/7}}$, with M_1 and M_2 positive constants. At this point we can apply the following lemma of [2], and obtain the final result. □

Lemma 11 *Let be $\lambda > 0$, $\varepsilon > 0$, $\beta > 0$. For any map $\varepsilon \mapsto \psi(\varepsilon) = o(1)$ when $\varepsilon \to 0$ and satisfying $\lambda + \psi(\varepsilon) > 0$, we put*

$$N(\varepsilon, \lambda, \beta) = \int_0^{\frac{\lambda + \psi(\varepsilon)}{\varepsilon^\beta}} \frac{du}{\sqrt{(1 + u^2)(1 + \varepsilon u^2)}} \quad \textit{and} \quad D(\varepsilon) = \int_0^{+\infty} \frac{du}{\sqrt{(1 + u^2)(1 + \varepsilon u^2)}}.$$

Then, when $\varepsilon \to 0$, we have $D(\varepsilon) \sim (1/2)\ln(1/\varepsilon)$, and, if $\beta < 1/2$, $N(\varepsilon, \lambda, \beta) \sim \beta \ln(1/\varepsilon)$.

4.2 The Limit of $\theta_d(K)$ at K_m

It is known (see [4] or [12]) that we have the formula

$$\lim_{K \to K_m} \theta_d(K) = \frac{1}{2\pi} \arccos\left(\frac{1}{2}\text{trace}\,(DF(L))\right). \tag{28}$$

where DF is the jacobian matrix of F. With formulas (2) and (8), it is easy to find the following result.

Proposition 12 *We have*

$$\theta_m(d) := \lim_{K \to K_m} \theta_d(K) = \frac{1}{2\pi} \arccos\left(\frac{1 - 2\ell^2 - \ell^4}{2\ell^4}\right) = \frac{1}{\pi} \arccos\left(\frac{\ell^2 - 1}{2\ell^2}\right), \tag{29}$$

and the function $d \mapsto \theta_m(d)$ is continuous on $]0, +\infty[$ and decreasing from $\frac{1}{2}$ to $\frac{1}{3}$ when d increases from 0 to $+\infty$. We have

$$Im(\theta_d) \supset \langle \frac{3}{7}, \theta_m(d) \rangle := \left]\min\left(\frac{3}{7}, \theta_m(d)\right), \max\left(\frac{3}{7}, \theta_m(d)\right)\right[. \tag{30}$$

One has the equivalence

$$\theta_m(d) = [\textit{resp.} < \textit{or} >]\frac{3}{7} \iff d = [\textit{resp.} > \textit{ or } <]d_0 := \frac{2\sin(\pi/14)}{[1 - 2\sin(\pi/14)]^{3/2}} \approx 1{,}076 \tag{31}$$

When $d = d_0$, *we have* $\ell = \ell_0 := \dfrac{1}{\sqrt{1 - 2\sin\frac{\pi}{14}}}$.

If $d \neq d_0$, the function $K \mapsto \theta_d(K)$ is not constant, because its image contains the interval $\langle \frac{3}{7}, \theta_m(d)\rangle$. But if $d = d_0$, it would be possible that θ_d be constant, equal to 3/7, and in this case all the solutions of (6) would be 7-periodic. But it is not the case.

Proposition 13 *The number 7 is not a common period to all the solutions of (6).*

Proof If 7 would be a common period to every solution of (6), we would have $F^7 = Id$, and so by (4), since $F^{-3} = S \circ F^3 \circ S$, we would have the relation $F^4 \circ S = S \circ F^3$. If we take $u_0 = v_0$, we would have $(u_4, v_4) = (v_3, u_3)$, and this relation gives $u_3 v_3^2 - v_3 - d = 0$. We start with $(u_0, v_0) = (1, 1)$, we calculate (u_3, v_3), and the previous relation gives an equation for number d. With a computer, this equation has for positive solution $d \approx 1.073$, which is different from d_0. This contradiction proves that when $d = d_0$ the solution with initial point $(1, 1)$ is not 7-periodic. □

Corollary 14 *The map* $K \mapsto \theta_{d_0}(K)$ *is non constant and not one-to-one. There exists an open interval* I *containing* d_0 *such that for each* $d \in I$ *the map* θ_d *is not one-to-one and not constant.*

Proof The first assertion results from Proposition 13. For the second, suppose for example that $K \mapsto \theta_{d_0}(K)$ has a maximum $M_0 > 3/7$ attained at $K_0 > K_m$. Since $d \mapsto \theta_d(K_0)$ and $d \mapsto \theta_m(d)$ are obviously continuous, it exists $\eta > 0$ such that for every $d \in]d_0 - \eta, d_0 + \eta[$ we have $\theta_d(K_0) > (M_0 + 3/7)/2$ and $\theta_m(d) < (M_0 + 3/7)/2$. Since $\lim_{K \to +\infty} \theta_d(K) = 3/7 < (M_0 + 3/7)/2$, the function θ_d attains the value $(M_0 + 3/7)/2$ twice (at least), and is not one-to-one. □

Problem Is the function $K \mapsto \theta_d(K)$ one-to-one if $|d - d_0|$ is sufficiently large?

4.3 *The Possible Periods of Periodic Solutions*

We know from Theorem 7 that the restriction of F to $\mathscr{C}_K^+$ is conjugated to a rotation on the circle with angle $2\pi\theta_d(K)$. So, if $\theta_d(K)$ is rational and equal to $\frac{p}{q}$ irreducible fraction, then the solutions with starting points on $\mathscr{C}_K^+$ are q-periodic, with q as minimal period, and the reciprocal is true.

In the contrary, if $\theta_d(K)$ is irrational, then every point of $\mathscr{C}_K^+$ has its orbit under the action of F which is dense in $\mathscr{C}_K^+$, and the converse is true.

How are distributed this two types of points, for a given d? What periods can appear, for a given d? The answers are given by the following result.

Theorem 15 *Let d be positive.*
(1) It exists a partition of $\mathbb{R}_*^{+^2}\setminus\{L\}$ *in two dense sets A and B, each of them union of invariant curves* $\mathscr{C}_K^+$*, such that every point in A is periodic and every point in B has a dense orbit in the positive part of the cubic which passes through it.*
(2) It exists an integer $N(d)$ *such that every integer* $q \geq N(d)$ *is the minimal period of some solution of (6).*

Proof **(1)** Put $]a, b[:=\mathrm{Im}(\overset{o}{\theta_d})$. This interval is not empty ($a < b$) if $d \neq d_0$ because $\theta_m(d) \neq 3/7$, and if $d = d_0$ because $K \mapsto \theta_{d_0}(K)$ is not constant. Since $K \mapsto \theta_d(K)$ is analytical, the function θ_d is constant on no not-empty interval of $]K_m, +\infty[$. So from the density of rational and irrational numbers in $]a, b[$ it results easily the density of the two sets $\theta_d^{-1}(]a, b[\cap\mathbb{Q})$ and $\theta_d^{-1}\big(]a, b[\cap(\mathbb{R}\setminus\mathbb{Q})\big)$ in $]K_m, +\infty[$. From this, one see that the two sets union of the curves $\mathscr{C}_K^+$ for K in the two previous dense sets are dense.
(2) The set of minimal periods of periodic solutions is exactly the set of integers q such that it exists a natural number p such that $\frac{p}{q}$ is in $\mathrm{Im}(\theta_d)$ and irreducible. So, if $\frac{p}{q}$ lies in $]a, b[$ and is irreducible, q is the minimal period of some solution of (6).
For finding such irreducible fractions, we fix the integer q (the eventual period) and search for p a *prime number* in the interval $]qa, qb[$, which is not a factor of q. It is known that the number of the distinct prime factors of a number q is majorized by $1.38402\,\frac{\ln q}{\ln(\ln q)}$ (see [9]). Denote $\pi(x)$ the cardinal of prime numbers not greater than x. So the number $P(q)$ of prime integers between qa and qb which do not divide q is at least

$$\pi(qb-1) - \pi(qa) - 1.38402\,\frac{\ln q}{\ln(\ln q)}.$$

From the prime number theorem (see [11]) we have $c(q)\frac{q}{\ln q} \leq \pi(q) \leq C(q)\frac{q}{\ln q}$, with $c(q) \leq 1 \leq C(q)$ and $\lim\limits_{q\to+\infty} c(q) = \lim\limits_{q\to+\infty} C(q) = 1$. So we have

$$\begin{aligned} P &\geq c(qb-1)\frac{qb-1}{\ln(qb-1)} - C(qa)\frac{qa}{\ln(qa)} - 1.38402\frac{\ln q}{\ln(\ln q)} \\ &= \big(c(qb-1)b - C(qa)a\big)\frac{q}{\ln q}(1+\eta(q)) \end{aligned}$$

where $\eta(q) \to 0$ when $q \to +\infty$. So it exists a number $N(d)$ sufficiently large such that for every $q \geq N(d)$ one has $P(q) \geq 1$, and so there exists a prime number p such that $\frac{p}{q} \in]a, b[$ and p does not divide q. Thus q is the minimal period of some solution of (6). □

Now we will search the set of integers which are minimal period of some solution of (6) for some value of $d > 0$. The principle is the same, and analogous to this of [2], but the interval is now explicit :]1/3, 1/2[, and so it is possible to improve the inequalities in the using of prime number theorem.

Theorem 16 *Every integer $q \geq 11$ is the minimal period of some solution of system (6) for some $d > 0$. Between 2 and 10, integers 2, 3, 4, 6, 10 are minimal period of no non-constant solution of (6), for no d, the others: 5, 7, 8, 9 are minimal periods.*

Proof The proof is long, and we split it in four steps.
Step (1) of the proof. First, we have, from Proposition 12, the relation

$$\bigcup_{d>0} \operatorname{Im}(\theta_d) \supset]1/3, 1/2[\setminus\{3/7\}. \tag{32}$$

Thus, for searching if an integer q is a minimal period of some solution of (6) for some $d > 0$, it suffices to find some prime number $p \in]q/3, q/2[$ which is not a factor of q. There is an exception if $p/q = 3/7$, but the only possibility for this is $q = 7$ and $p = 3$. So in the following we suppose $q \neq 7$.

We use an improvement of the prime number theorem, due to Rosser and Schoenfeld (see [11]):

$$\text{For } q \geq 52, \quad \frac{q}{\ln q} \leq \pi(q) \leq \left(1 + \frac{3}{2\ln q}\right)\frac{q}{\ln q}.$$

So if the function

$$f(q) := \frac{(q/2) - 1}{\ln((q/2) - 1)} - \frac{q/3}{\ln(q/3)}\left(1 + \frac{3}{2\ln(q/3)}\right) - 1.38402\frac{\ln q}{\ln(\ln q)} - 1 \tag{33}$$

is positive for some $q \geq 52$, it exists $p \in]q/3, q/2[$ which is prime and does not divide q. Thus q is a prime period of some solution of (6) for some $d > 0$. Of course the equivalent to $f(q)$ when $q \to +\infty$ is $(1/6)\frac{q}{\ln q}$, so it is true that $f(q) > 0$ for q sufficiently large. But we wish to have a quantitative version of this.

We put

$$f(x) := \frac{x/2 - 1}{\ln(x/2 - 1)} - \left(1 + \frac{3}{2\ln(x/3)}\right)\frac{x/3}{\ln(x/3)} - 1.38402\frac{\ln x}{\ln(\ln x)} - 1.$$

We have $f(780) < 0$ and $f(781) > 0$. From the graph of f on a computer, it seems that for every $x \geq 781$ we have $f(x) > 0$. We give a mathematical proof of this fact.

(a) We define the function $x \mapsto g_k(x) := k\frac{x}{\ln x} - 1.38402\frac{\ln x}{\ln(\ln x)} - 1$, for $k > 0$ to be chose. We study the monotonicity of g_k. We put $u := \ln x$, and so have

$$g_k(x) := h_k(u) = k\frac{e^u}{u} - 1.38402\frac{u}{\ln u} - 1,$$

and choose $u \geq u_0$, that is $x \geq x_0 := e^{u_0} \geq 52$ (and so $u_0 \geq 2$). We have

$$h_k' = ke^u\frac{u-1}{u^2} - 1.38402\frac{\ln u - 1}{\ln^2 u} \geq k(u_0 - 1)\frac{e^u}{u^2} - 1.38402\frac{1}{\ln u}$$

and so we have

$$\frac{1}{k} \leq \frac{(u_0 - 1)e^u \ln u}{1.38402u^2} := \phi(u) \Longrightarrow h_k'(u) > 0.$$

(b) Now we study the monotonicity of the function ϕ. First, $\left(\frac{e^u}{u^2}\right)' = \frac{e^u(u-2)}{u^3} \geq 0$ if $u \geq 2$. So ϕ is increasing and we have

$$\frac{1}{k} \leq \phi(u_0) \Longrightarrow \forall u \geq u_0,\ h_k'(u) > 0,$$

and so $g_k(x) > 0$ for $x \geq x_0$ if k satisfies the previous inequality and $g_k(x_0) > 0$.
(c) Now we search a condition which will imply the inequality

$$\forall x \geq x_0,\ \frac{x/2 - 1}{\ln(x/2 - 1)} - \left(1 + \frac{3}{2\ln(x/3)}\right)\frac{x/3}{\ln(x/3)} \geq k\frac{x}{\ln x} \Longleftrightarrow \forall x \geq x_0,\ f(x) \geq g_k(x).$$

Easy majorizations and minorations prove that a sufficient condition is to have, with $x_0 \geq 5$,

$$M(x_0) := \frac{1 - 2/x_0}{2} - \frac{1}{3}\left(1 + \frac{3}{2\ln(x_0/3)}\right)\frac{1}{1 - \frac{\ln 3}{\ln x_0}} \geq k.$$

(d) So the goal is to find an integer x_0 such that $M(x_0) > \frac{1}{\phi(u_0)}$ for $u_0 = \ln x_0$. We choose $x_0 = 2500$ and calculate the values $M(2500) \approx 0.0253\ldots$ and $\frac{1}{\phi(\ln 2500)} \approx 0.00241\ldots$ So we can take $k = 0.025$, and have to verify the sign of $g_k(x_0)$: $g_k(2500) = h_k(\ln 2500) \approx 2.7242 > 0$.
(e) *In fine*, for $x \geq 2500$ we have $f(x) > 0$.
(f) Now, using a computer, we see that $f(q) > 0$ for every integer $q \in [781, 2500]$. So our goal is to examine the integers in $[5, 780]$
First, remark the obvious inclusion:

$$\text{if } x \leq r \text{ and } I_q :=]q/3, q/2[, \quad]r/3, x/2[\subset \bigcap_{x \leq q \leq r} I_q. \tag{34}$$

We use this inclusion starting at $r = 780$: $780/3 = 260$; the smallest prime number majorizing 260 is $p = 263$, so we take $x/2 = 264$, and so $x = 528$. From (34), we have $p \in]780/3, 528/2[\subset]q/3, q/2[$ for every $q \in [528, 780]$, and p does not divide such a q, which is so a minimal period. Now, we use the same method starting from 527: $527/3 \approx 175.66$, and the smallest prime number majorizing this number is $p = 179$, so we take $x/2 = 180$, $x = 360$, and we can conclude that for all the integers $q \in [360, 527]$, the numbers p/q are in $]1/3, 1/2[$ and irreducible (because $p/q = p/pq' = 1/q' \in]1/3, 1/2[$ is not possible). So any integer in $[360, 527]$ is a minimal period.

We continue this procedure, and finish with the fact that all the integers $q \geq 24$ are minimal periods. The integers between 5 and 23 are examined one by one, and only for 6 and 10 there is no irreducible fraction $p/6$ nor $p/10$ in $]1/3, 1/2[$. Of course $3/7 \in]1/3, 1/2[$, but it is not *a priori* in the set $\bigcup_{d>0} \text{Im}(\theta_d)$.

Step (2) of the proof. We prove that 7 is a minimal period. We have to use the following result, which was previously already used.

Lemma 17 *For $K > K_m$, the places and the shapes of the different branches of the cubic $\mathscr{C}_K$ are the same as in Fig. 1.*

This Lemma will be proved after the end of the proof of Theorem 16.

We must also use the following result.

Lemma 18 *For every integer $n \geq 0$ we have the following formula for the group law on the curve $\overline{\mathscr{C}_K}$*

$$-nH = S[(n+1)H], \tag{35}$$

where S is the symmetry with respect to the diagonal.

Proof We denote (R_n) this relation. For $n = 0$, (R_0) is $V = S(H)$, which is true. Suppose that (R_{n-1}) is true: $-(n-1)H = S(nH)$. Since $F(M) = M \underset{V}{+} H$ and $F^{-1}(M) = M \underset{V}{-} H$, we have, using (R_{n-1}) and formula (4),

$$-nH = F^{-1}[-(n-1)H] = F^{-1}[S(nH)] = (S \circ F)(nH) = S[(n+1)H].$$

This recursion proves the lemma. □

Now the condition for having 7 as a period is $7H = V$ or $-3H = 4H$. But by Lemma 18, $-3H = S(4H)$, and so the condition is exactly $4H = S(4H)$, that is $4H = F_1$ or $4H = D$ (it is easy to see that each $nH \in \overline{\mathscr{C}_K} \backslash \mathscr{C}_K^+$). Let be N_0 the point of $\mathscr{S}^+$ with horizontal tangent (where y is minimum on $\mathscr{S}^+$), which exists because $\mathscr{S}^+$ is a convex curve, by Lemma 17 and Fig. 1. If $2H \neq N_0$, one see easily that $4H = (2H * 2H) * V$ is finite and on the branches $\mathscr{S}^+$ or $\mathscr{S}^-$, and so cannot be equal to F_1. So the unique possibility is $2H = N_0$, and then $4H = D$. Conversely, if $4H = D$, then $2H = N_0$, that is N_0 is in the vertical line through point A.

But it is easy to calculate the second coordinate of $2H$:

$$y_{2H} = \frac{d^2 + Kd + 1}{d}. \tag{36}$$

So the condition for having 7 as a period is $f'_x(-d, y_{2H}) = 0$, that is

$$Kd(1-d^2) = d^4 - d^2 - 1. \tag{37}$$

So 7 is a period if and only if (37) is true for some $d > 0$ and $K > K_m$. So we must have

$$1 < d < \sqrt{\frac{1+\sqrt{5}}{2}}. \tag{38}$$

And now we have $K = \dfrac{d^4 - d^2 - 1}{d(1-d^2)}$. So 7 is a period iff

$$\frac{d^4 - d^2 - 1}{d(1-d^2)} > K_m = 3\ell + \frac{1}{\ell}.$$

Since the map $d \mapsto \ell$ is bijective increasing from 1 to $+\infty$ when d increases from 0 to $+\infty$, this condition can be translated as

$$\frac{(\ell^3-\ell)^4 - (\ell^3-\ell)^2 - 1}{(\ell^3-\ell)[1-(\ell^3-\ell)^2]} > 3\ell + \frac{1}{\ell}.$$

We put $\ell^2 := x$ and obtain the condition $u(x) := x^3 - x^2 - 2x + 1 < 0$. It is easy to see that the roots of u satisfy the inequalities $x_2 < 0 < x_1 < 1 < x_m$. So the condition $u(x) < 0$, for $x > 1$, is $x < x_m = 1.80193377$ and then $\ell < \ell_m = 1.34236$, and so

$$1 < d < d_m = 1.07649\cdots < \sqrt{\frac{1+\sqrt{5}}{2}} = 1.2220\cdots$$

Then it is exactly for $1 < d < d_m$ that it exists a (unique) $K(d) := \dfrac{d^4 - d^2 - 1}{d(1-d^2)} > K_m$ such that 7 is a period for the solutions of (6) which are on the curve $\mathscr{C}_{K(d)}$.

Remark 19 With some calculation it is possible to see that one has

$$d_m = d_0 \quad \text{and} \quad \ell_m = \ell_0. \tag{39}$$

In fact, look at the equation $w^7 + 1 = 0$, with $w \neq -1$. By setting $X := w + \dfrac{1}{w}$, this equation is equivalent to the new equation $X^3 - X^2 - 2X + 1 = 0$, whose roots are

$$2\cos\frac{5\pi}{7} = -2\sin\frac{3\pi}{14} < 0 < 2\cos\frac{3\pi}{7} = 2\sin\frac{\pi}{14} < 1 < 2\cos\frac{\pi}{7} = 2\sin\frac{5\pi}{14}.$$

So $\ell_m^2 = 2\sin\frac{5\pi}{14}$, and we have easily the relation $2\sin\frac{5\pi}{14} = \frac{1}{1-2\sin\frac{\pi}{14}} = \ell_0^2$ (see Proposition 12), because $2\sin\frac{5\pi}{14} - 2\sin\frac{3\pi}{14} + 2\sin\frac{\pi}{14} = 1$, sum of the roots of the equation $X^3 - X^2 - 2X + 1 = 0$.

Remark 20 There is also a direct simple analytic proof that if d is near d_0 the function $K \mapsto \theta_d(K)$ attains the value 3/7, by using the method of the proof of Corollary 14 (with the intermediate value theorem). The comparizon of this method with this previous geometric one and Remark 19 gives the following qualitative result:

$$\forall\, K > K_m,\ \ \theta_{d_0}(K) \leq 3/7. \tag{40}$$

Step (3) of the proof. We solve the case $q = 6$ by geometric method. First, we remark that $2n$ is a period of solutions on $\overline{\mathscr{C}_K}$ iff $2nH = V$, or $nH = -nH$ in the group law, that is $nH = (nH) * B$ (see formula (14)), and this is equivalent to the fact that B belongs to the tangent to $\overline{\mathscr{C}_K}$ at the point nH. So, in the case of $q = 6 = 2 \times 3$, we will see that B does not belong to the tangent at the point $3H$.

First we localize $3H$. Since $H * H = A = (-d, 0)$, we have $2H = A * V \in \mathscr{S}^+$ (see the proof of Lemma 6). So $(2H) * H \in \mathscr{S}^+$, and $3H = (2H) + H = [(2H) * H] * V$ is the vertical projection of $(2H) * H$ on $\mathscr{S}$, and so $3H$ is on the arc $\widehat{HAB}$ of $\mathscr{S}$. Now we use the following lemma.

Lemma 21 *There is exactly (for $K > K_m$) two real and finite inflection points I, J on $\overline{\mathscr{C}_K}$, symmetric with respect to the diagonal, situated on $\mathscr{S}$, in the second and the fourth quadrant, I on the arc $\widehat{HA}$ and J on the arc $\widehat{BV}$. The arc $\widehat{IABJ}$ of $\mathscr{S}$ is strictly convex, the arcs $\widehat{HI}$ and $\widehat{JV}$ also, and it is also the case for $\mathscr{S}^+$ and $\mathscr{S}^-$.*

With this lemma, we see that if $3H \in \widehat{IB}$, since $3H \neq B$, the tangent at $3H$ does not cut, by convexity, the arc $\widehat{IB}$, and so B does not belong to this tangent. If now $3H \in \widehat{HI}$, by convexity the tangent at $3H$ has positive slope, and cannot pass to B. So in each case we conclude that $3H \neq -3H$, and so 6 is not a minimal period.

Proof of Lemma 21. The transformation ϕ of $\overline{\mathscr{C}_K}$ onto Γ_K is real and projective, and so preserves the real inflection points. So we search the inflection points of Γ_K.

The positive part ($y \geq 0$) of this curve is defined by $y = \sqrt{P(x)}$, where $P(x) = 4x^3 - g_2x - g_3 = 4(x-e_1)(x-e_2)(x-e_3)$. One shows easily that the roots of $(\sqrt{P})''$ are those of $2PP'' - P'^2$. Let h be this function. It is a degree 4 polynomial, and $h(e_i) = -P'^2(e_i) < 0$. But $h' = 2PP''' = 48P$, which is zero at the e_i's. So the function h is negative on $[e_3, e_1]$ and has only two real roots. The smallest gives no real inflection point, but the greatest gives two inflection points on the unbounded branch of Γ_K, symmetric with respect to the x-axis. When we transform these two points by ϕ^{-1}, we obtain exactly two real finite inflection points I and J, symmetric with respect to the diagonal. But, with Lemma 17, there are necessary at least one inflection point on the arc $\widehat{HA}$ and on the arc $\widehat{BV}$. So these points are exactly I and

J. The unicity of this inflection points and Lemma 17 give the results on concavity and convexity asserted in Lemma 21. □

Step (4) of the proof. Now we study if 10 is a minimal period (it is a period because 5 is period). So we suppose that $5H \neq V$ and $10H = V$, that is $5H = -5H = (5H) * B$, or: B belongs to the tangent to $\overline{\mathscr{C}_K}$ at $5H$. We examine two cases.

(4a) $3H = A$, that is $(2H) * H$ is on the vertical line through A, or $2H = N_0$ (see step (2) of the proof). In this case we have $5H = \big((3H) * (2H)\big) * V = V * V = B$ and then $B * B = B * 5H = 5H = B$. Hence B is an inflection point, and this is impossible by Lemmas 17 and 21.

(4b) $3H \neq A$, that is $2H \neq N_0$. We look at two subcases.

(i) The point $2H$ is at the left side of N_0. In this case $(3H) * (2H) \in \widehat{BV}$, and then $5H \in \widehat{BV}$. But by hypothesis $5H \neq V$ and $5H$ cannot be equal to B by (4a). So $5H$ belongs to the *open* arc $\widehat{BJV}$. If $5H \in \widehat{JV}$, the tangent at $5H$ has a positive slope and cannot pass through B, and if $5H \in \widehat{BJ}$, the strict convexity of this arc proves that the tangent at $5H$ cannot pass through B.

(ii) The point $2H$ is at the right side of N_0. In this case $5H \in \mathscr{S}^+$, and no tangent at this branch passes through B, except if $5H = V$, which is excluded by hypothesis. So in every case of (4b) the tangent at $5H$ cannot pass through B, and then 10 is not a minimal period.

This achieves the proof of Theorem 16. □

Proof of Lemma 17. We will use the form of the curve Γ_K, or more exactly of the curve Γ_K^* whose horizontal translated (by $\frac{A}{12}$, see formulas (14), (16), (19)) is Γ_K, that is $\Gamma_K^* = \mathscr{T}_2 \circ \mathscr{T}_1(\overline{\mathscr{C}_K})$. We denote $V^* = \mathscr{T}_2 \circ \mathscr{T}_1(V)$, and with the same notation the image in Γ_K^* of other remarkable points of $\overline{\mathscr{C}_K}$. With previous (see formula (22)) and new easy calculations, we find:

$$\begin{cases} A^* = (d, -dK^{3/2}, 1), \\ B^* = (d, dK^{3/2}, 1), \\ E_1^* = (f_1^*, 0, 1), \\ H^* = (K + d, -(K + d)K^{3/2}, 1), \\ V^* = (K + d, (K + d)K^{3/2}, 1), \\ D^* = (0, 1, 0), \end{cases} \tag{41}$$

with $0 < f_1^* < d$ (use the inequalities of Lemma 8 and formula (26)). So these points are easy to place on the unbounded branch of the curve Γ_K^*, because $K > 0$: see Fig. 3a.

Now we will locate the two inflection points J^* (on the positive part of the branch) and I^* (on the negative part), which have the same abscissa and opposite second coordinates (see the proof of Lemma 9). We know that the three arcs $\widehat{D^*I^*}$, $\widehat{I^*J^*}$ and $\widehat{J^*D^*}$ are strictly convex, and that we have the relation $B^* = V^* * V^*$ which

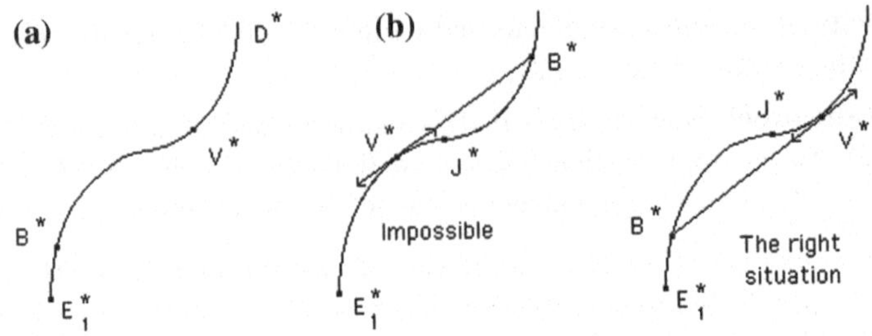

Fig. 3 Inflection points

means that the tangent to the curve at V^* cuts again the curve at the point B^*. If we suppose that J^* was on the arc $\widehat{V^*D^*}$, by the convexity of the arc $\widehat{E_1^*J^*}$, the tangent at V^* would cut the curve at the point $B^* \in \widehat{J^*D^*} \subset \widehat{V^*J^*D^*}$, which is impossible, because we have $x_{B^*} < x_{V^*}$ by (41) (see Fig. 3b).

So the order of the points on the arc $\widehat{E_1^*D^*}$ is strict and is $E_1^* < B^* < J^* < V^* < D^*$, where "$P < Q$" means that P is at left of Q and $P \neq Q$. For transporting the different connected arcs into the initial curve $\overline{\mathscr{C}_K}$, we use an easy fact:

Fact. The slope p of the tangent to $\overline{\mathscr{C}_K}$ at the point $B = (0, -d, 1)$ is, for $K > 0$,

$$p = -(d^2 + Kd + 1) < -1. \tag{42}$$

Now, with this tangent, it is clear that the arc $\widehat{E_1BJV}$ is as in Fig. 1, and also the arc $\widehat{E_1AIH}$ by symmetry.

For the two symmetric and convex arcs $\widehat{VD}$ and $\widehat{DH}$, we make a little calculation: we put, in the equation of $\mathscr{C}_K$, $x + y - K = t$; we obtain $ty^2 + t(t-K)y - (t + K + d) = 0$. So, if $t \to 0$, we have

$$y = \frac{\sqrt{K+d}}{\sqrt{t}}(1 + o(1)),$$

which is defined only for $t > 0$, with $y \to \infty$ when $t \to 0_+$. So the two connected convex arcs with asymptotes $x + y - K = 0$, $x = 0$ or $y = 0$ are above the line $x + y - K = 0$: it is the case at $\pm\infty$, and this line does not cut the curve. So we have the form of Fig. 1. □

Remark 22 The necessity of Lemma 17 is justified by the other forms of the curve in some cases (in fact when $K \leq -d$), see for example the following Fig. 4.

Remark 23 It results of the relation $\widetilde{B} = \widetilde{V} * \widetilde{V}$ (which comes by translation from the formula $B^* = V^* * V^*$) that we have

$$\widetilde{\widetilde{H}} = \psi(\widetilde{H}) = \widetilde{H} \underset{\widetilde{V}}{+} \omega = (\widetilde{H} * \omega) * \widetilde{V} = \widetilde{V} * \widetilde{V} = \widetilde{B} = (\frac{K^3}{12} - \frac{2K}{3}, dK^{3/2}, 1)$$

(see (41) and the proof of Proposition 10).

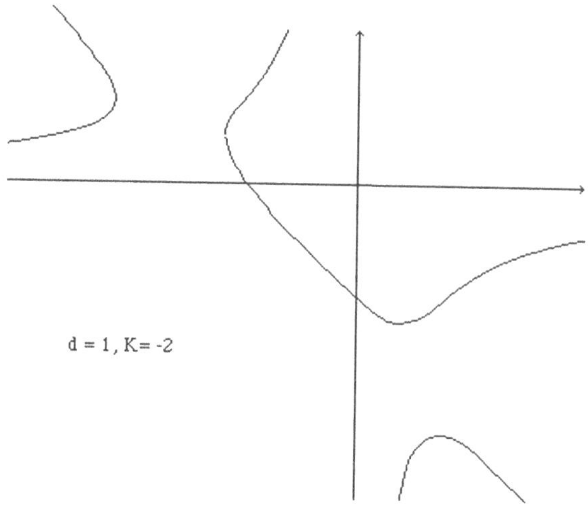

Fig. 4 An other form of the cubic curve

5 Chaotic Behaviour of the Dynamical System (2)

In this part we will see that the dynamical system (2) in $\mathbb{R}_*^{+^2}$ associated to the homographic system of difference equations (6) has uniform sensitiveness to initial conditions on every compact set not containing the equilibrium L.

Theorem 24 *For every compact set $\mathcal{K} \subset \mathbb{R}_*^{+^2}$ not containing the equilibrium L, it exists a number $\delta(\mathcal{K}) > 0$ such that for every point $M \in \mathcal{K}$ and every neighborhood W of M it exists $M' \in W$ such that $dist(F^n(M), F^n(M') \geq \delta(\mathcal{K})$ for infinitely many integers n.*

Of course, in this assertion, "*dist*" denote the euclidean distance in $\mathbb{R}_*^{+^2}$.

The proof will use a general topological result (probably well known) on "dynamical systems fibring in rotations on $\mathbb{T}$", the critical point for using this result being the proof of some uniformity in the inverse conjugation of previous Sect. 3.

Proposition 25 *Let X be a metric space. Let be also $\theta : X \to \mathbb{T} = \mathbb{R}/\mathbb{Z}$ a continuous map such that for every non-empty open set U, the set $\theta(U)$ contains a non-empty open set. Define the map*

$$g : X \times \mathbb{T} \to X \times \mathbb{T} : (x, \alpha) \to (x, \alpha + \theta(x)).$$

Then the dynamical system $(X \times \mathbb{T}, g)$ has δ-sensitiveness to initial conditions for every $\delta \in]0, 1/2[$.

Proof Although this is probably a classical result, we give a short proof for the completeness. Let be $M := (x_0, \alpha_0) \in X$, and let be $W := U \times I$ an open neighborhood of (x_0, α_0). Let be J a non-empty open interval contained in $\theta(U)$. It is possible

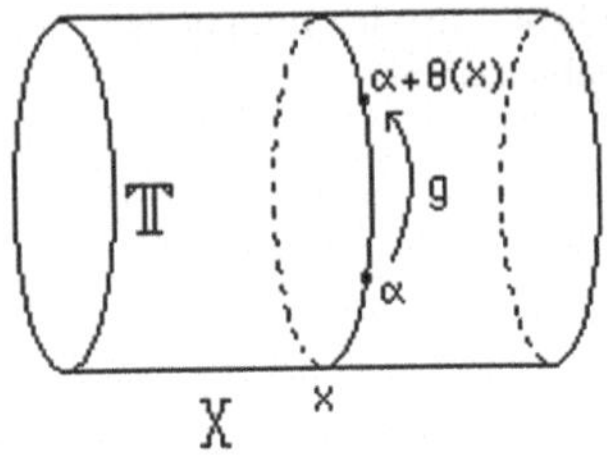

Fig. 5 Fibring in rotations

to find in J a point θ_1 such that in $\mathbb{T} = \mathbb{R}/\mathbb{Z}$ the number $|\theta_1 - \theta(x_0)|$ is irrational and in $]0,1[$. It exists $x' \in U$ such that $\theta(x') = \theta_1$. We put $M' = (x', \alpha_0) \in W$, and calculate the distance between the points $g^n(M)$ and $g^n(M')$: $\widetilde{d}\big(g^n(M), g^n(M')\big) = d_X(x, x') + ||n(\theta_1 - \theta(x_0))||$ ($\widetilde{d}$ denote the ℓ^1-distance in the product $X \times \mathbb{T}$, and $||.||$ the distance to the point 0 in $\mathbb{R}/\mathbb{Z}$). Let be $\delta \in]0, 1/2[$. Since $\theta_1 - \theta(x_0)$ is irrational, for infinitely many values of n we have $||n(\theta_1 - \theta(x_0))|| > \delta$, and so $\widetilde{d}\big(g^n(M), g^n(M')\big) \geq \delta$ for infinitely many n: this is the δ-sensitiveness to initial conditions of the dynamical system $(X \times \mathbb{T}, g)$ (Fig. 5). □

Now it is easy to prove the following fact:

Fact. Let be (X, g) a dynamical system which has δ-sensitiveness to initial conditions, where X is a *compact* metric space. Let be Y another compact metric space, and $h : X \to Y$ a homeomorphism. Then the conjugated dynamical system $(Y, \widetilde{g})$ by h ($\widetilde{g} = h \circ g \circ h^{-1}$) has η-sensitiveness to initial conditions for some $\eta > 0$.

The last argument for the proof of Theorem 24 will be the following result.

Lemma 26 *Let be $\wp_K$ the Weierstrass' function built with the numbers $g_2(K)$ and $g_3(K)$ of formulas (20) and (21), and defined by the formula*

$$\wp_K(z) = \frac{1}{z^2} + \sum_{\lambda \in \Lambda_K,\, \lambda \neq (0,0)} \left(\frac{1}{(z-\lambda)^2} - \frac{1}{\lambda^2} \right), \tag{43}$$

where Λ_K is the lattice of the points of $\mathbb{C}$ defined by

$$\Lambda_K := \{2p\omega_1(K) + 2iq\omega_2(K) \,|\, (p, q) \in \mathbb{Z}^2\},$$

with

$$\omega_1(K) = \frac{1}{\sqrt{e_1 - e_3}} \int_0^{+\infty} \frac{du}{\sqrt{(1+u^2)(1+\varepsilon u^2)}} \tag{44}$$

and

$$\omega_2(K) = \frac{1}{\sqrt{e_1 - e_3}} \int_0^{\pi/2} \frac{du}{\sqrt{1 - \varepsilon \sin^2 u}}, \tag{45}$$

where e_1, e_2, e_3, ε are as in formula (24). Then we have the property:

$$\wp_K(2x\omega_1(K) + i\omega_2(K)) \to \wp_{K_0}(2x\omega_1(K_0) + i\omega_2(K_0)) \text{ uniformly for } x \in [0, 1]$$

when $K \to K_0$, and the same property for the derivative $\wp'_K$.

Sketch of the proof of the lemma. For formulas (44) and (45) we refer to [1] or [2]. The segments $H_K := \{2x\omega_1(K) + i\omega_2(K) \mid x \in [0, 1]\}$, when $K \to K_0$, are contained in a fixed compact set disjoint from Λ_{K_0} and from the Λ_K if K is near to K_0. So, by some calculations, one can easily have uniformly for $x \in [0, 1]$ an upper bound for the remainder of the series (43). For the finite part of the sum, we use the continuity in K, uniformly with respect to x, of each of its terms, by using the continuity of the functions $K \mapsto e_j(K)$, $K \mapsto \omega_i(K)$. □

Sketch of the proof of the theorem. It is well known (see [1]) that for x varying in $[0, 1]$ the formula

$$\mathscr{P}_K(x) := \left(\wp_K(2x\omega_1(K) + i\omega_2(K), \wp'_K(2x\omega_1(K) + i\omega_2(K))\right)$$

gives a one-to-one parametrization of the bounded component Γ_K^+ of Γ_K. From now on we denote ϕ_K and ψ_K the maps defined by formulas (15), (16), (18) and (22), for marking their dependance on K. Now, the map $z \mapsto (\psi_K \circ \phi_K)^{-1}(z) = (\psi_K \circ \mathscr{T}_3 \circ \mathscr{T}_2 \circ \mathscr{T}_1)^{-1}(z)$ depends continuously on K, uniformly for z in a given compact in $\mathbb{R}_*^{+^2}$ (where the Γ_K^+ are situated for K near K_0).

So, if we choose K_1, K_2 such that $K_m < K_1 < K_2$, the map h defined on $[K_1, K_2] \times \mathbb{T}$ by

$$h(K, x) = (\psi_K \circ \phi_K)^{-1} \circ \mathscr{P}_K(x)$$

is a homeomorphism of $[K_1, K_2] \times \mathbb{T}$ onto the set

$$\Delta(K_1, K_2) := \{K_1 \le G \le K_2\} = \bigcup_{K_1 \le K \le K_2} \mathscr{C}_K^+.$$

But the map $F : \Delta(K_1, K_2) \to \Delta(K_1, K_2)$ is conjugated by h to the map $g : [K_1, K_2] \times \mathbb{T} \to [K_1, K_2] \times \mathbb{T} : (K, x) \mapsto (K, x + \theta_d(K))$ (see Theorem 7 and Fig. 6). So by Lemma 26, the fact and Proposition 25, the dynamical system $\left(\Delta(K_1, K_2), F\right)$ has a δ-sensitiveness to initial conditions for some $\delta > 0$. Now, if a compact set $\mathscr{K} \subset \mathbb{R}_*^{+^2}$ does not contain the equilibrium L, put $K_1 = \min_{\mathscr{K}} G$ and $K_2 = \max_{\mathscr{K}} G$. Then $\mathscr{K} \subset \Delta(K_1, K_2)$, and $F_{|\mathscr{K}}$ has a δ-sensitiveness to initial conditions. □

Remark 27 The proof of Theorem 24 gives in fact an improvement to the assertion on "pointwise" chaotic behavior of the dynamical systems studied in [2–4].

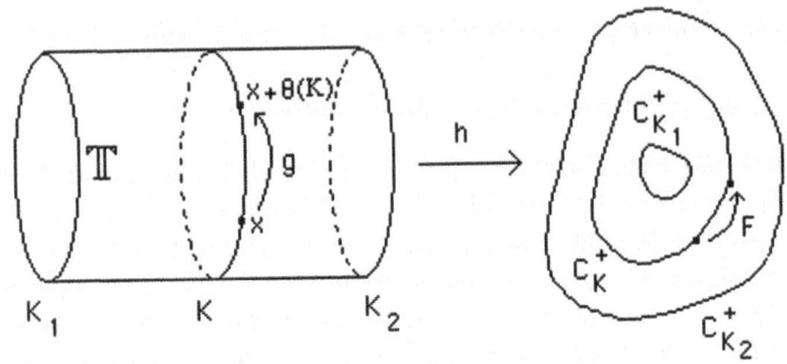

Fig. 6 Homeomorphism

Appendix: An Assisted Computer Proof that the Map ψ Defined by Formula (22) Is an Isomorphism Between the Two Chord-Tangent Group Laws

We start with a standard regular cubic $\mathscr{C}$ in Weierstrass' form: $y^2 = 4x^3 + ax + b$, with a given point Z on it, we denote ω the infinite point of $\mathscr{C}$ in vertical direction, and suppose $Z \neq \omega$. We denote $\underset{Z}{+}$ the addition for the chord-tangent law on $\mathscr{C}$ with zero element Z. So we have the following result.

Proposition 28 *The map $\psi : \mathscr{C} \to \mathscr{C} : M \mapsto M \underset{Z}{+} \omega$ is an isomorphism of the chord-tangent law on $\mathscr{C}$ with unit element Z on the standard chord-tangent law on $\mathscr{C}$ with unit element ω.*

The following result is asserted without proof in [10, p. 21].

Corollary 29 *Let be A and Z two different points on the regular cubic $y^2 = 4x^3 + ax + b$. Then the two chord-tangent group laws with unit elements Z and A are isomorphic.*

A computer assisted proof of Proposition 28 First, we remark that the map ψ is obviously an isomorphism of the group $(\mathscr{C}, \underset{Z}{+}, Z)$ onto the group $(\mathscr{C}, \times, \omega)$, where

$$P \times Q := P \underset{Z}{+} Q \underset{Z}{+} \big(\underset{Z}{-}\,\omega\big). \tag{46}$$

We remark also that we have

$$\big(\underset{Z}{-}\,\omega\big) = (Z * Z) * \omega := W. \tag{47}$$

Now we have to show that the law $\times$ coincides with the law $\underset{\omega}{+}$, that is

$$(P * Q) * \omega = \{[(P * Q) * Z] * W\} * Z$$

for every P, Q. This is equivalent to the relation

$$\forall R \in \mathscr{C}, \quad R * \omega = [(R * Z) * W] * Z. \tag{48}$$

For proving relation (48) we use Maple, and present a sequence of calculation instructions which gives the result, with comments.

We put $Z = (u, v)$, with $v \geq 0$ (this is possible by the symmetry of the curve). We put $W = (U, V)$, and denote $R = (x, y)$. We have $v = \sqrt{4u^3 + au + b}$, and two possibilities $y = \sqrt{4x^3 + ax + b}$ or $y = -\sqrt{4x^3 + ax + b}$, so we write two lists of instructions, depending on the sign of y. We denote

$$R * Z := R_1 = (x_1, y_1), \quad R_1 * W := R_2 = (x_2, y_2), \quad R_2 * Z := R_3 = (X, Y).$$

Our goal is to prove that $R_3 = R * \omega$, that is $X = x$ and $Y = -y$. We remark that if a line $y = px + q$ cuts the curve at three points, their abscisses satisfy the relation $x_1 + x_2 + x_3 = p^2/4$ (see the proof of Proposition 10).

The following Fig. 7 shows the geometric construction.

First list of Maple instructions

$v := \text{sqrt}(4 * u^3 + a * u + b);$ $\qquad [Z = (u, v) \in \mathscr{C}]$

$y := \text{sqrt}(4 * x^3 + a * x + b);$ $\qquad [R = (x, y) \in \mathscr{C}]$

Fig. 7 Group isomorphism

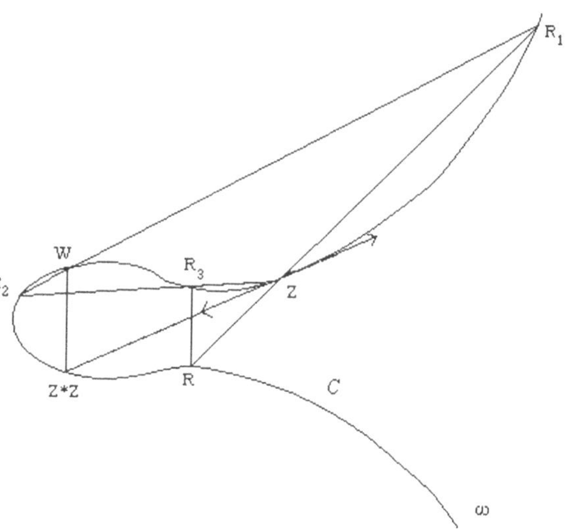

$p := \text{simplify}((12 * u^2 + a)/(2v));$
$q := \text{simplify}(v - p * u);$ [$y = px + q$ is the tangent to $\mathscr{C}$ at Z]

$U := \text{simplify}(p^2/4 - 2 * u);$
$V := \text{simplify}(-(p * U + q));$ [coordinates of W]

$p[1] := \text{simplify}((y - v)/(x - u));$
$q[1] := \text{simplify}(v - p[1] * u);$ [$y = p_1 x + q_1$ is the line (RZ)]

$x[1] := \text{simplify}(p[1]^2/4 - x - u);$
$y[1] := \text{simplify}(p[1] * x[1] + q[1]);$ [coordinates of R_1]

$p[2] := \text{simplify}((y[1] - V)/(x[1] - U));$
$q[2] := \text{simplify}(V - p[2] * U);$ [$y = p_2 x + q_2$ is the line $(R_1 W)$]

$x[2] := \text{simplify}(p[2]^2/4 - x[1] - U);$
$y[2] := \text{simplify}(p[2] * x[2] + q[2]);$ [coordinates of R_2]

$p[3] := \text{simplify}((y[2] - v)/(x[2] - u));$
$q[3] := \text{simplify}(v - p[3] * u);$ [$y = p_3 x + q_3$ is the line $(R_2 Z)$]

$X := \text{simplify}(p[3]^2/4 - u - x[2]);$
$Y := \text{simplify}(p[3] * X + q[3]);$ [coordinates of R_3]

At this point, we see that X has the form $X = \dfrac{N.x}{D^2}$. So we continue with

H := simplify (D^2);
A := simplify (X*H-H*x);

and obtain $A = 0$, that is $X = x$. Now we substitute x to X in the formula giving Y, obtain Y', and then make

B := simplify (Y'*sqrt(4*x^3+a*x+b)+4*x^3+a*x+b);

We obtain $B = 0$, that is $Y = Y' = -\sqrt{4x^3 + ax + b} = -y$, and so $X = x$ and $Y = -y$, that is the result.

Now we write the same list of instructions, but with $y = -\sqrt{4x^3 + ax + b}$ and an obvious modification in B, and obtain also the same result. □

References

1. Appel et, P., Lacour, E.: Principes de la théorie des fonctions elliptiques. Gauthier-Villars, Paris (1897)
2. Bastien, G., Rogalski, M.: Global behaviour of the solutions of Lyness' difference equations. J. Differ. Equ. Appl. **10**, 977–1003 (2004)
3. Bastien, G., Rogalski, M.: On some algebraic difference equations $u_{n+2}u_n = \psi(u_{n+1})$ in $\mathbb{R}_*^+$, related to families of conics or cubics: generalization of the Lyness' sequences. J. Math. Anal. Appl. **300**, 303–333 (2004)
4. Bastien, G., Rogalski, M.: On the algebraic difference equations $u_{n+2} + u_n = \psi(u_{n+1})$ in $\mathbb{R}$, related to a family of elliptic quartics in the plane. J. Math. Anal. Appl. **326**, 822–844 (2007)
5. Bastien, G., Rogalski, M.: Level sets lemmas and unicity of critical point of invariants, tools for local stability and topological properties of dynamical systems. Sarajev. J. Math. **8**(21), 273–282 (2012)
6. Duistermaat, J.: Discrete Integrable Systems. QRT Maps and Elliptic Surfaces. Springer, New York (2010)
7. Husemoller, D.: Elliptic Curves. Springer, New-York (1987)
8. Quispel, G.R.W., Roberts, J.A.G., Thompson, C.J.: Integrable mappings and soliton equations. Phys. Lett. A **126**, 419–421 (1988)
9. Robin, G.: Estimation de la fonction de Tchebychef Θ sur le k-ième nombre premier et grandes valeurs de la fonction $\omega(n)$ nombre de diviseurs premiers de n. Acta Arithmetica **XLII**, 367–389 (1983)
10. Silverman, J.H., Tate, J.: Rational Points on Elliptic Curves. Springer, New York
11. Tenenbaum, G.: Introduction à la théorie analytique et probabiliste des nombres, Cours spécialisé n°1. Collection Société Mathématique de France, Paris (1995)
12. Zeeman, E.C.: Geometric unfolding of a difference equation. unpublished paper (1996)

References

[illegible]

Continuous Limit in Dynamics with Choice

Lev Kapitanski and Sanja Živanović Gonzalez

Abstract We are interested in time evolution of systems that switch their modes of operation at discrete moments of time. The intervals between switching may, in general, vary. The number of modes may be finite or infinite. The mathematical setting for such systems is variable time step dynamics with choice. We have used this setting previously to study the long term behavior of such systems. In this paper, we define and study the continuous time dynamics whose trajectories are limits of trajectories of discrete systems as time step goes to zero. The limit dynamics is multivalued. In the special case of a switched system, when the dynamics is generated by switching between solutions of a finite number of systems of ODEs, we show that our continuous limit solution set coincides with the solution set of the relaxed differential inclusion.

Keywords Switched systems · Symbolic dynamics · Differential inclusions

1 Introduction

Many if not most of natural (real life) and man-made (engineered) systems switch their regimes of operation at discrete moments of time. The intervals between switching—the dwell times—may, in general, vary. The number of regimes may be finite or infinite. It is important to understand the workings of such systems. In control theory, the systems with both continuous and discrete behavior are known as hybrid systems. There is a large literature on control, stability, etc. of such systems, see [2, 4, 10, 11]. A special class of hybrid systems is formed by the so-called

L. Kapitanski
Department of Mathematics, University of Miami, Coral Gables, FL 33124, USA
e-mail: levkapit@math.miami.edu

S. Živanović Gonzalez (✉)
Department of Mathematics and Computer Science, Barry University,
Miami Shores, FL 33161, USA
e-mail: szivanovic@mail.barry.edu

L. Alsedà i Soler et al. (eds.), *Difference Equations, Discrete Dynamical Systems and Applications*, Springer Proceedings in Mathematics & Statistics 180, DOI 10.1007/978-3-662-52927-0_19

switched systems, [9]. A switched system (with time-dependent switching) is governed by a finite set of (systems of) ordinary differential equations:

$$\dot{x} = F_n(x) \quad \text{on the interval } [t_n, t_{n+1}), \quad j = 0, 1, \dots, \tag{1}$$

where each F_n is taken from a finite set of (vector-)functions $\{f_1, \dots, f_N\}$, see [9]. If we know in advance the sequence of dwell times, $\tau_0, \tau_1, \dots, \tau_n, \dots$ and the sequence of the corresponding regimes, $j_0, j_1, \dots, j_n, \dots \in \{1, 2, \dots, N\}$, then the values of $x(t)$ at the switching instances, $x_n = x(t_n)$, can be determined successively starting from $x(0) = x_0$ as

$$x_{n+1} = S_{j_n}^{\tau_n}(x_n), \tag{2}$$

where S_j^τ is the evolution map for the equation in the regime number j, i.e., $S_j^\tau(\eta)$ is the value at time $t = \tau$ of the solution $y(t)$ of the jth constituent equation $\dot{y} = f_j(y)$ with the initial condition $y(0) = \eta$.

Control theorists also study the discrete-time switched systems where the continuous phase is missing or not important. The discrete-time switched systems are described directly by Eq. (2) where the maps S_j^τ are given in advance. Most of the studies of such systems deal with the (already difficult) case when all S_j^τ are constant matrices, see e.g., [10]. However, the nonlinear equations (2) appear naturally in a quantized setting when discretized equations (1) are used (or being implemented),

$$x_{n+1} = x_n + \tau_n f_{j_n}(x_n). \tag{3}$$

In this example, $S_j^\tau(x) = x + \tau f_j(x)$, and the "dwell-time" τ is interpreted as the (variable) time-step. The occurrence of τ_n may be more intricate as in the following example:

$$x_{n+1} = x_n + \sin(\tau_n x_n). \tag{4}$$

There is only one regime, while τ_n may vary over a finite set or some interval, and the evolution map depends on τ: $S^\tau(x) = x + \sin(\tau x)$.

In [7] we propose a convenient mathematical setting to work with systems like (2). Let X be the state space, let $\mathcal{J}$ be the set of labels for all possible regimes, and let $\mathcal{I}$ be the set of possible dwell-times. Suppose that to every choice of $j \in \mathcal{J}$ and $\tau \in \mathcal{I}$ corresponds a map $S_j^\tau : X \to X$. This makes it possible to consider the switched trajectories generated as in (2). We have trajectories, but we don't have dynamical system. However, it is not hard to build a dynamical system of which (2) is a part. Denote by $\Sigma_{\mathcal{J}}$ the set of all infinite one-sided strings (sequences) of symbols from $\mathcal{J}$, and denote by $\Sigma_{\mathcal{I}}$ the set of all infinite one-sided strings of numbers from $\mathcal{I}$. In the lingo of symbolic dynamics, $\Sigma_{\mathcal{J}}$ and $\Sigma_{\mathcal{I}}$ are the full one-sided shifts over the alphabets $\mathcal{J}$ and $\mathcal{I}$. The strings from $\Sigma_{\mathcal{J}}$ encode the sequences of regimes, and the strings from $\Sigma_{\mathcal{I}}$ encode the sequences of the dwell-times during which the corresponding regimes are active. It is convenient to identify the infinite strings in $\Sigma_{\mathcal{J}}$ with the maps $w : [0, 1, 2, \dots] \to \mathcal{J}$ and the infinite strings in $\Sigma_{\mathcal{I}}$ with the maps $h : [0, 1, 2, \dots] \to \mathcal{I}$. Thus, $w(n-1)$ is the label of the nth regime in the sequence w, and $h(n-1)$ is the nth time interval in the sequence h. Often we think of w as a sequence and write down the consecutive symbols as $w(0)w(1)w(2)\ldots$. On the sets $\Sigma_{\mathcal{J}}$ and $\Sigma_{\mathcal{I}}$ acts the shift operator, σ, that erases the first symbol in sequences. Thus, $\sigma(w) = w(1)w(2)\ldots$, or we can write $\sigma(w)(n) = w(n+1)$; in the same fashion σ acts on $\Sigma_{\mathcal{I}}$. Now, to include the trajectories (2) into a semi-dynamical system, we use the skew-product construction. Consider the product space $\mathfrak{X} = X \times \Sigma_{\mathcal{J}} \times \Sigma_{\mathcal{I}}$ and define the map $\mathfrak{S} : \mathfrak{X} \to \mathfrak{X}$ acting by the rule

$$\mathfrak{S}(x, w, h) = \left(S_{w(0)}^{h(0)}(x), \sigma(w), \sigma(h) \right).$$

The iterations of the map $\mathfrak{S}$ define on the space $\mathfrak{X}$ a semi-dynamical system which we call the *variable time-step dynamics with choice* associated with the maps S_j^τ. The switched trajectories (2) appear as the first coordinates of the trajectories of the points in $\mathfrak{X}$ under the iterations of the map $\mathfrak{S}$.

We used this setting in [7] to study the long-term behavior in variable time-step dynamics with choice. We were interested, in particular, in global compact attractors for such systems. Our assumptions on the spaces X, $\mathcal{J}$, $\mathcal{I}$, and on the maps S_j^τ were quite general: X could be any complete metric space, $\mathcal{J}$ was a metric compact, and $\mathcal{I}$ was a compact subset of $(0, +\infty)$. The assumptions on the maps S_j^τ were basically minimal to guarantee that for fixed j and τ the discrete-time system (S_j^τ, X) would possess a global compact attractor.

In this paper we use the same setting but for a completely different purpose. We want to understand what happens with the dynamics (2) when the intervals between switching the regimes shrink to 0.

Prior to [7] we studied global attractors in dynamics with choice when the time-step, or the dwell-time, is fixed, [5, 6, 12]. One of the numerical examples involved the following discrete version of the Ross-Macdonald type model of malaria transmission:

$$\begin{aligned} x_{n+1} &= x_n + \tau\,[a_n\, y_n(1-x_n) - r_n\, x_n] \\ y_{n+1} &= y_n + \tau\,[b_n\, x_n(1-y_n) - m_n\, y_n.] \end{aligned} \tag{5}$$

For computational purposes, the coefficients (a_n, b_n, r_n, m_n) were chosen from a set of two quadruples, $(4, 6, 1, 2)$ and $(2, 10, 3, 2)$ (just two different regimes, $\mathcal{J} = \{1, 2\}$). We have observed that the appearance of the global attractor depends on the size of τ. If τ is relatively large, a certain part of the boundary of the attractor looks like the feathery end of a brush stroke. For smaller τ, the same part of the boundary becomes smoother, approaching a smooth curve, it seems, as τ goes to zero. It was then that we started thinking about the limit $\tau \to 0$ in dynamics with choice.

The question of continuous limit itself suggests that $\tau = 0$ should be included in the set of admissible dwell-times, so we take $\mathcal{I} = [0, \tau_*]$, some $\tau_* > 0$. It is natural to assume that S_j^0 is the identity map for every j. However, we do not assume that S_j^τ form a semi-group with respect to τ, because we want to include examples (3) and (4). To handle various limits, we need some continuity and compactness from the maps S_j^τ. However, since the identity map is among them, we are forced to impose the assumption on X, that closed, bounded subsets of X are compact. Since we always assume that X is a complete metric space, the latter restriction still permits X to be a finite-dimensional Riemannian manifold, or a Frechét-Montel vector space. Our other assumptions on the maps S_j^τ refer to reasonable properties of the switched trajectories, and in examples can be checked with a priori estimates. We try to keep the number of assumptions to the minimum.

The main results can be stated informally as follows. The continuous limit dynamics associated with the maps S_j^τ is inherently multi-valued. For every initial state x_0, the set $\mathscr{F}_t(x_0)$ of the limits of the values at time t of the switched trajectories with the maximal dwell-time going to 0 is compact. For any closed, bounded set $B \subset X$, the set $\mathscr{F}_t(B) = \cup_{x \in B} \mathscr{F}_t(x)$ is compact. On the hyperspace $X^\sharp$ of all closed, bounded (= compact) subsets of X with the Hausdorff metric, the map $\mathscr{F}_t : X^\sharp \to X^\sharp$ is continuous, satisfies the semigroup property, $\mathscr{F}_{t_2} \circ \mathscr{F}_{t_1} = \mathscr{F}_{t_1+t_2}$, and thus defines a semi-dynamical system. If we assume in addition that the points of X that are close to each other can be connected by a unique "shortest curve" (with reasonable properties), then each set

$$\bigcup_{0 \le t \le T} \mathscr{F}_t(B)$$

is lined (filled) by the integral curves. We say that $\gamma(t)$ is an integral curve of the continuous limit dynamics if $\gamma : [0, T] \to X$ is a continuous map and $\gamma(t_1 + t_2) \in \mathscr{F}_{t_2}(\gamma(t_1))$ for all $t_1, t_2 \in [0, T]$ with $t_1 + t_2 \in [0, T]$.

This result can be viewed as an extension of the Filippov–Ważewski relaxation theorem. An application to the switched system (1) shows that in that case the continuous limit dynamics is the same as the dynamics generated by the differential inclusion

$$\dot{x} \in \overline{co}\,(f_1(x), \ldots, f_N(x))\,. \tag{6}$$

The continuous limit dynamics for the Euler approximations (3) is also given by (6). However, our abstract results are much more general and we can apply them in a much greater variety of situations.

The paper is organized as follows. In Sect. 2 we set the notation for the variable time-step dynamics with choice. In Sect. 3 we prove all the abstract results on the properties of the maps $\mathscr{F}_t$ and define the Continuous Limit Dynamics. In Sect. 4 we show that the Continuous Limit Dynamics for the switched system (1) is governed by the differential inclusion (6) thus making a link with the Filippov–Ważewski theorem.

2 Abstract Setting

State space. Throughout this paper (X, d) is a complete metric space with the Heine–Borel property (i.e., every closed, bounded subset of X is compact). For example, X can be a finite-dimensional Riemannian manifold, or X can be a Montel space.

The distance from a point, x, to a set, B, is $d(x, B) = \inf_{y \in B} d(x, y)$. We denote by $e(A, B) = \sup_{x \in A} d(x, B)$ the excess of the set A over the set B. The Hausdorff distance between the sets A and B is the maximum of the two numbers, $e(A, B)$ and $e(B, A)$; we write

$$d_H(A, B) = \max\{e(A, B), e(B, A)\}.$$

Regime switching functions. Defined on X is a two-parameter family of maps S_j^τ. Here j represents the regime. The space of available regimes will be denoted $\mathcal{J}$. We assume that $\mathcal{J}$ is a compact metric space with the metric $d_{\mathcal{J}}$. The superscript τ shows the dwell time (along which the regime j is engaged). We take τ from the interval $\mathcal{I} = [0, \tau_*]$, some $\tau_* > 0$. The metric on $\mathcal{I}$ is induced from $\mathbb{R}$.

A regime switching function is an infinite string of symbols from $\mathcal{J}$. Denote by $\Sigma_{\mathcal{J}}$ the space of all strings $w = w(0)w(1)w(2)\dots$ with $w(n) \in \mathcal{J}$. This is the full one-sided shift on the alphabet $\mathcal{J}$. Equipped with the metric

$$d_{\Sigma_{\mathcal{J}}}(u, v) = \sum_{i=0}^{\infty} \frac{1}{2^i}\, d_{\mathcal{J}}(u(i), v(i)),$$

$\Sigma_{\mathcal{J}}$ is a compact metric space. The shift operator σ, acting according to the rule

$$\sigma(w(0)w(1)w(2)w(3)\dots) = w(1)w(2)w(3)\dots,$$

is continuous on $\Sigma_{\mathcal{J}}$. We use the notation $w[n] = w(0)w(1)w(2)\dots w(n-1)$.

Similarly, we define $\Sigma_{\mathcal{I}}$ with the metric $d_{\Sigma_{\mathcal{I}}}$. This space is a metric compact as well.

Along with infinite strings we need finite strings, or words. For a finite word, u, its length is the number of letters (symbols) in it; the length is denoted by $|u|$. Sometimes we put an asterisk after a word to indicate that the word is finite, e.g., $u*$. The space of all finite words over $\mathcal{J}$ will be denoted $\Sigma^*_{\mathcal{J}}$. If w is an infinite word, the notation $w[n]$ will be used for the finite word formed by the first n symbols of w, i.e., $w[n] = w(0)\dots w(n-1)$. Clearly, $|w[n]| = n$. For infinite words, $|w| = \infty$. The concatenation of u, a finite word, and v, a finite or infinite word, is denoted $u.v$.

For the words h in $\Sigma^*_{\mathcal{I}}$ and $\Sigma_{\mathcal{I}}$ we need to measure the total time span. The notation is this:

$$\|h\| = \sum_{n=0}^{|h|} h(n).$$

Evolution maps. For every choice of $\tau \in \mathcal{I}$ and $j \in \mathcal{J}$, a continuous map $S_j^\tau : X \to X$ is given. Thanks to our assumptions on the state space X, each S_j^τ is a bounded map. The maps S_j^τ may not satisfy the semi-group property, i.e., $S_j^{\tau_1} \circ S_j^{\tau_2}$ is not assumed to be equal to $S_j^{\tau_1+\tau_2}$, but we assume that S_j^0 is the identity map for any j. Starting with a point x and acting successively with different or the same maps S_j^τ, one obtains a switched trajectory.

Definition 1 The switched trajectory corresponding to the regime switching $(h, w) \in \Sigma_{\mathscr{I}} \times \Sigma_{\mathscr{J}}$ and starting at $x \in X$ is the sequence of points defined recursively as

$$y_0 = x, \quad y_{n+1} = S_{w(n)}^{h(n)}(y_n), \ n \geq 0.$$

As explained in our paper [7], a convenient way to study the switched trajectories is to use the scew-product construction and work in the product space $X \times \Sigma_{\mathscr{J}} \times \Sigma_{\mathscr{I}}$ with the metric of the product space:

$$d_{\mathfrak{X}}((x, w, h), (x', w', h')) = d_X(x, x') + d_{\Sigma_{\mathscr{J}}}(w, w') + d_{\Sigma_{\mathscr{I}}}(h, h').$$

Here is the relevant definition.

Definition 2 The variable time step dynamics with choice on X associated with the maps S_j^τ, $j \in \mathscr{J}$, $\tau \in \mathscr{I}$, is the discrete time dynamics generated on the product space $\mathfrak{X} = X \times \Sigma_{\mathscr{J}} \times \Sigma_{\mathscr{I}}$ by the iterations of the map

$$\mathfrak{F} : (x, w, h) \mapsto \left(S_{w(0)}^{h(0)}(x), \sigma(w), \sigma(h)\right). \tag{7}$$

Remark 1 This definition makes sense without any assumptions of metricity and compactness and such. But we are going to use this definition only when those assumptions are satisfied.

Note that after n steps the point (x, w, h) is moved to the point

$$\mathfrak{F}^n(x, w, h) = \left(S_{w[n]}^{h[n]}(x), \sigma^n(w), \sigma^n(h)\right),$$

where we use a convenient shorthand

$$S_{w[n]}^{h[n]} = S_{w(n-1)}^{h(n-1)} \circ S_{w(n-2)}^{h(n-2)} \circ \cdots \circ S_{w(0)}^{h(0)}.$$

3 Continuous Limit Dynamics

In this section we define the continuous limit of a variable time step dynamics with choice. The resulting *continuous limit dynamics* (CLD for short) will be multivalued. We start by listing all the assumptions on the maps S_j^τ.

Assumptions on S_j^τ:

A0 Each $S_j^\tau : X \to X$ is a continuous map; S_j^0 is the identity map.
A1 For any closed, bounded set $B \subset X$, the maps S_j^τ, restricted to B, depend uniformly continuously on $j \in \mathscr{J}$ and $\tau \in \mathscr{I}$. More precisely, given a closed, bounded B, for every $\varepsilon > 0$ there exist $\delta_{\mathscr{J}} > 0$ and $\delta_{\mathscr{I}} > 0$ such that

$$\sup_{x \in B} d_X(S_{j_1}^{\tau_1}(x), S_{j_2}^{\tau_2}(x)) \leq \varepsilon$$

provided $d_{\mathcal{J}}(j_1, j_2) \leq \delta_{\mathcal{J}}$ and $|\tau_1 - \tau_2| = d_{\mathcal{I}}(\tau_1, \tau_2) \leq \delta_{\mathcal{I}}$.

A2 For every $x \in X$ and for every $T > 0$, there is a constant $C = C(x, T)$ such that, for any finite word $s* \in \Sigma^*_{\mathcal{I}}$ and for any finite word $w* \in \Sigma^*_{\mathcal{J}}$ of the same length as $s*$, if $\|s*\| \leq T$ then

$$d(x, S_{w*}^{s*}(x)) \leq C.$$

A3 For every $x \in X$ and for every $\varepsilon > 0$ there exists $\delta = \delta(x, \varepsilon) > 0$ such that, for any finite word $s* \in \Sigma^*_{\mathcal{I}}$ and for any finite word $w* \in \Sigma^*_{\mathcal{J}}$ of the same length as $s*$,

$$d(S_{w*}^{s*}(x), x) < \varepsilon,$$

provided $\|s*\| \leq \delta$.

A4 Suppose $x_n \to x$. Suppose s_n is a sequence of finite words of symbols in $\mathcal{I}$ and u_n is a sequence of finite words of symbols in $\mathcal{J}$ of the same length as s_n. Suppose $\sup_n \|s_n\| \leq C$. Then for every $\varepsilon > 0$ there exists $N = N(C, \varepsilon)$ such that for all $n \geq N$,

$$d(S_{u_n}^{s_n}(x_n), S_{u_n}^{s_n}(x)) \leq \varepsilon.$$

The first two assumptions refer to the properties of individual maps, while the last three assumptions refer to switched trajectories. In applications with differential or difference equations, verification of these assumptions relies on a priori estimates. Assumption **A2** says that every switched trajectory starting from (any point) x stays bounded over any finite time interval. Assumptions **A3** and **A4** say that over short time intervals the switched trajectories remain close to the initial position.

Lemma 1 *(a) If $\tau_n \to \tau$ in $\mathcal{I}$, $j_n \to j$ in $\mathcal{J}$ and $x_n \to x$ in X, then $S_{j_n}^{\tau_n}(x_n) \to S_j^{\tau}(x)$.*

(b) For every bounded set A, $S_{j_n}^{\tau_n}(A)$ converges to $S_j^{\tau}(A)$ in the Hausdorff metric when $\tau_n \to \tau$ and $j_n \to j$.

(c) For every bounded A, for every integer $n > 0$, and for every $\varepsilon > 0$, there exists $\delta = \delta(A, n, \varepsilon) > 0$ such that

$$\sup_{h:\, \|h[n]\| \leq \delta} \sup_{w} d_H(S_{w[n]}^{h[n]}(A), A) \leq \varepsilon.$$

Proof To prove (a) consider

$$d(S_{j_n}^{\tau_n}(x_n), S_j^{\tau}(x)) \leq d(S_{j_n}^{\tau_n}(x_n), S_j^{\tau}(x_n)) + d(S_j^{\tau}(x_n), S_j^{\tau}(x)).$$

The first term on the right will become small due to **A1**, and the second term will become small due to continuity of S_j^{τ}.

For (b), observe that

$$\sup_{x\in A}\inf_{y\in A} d(S_{j_n}^{\tau_n}(x), S_j^\tau(y)) \le \sup_{x\in A} d(S_{j_n}^{\tau_n}(x), S_j^\tau(x))$$

and the right side goes to zero due to **A1**. Swapping x and y completes the proof of (b).

To prove (c) use induction on n. For $n = 1$ we have $d_H(S_j^\tau(A), S_j^0(A)) \to 0$ as $\tau \to 0$ due to **A1**. Indeed,

$$\sup_{x\in A}\inf_{y\in A} d(S_j^\tau(x), y) \le \sup_{x\in A} d(S_j^\tau(x), x) \le \varepsilon$$

and

$$\sup_{x\in A}\inf_{y\in A} d(S_j^\tau(y), x) \le \sup_{x\in A} d(S_j^\tau(x), x) \le \varepsilon$$

by **A1**.

Suppose the statement is true for $n-1$ and suppose that it is not true for n. Then there exist a bounded A, an $\varepsilon > 0$, and the sequences h_k and w_k such that $\|h_k[n]\| < 1/k$, but

$$d_H(S_{w_k[n]}^{h_k[n]}(A), A) > \varepsilon.$$

Each of $h_k(r)$ converges to zero, and we may assume that each of $w_k(r)$, $0 \le r \le n-1$, converges. We have

$$d_H(S_{w_k[n]}^{h_k[n]}(A), A) \le d_H(S_{j_k}^{\tau_k} \circ S_{w_k[n-1]}^{h_k[n-1]}(A), S_{w_k[n-1]}^{h_k[n-1]}(A)) + d_H(S_{w_k[n-1]}^{h_k[n-1]}(A), A).$$

The second term on the right goes to 0 by the induction hypothesis, so we may assume that

$$d_H(S_{j_k}^{\tau_k} \circ S_{w_k[n-1]}^{h_k[n-1]}(A), S_{w_k[n-1]}^{h_k[n-1]}(A)) \ge \frac{\varepsilon}{2}.$$

The Hausdorff distance can be estimated as follows.

$$\sup_{x\in A}\inf_{y\in A} d(S_{j_k}^{\tau_k} \circ S_{w_k[n-1]}^{h_k[n-1]}(x), S_{w_k[n-1]}^{h_k[n-1]}(y)) \le \sup_{x\in A} d(S_{j_k}^{\tau_k} \circ S_{w_k[n-1]}^{h_k[n-1]}(x), S_{w_k[n-1]}^{h_k[n-1]}(x))$$

and

$$\sup_{y\in A}\inf_{x\in A} d(S_{j_k}^{\tau_k} \circ S_{w_k[n-1]}^{h_k[n-1]}(x), S_{w_k[n-1]}^{h_k[n-1]}(y)) \le \sup_{y\in A} d(S_{j_k}^{\tau_k} \circ S_{w_k[n-1]}^{h_k[n-1]}(y), S_{w_k[n-1]}^{h_k[n-1]}(y))$$

Thus, for every sufficiently large k there is an $x_k \in A$ such that

$$d(S_{j_k}^{\tau_k} \circ S_{w_k[n-1]}^{h_k[n-1]}(x_k), S_{w_k[n-1]}^{h_k[n-1]}(x_k)) \ge \frac{\varepsilon}{4}.$$

Because bounded sequences are relatively compact in X we may assume $x_k \to x_*$, and at the same time, because $\mathscr{J}$ and $\mathscr{I}$ are compact, we may assume that each of the sequences τ_k, j_k, $h_k[n-1]$ and $w_k[n-1]$ converges. We must have $\tau_k \to 0$ and $h_k[n-1] \to 0\ldots0$. By part (a), the left side must go to 0, a contradiction. □

The following observation is a simple exercise.

Lemma 2 *The map $\mathfrak{F}$ is continuous and compact.*

3.1 Switched Trajectories

Starting the switched trajectories from the initial point $x \in X$, we would like to know where we can end up at time $T > 0$ if we use all possible regime switching. This is the set reachable from x at time T,

$$R_T(x) = \{y \in X : y = S^{h[n]}_{w[n]}(x) \text{ for some } n \geq 0,\ (h, w) \in \Sigma_{\mathscr{I}} \times \Sigma_{\mathscr{J}} \text{ with } \|h[n]\| = T\} \tag{8}$$

Due to Assumption **A2**, the set $R_T(x)$ is bounded. In fact, the set $\bigcup_{t\in[0,T]} R_T(x)$ is bounded. It makes sense to consider the closure $\overline{R_T(x)}$. This is the set of points that can be if not reached then approximated by switched trajectories. However, we propose to consider the following set instead.

The sets $\mathscr{F}_T(x)$. For every $x \in X$ and every $T > 0$ define $\mathscr{F}_T(x)$ as the set of all possible limits of sequences of the form $S^{h_k[n_k]}_{w_k[n_k]}(x)$, where $n_k \nearrow \infty$, $w_k \in \Sigma_{\mathscr{J}}$, and $h_k \in \Sigma_{[0,\varepsilon_k]}$ with $\|h_k[n_k]\| = T$ and $\varepsilon_k \searrow 0$. For $T = 0$ define $\mathscr{F}_T(x) = \{x\}$.

The difference between $\overline{R_T(x)}$ and $\mathscr{F}_T(x)$ is that $\mathscr{F}_T(x)$ may not contain some of the limits corresponding to the partitions where some intervals are not getting smaller in size.

Partitions. Consider the finite partitions of the interval $[0, T]$ into sufficiently small subintervals. If p is a partition, $|p|$ denotes the number of elements of p. Each partition into r subintervals is encoded by a finite word, which we can write as $h[r]$. Define the max-mesh size of the word $h[r]$ as *max-mesh* $(h[r]) = \max_{0\leq k\leq r-1} h(k)$, and define the min-mesh as *min-mesh* $(h[r]) = \min_{0\leq k\leq r-1} h(k)$. The strings $h_k[n_k]$ in the definition of $\mathscr{F}_T(x)$ have max-mesh going to 0.

Denote by $P(T; n, m)$ the set of all partitions p of $[0, T]$ with *max-mesh* $(p) \leq 2^{-n}$ and *min-mesh* $(p) \geq 2^{-m}$. We identify $p \in P(T; n, m)$ with the corresponding word in $\mathscr{I}^{|p|}$.

Define the sets

$$A_N = \bigcup_{n\geq N} \bigcup_{m\geq n} \bigcup_{p\in P(T;n,m)} \bigcup_{w*\in \mathscr{J}^{\|p\|}} S^p_{w*}(x).$$

Notice that $A_N \supset A_{N+1} \supset \dots$. Moreover,

$$\mathscr{F}_T(x) = \bigcap_{N\geq 0} \overline{A_N}. \tag{9}$$

Thus, $\mathscr{F}_T(x)$ is closed in addition to being bounded, and hence it is compact. This and other useful properties of $\mathscr{F}_T(x)$ are collected in Proposition 1 below. Here is a useful auxiliary result that exploits Assumption **A4**.

Lemma 3 *Suppose s_n is a sequence of finite words of symbols in $\mathscr{I}$ and u_n is a sequence of finite words of symbols in $\mathscr{J}$ of the same length as s_n. Suppose $\|s_n\| \to 0$. Let x_n be a bounded sequence. Then*

$$\lim_{n\to\infty} d\left(S_{u_n}^{s_n}(x_n), x_n\right) = 0.$$

Proof Suppose the assertion is wrong and there exist s_n, u_n, and x_n with the properties described in the statement but

$$d\left(S_{u_n}^{s_n}(x_n), x_n\right) > \varepsilon$$

for some $\varepsilon > 0$. In our settings, the sequence x_n has a convergent subsequence, and we call this subsequence x_n and work with it. Let x be the limit of x_n. Consider the inequality

$$d\left(S_{u_n}^{s_n}(x_n), x_n\right) \leq d\left(S_{u_n}^{s_n}(x_n), S_{u_n}^{s_n}(x)\right) + d\left(S_{u_n}^{s_n}(x), x\right) + d(x, x_n).$$

Each term on the right goes to 0 due to **A4** (the first term), to **A3** (the second term), and because $x_n \to x$. A contradiction. This proves the lemma. □

Proposition 1 *(a) The set $\mathscr{F}_T(x)$ is compact.*
(b) If $y_k = S_{w_k[n_k]}^{h_k[n_k]}(x) \to y$ as $k \to +\infty$ with $\|h_k[n_k]\| = T_k \to T$, max-mesh $(h_k[n_k]) \searrow 0$, and $n_k \nearrow +\infty$, then $y \in \mathscr{F}_T(x)$.
(c) $\mathscr{F}_T(x)$ is continuous with respect to T, i.e., if $T_k \to T$, then $d_H\left(\mathscr{F}_{T_k}(x), \mathscr{F}_T(x)\right) \to 0$.
(d) If $T_k \to T$ and $x_k \to x$, then $d_H\left(\mathscr{F}_{T_k}(x_k), \mathscr{F}_T(x)\right) \to 0$.
(e) The set $\bigcup_{0\leq t\leq T} \mathscr{F}_t(x)$ is compact.

Proof We have already established compactness of $\mathscr{F}_T(x)$. To prove part (b), first assume that there are infinitely many k for which $T_k \leq T$ and work with those k only. Pick your favorite $j \in \mathscr{J}$ and let $z_k = S_j^{T-T_k}(y_k) = S_j^{T-T_k} \circ S_{w_k[n_k]}^{h_k[n_k]}(x)$. By Lemma 1(a), $z_k \to y$. On the other hand, the limit of z_k belongs to $\mathscr{F}_T(x)$. Hence, $y \in \mathscr{F}_T(x)$.

Next, assume that there are infinitely many k for which $T_k > T$ and work with those k. Denote by m_k the integer such that $\|h_k[m_k]\| \leq T$, but $\|h_k[m_k+1]\| > T$. The sequence $z_k = S_{w_k[m_k]}^{h_k[m_k]}(x)$ is bounded, hence it has a convergent subsequence. Let z_* be its limit. Define $s_k = T - \|h_k[m_k]\|$ and consider the sequence $S_{j.w_k[m_k]}^{s_k.h_k[m_k]}(x)$.

This sequence is bounded by Lemma 1(c), hence it has a convergent subsequence. The limit is in $\mathcal{F}_T(x)$ by construction, but this limit is z_* due to Lemma 3. Now we apply Lemma 3 to the sequence $S^{h_k[n_k]}_{w_k[n_k]}(x)$ to conclude that its limit is z_*, which then is equal to y and, hence, $y \in \mathcal{F}_T(x)$.

Now turn to the proof of continuity of $\mathcal{F}_T(x)$ with respect to T. We argue by contradiction. Suppose there is no continuity. Then there is a sequence $T_\ell \to T$ and there exists an $\varepsilon > 0$ such that either

$$e(\mathcal{F}_{T_\ell}(x), \mathcal{F}_T(x)) > \varepsilon \tag{10}$$

for all ℓ, or

$$e(\mathcal{F}_T(x), \mathcal{F}_{T_\ell}(x)) > \varepsilon \tag{11}$$

for all ℓ, or both. First assume (10). Because each $\mathcal{F}_{T_\ell}(x)$ is compact, there exist $y^\ell \in \mathcal{F}_{T_\ell}(x)$ such that

$$d(y^\ell, \mathcal{F}_T(x)) = \sup_{y \in \mathcal{F}_{T_\ell}(x)} d(y, \mathcal{F}_T(x))$$

for every ℓ. Each y^ℓ is a limit of the form

$$y^\ell = \lim_k S^{h^\ell_k[n^\ell_k]}_{w^\ell_k[n^\ell_k]}(x),$$

where $\|h^\ell_k[n^\ell_k]\| = T_\ell$ and *max-mesh* $(h^\ell_k[n^\ell_k]) \to 0$ as $k \to \infty$.

If there are infinitely many ℓ for which $T_\ell > T$, then use only this subsequence and do the following. There are integers $m^\ell_k < n^\ell_k$ such that $\|h^\ell_k[m^\ell_k]\| \leq T$, but $\|h^\ell_k[m^\ell_k + 1]\| > T$. The sequence $z^\ell_k = S^{h^\ell_k[m^\ell_k]}_{w^\ell_k[m^\ell_k]}(x)$ has a convergent subsequence as $k \to \infty$, and its limit, z^ℓ_*, belongs to $\mathcal{F}_T(x)$ by part (b). Again, let s^ℓ_k and u^ℓ_k be such that $s^\ell_k.h^\ell_k[m^\ell_k] = h^\ell_k[n^\ell_k]$ and $u^\ell_k.w^\ell_k[m^\ell_k] = w^\ell_k[n^\ell_k]$. Consider the right hand side of the inequality

$$d(y^\ell, z^\ell_*) \leq d(y^\ell, S^{s^\ell_k}_{u^\ell_k}(z^\ell_k)) + d(S^{s^\ell_k}_{u^\ell_k}(z^\ell_k), S^{s^\ell_k}_{u^\ell_k}(z^\ell_*)) + d(S^{s^\ell_k}_{u^\ell_k}(z^\ell_*), z^\ell_*)$$

written for $k \geq k_\ell$ large enough so that *max-mesh* $(h^\ell_k[n^\ell_k]) \leq 2^{-\ell}$. Note that $\|s^\ell_k\| \leq |T_\ell - T| + 2^{-\ell}$. By Lemma 3, $d(S^{s^\ell_k}_{u^\ell_k}(z^\ell_*), z^\ell_*)$ can be made arbitrarily small by choosing ℓ large enough. The remaining terms can be made small by choosing k large enough. Thus, $e(\mathcal{F}_{T_\ell}(x), \mathcal{F}_T(x)) \to 0$ as $\ell \to \infty$.

If there are at most finitely many ℓ for which $T_\ell > T$ but infinitely many ℓ for which $T_\ell < T$, then use only this subsequence and do the following. Fix your favorite $j \in \mathcal{J}$ and define $s_\ell = T - T_\ell$. The sequence y^ℓ is bounded as a subset of the bounded set

$$\bigcup_{0\le t\le 2T} \mathscr{F}_t(x) \subset \overline{\bigcup_{0\le t\le 2T} R_t(x)}. \tag{12}$$

By Lemma 3, we can find an ℓ such that

$$d(y^\ell, S_j^{s_\ell}(y^\ell)) \le \frac{\varepsilon}{4}.$$

For this ℓ there exists a $k = k_\ell$ such that

$$d\left(S_j^{s_\ell} \circ S_{w_k^\ell[n_k^\ell]}^{h_k^\ell[n_k^\ell]}(x),\ S_j^{s_\ell}(y^\ell)\right) \le \frac{\varepsilon}{4}.$$

The sequence $z^\ell = S_j^{s_\ell} \circ S_{w_k^\ell[n_k^\ell]}^{h_k^\ell[n_k^\ell]}(x)$ (where $k = k_\ell$) has a subsequence converging to some point $z \in \mathscr{F}_T(x)$. Hence, $d(y^\ell, \mathscr{F}_T(x)) < \varepsilon$ for large ℓ, a contradiction.

Now, assume (11). Again, for every ℓ there exist an $x^\ell \in \mathscr{F}_T(x)$ and a $y^\ell \in \mathscr{F}_{T_\ell}(x)$ such that

$$e(\mathscr{F}_T(x), \mathscr{F}_{T_\ell}(x)) = d(x^\ell, \mathscr{F}_{T_\ell}(x)) = d(x^\ell, y^\ell).$$

Again, consider first the case $T_\ell \searrow T$. Every x^ℓ is a limit, $x^\ell = \lim_k x_k^\ell$,

$$x^\ell = \lim_k x_k^\ell, \quad x_k^\ell = S_{w_k^\ell[n_k^\ell]}^{h_k^\ell[n_k^\ell]}(x), \tag{13}$$

where $\|h_k^\ell[n_k^\ell]\| = T$. Denote by p_k^ℓ any partition of the interval $[T, T_\ell]$ into pieces of size not greater than *max-mesh* $(h_k^\ell[n_k^\ell])$. Pick your favorite j and denote by u_k^ℓ the word $j \dots j$ with as many j's as the length of p_k^ℓ. For every fixed ℓ, the sequence

$$\{S_{u_k^\ell}^{p_k^\ell}(x_k^\ell)\}_{k=1}^\infty$$

has a convergent subsequence, and its limit, call it z^ℓ, belongs to $\mathscr{F}_{T_\ell}(x)$. We have

$$d(z^\ell, x^\ell) \le d\left(z^\ell, S_{u_k^\ell}^{p_k^\ell}(x_k^\ell)\right) + d\left(S_{u_k^\ell}^{p_k^\ell}(x_k^\ell), x_k^\ell\right) + d(x_k^\ell, x^\ell). \tag{14}$$

Choose $k = k_\ell$ to make the first and the third terms on the right less than $\varepsilon/4$. The second term then is small for large ℓ thanks to Lemma 3.

If $T_\ell \nearrow T$, then use the same approximation (13) to x^ℓ and split $h_k^\ell[n_k^\ell]$ into $h_k^\ell[n_k^\ell] = p_k^\ell.h_k^\ell[m_k^\ell]$, where $\|h_k^\ell[m_k^\ell]\| \le T_\ell$ and $\|h_k^\ell[m_k^\ell + 1]\| > T_\ell$. Thus,

$$x_k^\ell = S_{u_k^\ell}^{p_k^\ell} \circ S_{w_k^\ell[m_k^\ell]}^{h_k^\ell[m_k^\ell]}(x).$$

The sequence $z_k^\ell = S_{w_k^\ell[m_k^\ell]}^{h_k^\ell[m_k^\ell]}(x)$ has a convergent subsequence, and we work with this subsequence and let $z^\ell = \lim_k S_{w_k^\ell[m_k^\ell]}^{h_k^\ell[m_k^\ell]}(x)$. By part (b), $z^\ell \in \mathscr{F}_{T_\ell}(x)$. Again, similar to (14), we have inequality

$$d(z^\ell, x^\ell) \le d(z^\ell, z_k^\ell) + d\left(z_k^\ell, S_{u_k^\ell}^{p_k^\ell}(z_k^\ell)\right) + d(x_k^\ell, x^\ell).$$

There exists k_ℓ such that $d(z^\ell, z_k^\ell) + d(x_k^\ell, x^\ell) \le \varepsilon/2$ for all $k \ge k_\ell$. With $k = k_\ell$, the middle term is small for large ℓ by Lemma 3. This contradicts (11) and completes the proof of part (c).

Now assume $T_\ell \to T$ and $x^\ell \to x$. In order to prove part (d) it suffices to prove that

$$\lim_\ell e(\mathscr{F}_{T_\ell}(x^\ell), \mathscr{F}_{T_\ell}(x)) = 0 \text{ and } \lim_\ell e(\mathscr{F}_{T_\ell}(x), \mathscr{F}_{T_\ell}(x^\ell)) = 0.$$

For the first limit, notice that every point $y^\ell \in \mathscr{F}_{T_\ell}(x^\ell)$ can be approximated by a sequence of the form $S_{w_k^\ell[n_k^\ell]}^{h_k^\ell[n_k^\ell]}(x^\ell)$ and that

$$d(S_{w_k^\ell[n_k^\ell]}^{h_k^\ell[n_k^\ell]}(x^\ell), \mathscr{F}_{T_\ell}(x)) \le d(S_{w_k^\ell[n_k^\ell]}^{h_k^\ell[n_k^\ell]}(x^\ell), S_{w_k^\ell[n_k^\ell]}^{h_k^\ell[n_k^\ell]}(x)) + d(S_{w_k^\ell[n_k^\ell]}^{h_k^\ell[n_k^\ell]}(x), \mathscr{F}_{T_\ell}(x))$$

Choose $k = k_\ell$ so that the distance between $S_{w_k^\ell[n_k^\ell]}^{h_k^\ell[n_k^\ell]}(x^\ell)$ and y^ℓ be less than $2^{-\ell}$. Then, for large ℓ, the first term on the right becomes small due to Assumption **A4**, while the second becomes small because $S_{w_k^\ell[n_k^\ell]}^{h_k^\ell[n_k^\ell]}(x)$ converges to a point in $\mathscr{F}_T(x)$ and $\mathscr{F}_{T_\ell}(x)$ converges to $\mathscr{F}_T(x)$ by part (c). In a similar fashion, for the second limit, we represent any $y^\ell \in \mathscr{F}_{T_\ell}(x)$ as $y^\ell = \lim_k S_{w_k^\ell[n_k^\ell]}^{h_k^\ell[n_k^\ell]}(x)$ and estimate

$$d(S_{w_k^\ell[n_k^\ell]}^{h_k^\ell[n_k^\ell]}(x), \mathscr{F}_{T_\ell}(x^\ell)) \le d(S_{w_k^\ell[n_k^\ell]}^{h_k^\ell[n_k^\ell]}(x), S_{w_k^\ell[n_k^\ell]}^{h_k^\ell[n_k^\ell]}(x^\ell)) + d(S_{w_k^\ell[n_k^\ell]}^{h_k^\ell[n_k^\ell]}(x^\ell), \mathscr{F}_{T_\ell}(x^\ell))$$

Now, choose $k = k_\ell$ so that

$$d(y^\ell, S_{w_k^\ell[n_k^\ell]}^{h_k^\ell[n_k^\ell]}(x^\ell)) < \varepsilon/3$$

and

$$d(S_{w_k^\ell[n_k^\ell]}^{h_k^\ell[n_k^\ell]}(x^\ell), \mathscr{F}_{T_\ell}(x^\ell)) < \varepsilon/3$$

and then use Assumption **A4** to choose ℓ large enough to have

$$d(S_{w_k^\ell[n_k^\ell]}^{h_k^\ell[n_k^\ell]}(x), S_{w_k^\ell[n_k^\ell]}^{h_k^\ell[n_k^\ell]}(x^\ell)) < \varepsilon/3.$$

This show that for every $\varepsilon > 0$, $e(\mathscr{F}_{T_\ell}(x), \mathscr{F}_{T_\ell}(x^\ell)) < \varepsilon$ for all large enough ℓ. This proves (d).

Next we show that the set $\bigcup_{0\le t\le T} \mathscr{F}_t(x)$ is closed. Suppose $x^\ell \in \mathscr{F}_{t_\ell}(x)$ and $x^\ell \to x_*$. We may assume that $t_\ell \to t_*$. Since $\mathscr{F}_{t_\ell}(x) \to \mathscr{F}_{t_*}(x)$ in the Hausdorff metric, $d(x^\ell, \mathscr{F}_{t_*}(x)) \to 0$. There exist $z^\ell \in \mathscr{F}_{t_*}(x)$ such that $d(x^\ell, \mathscr{F}_{t_*}(x)) = d(x^\ell, z^\ell)$. We see that z^ℓ must converge to x_*, and hence $x_* \in \mathscr{F}_{t_*}(x)$. This proves part (e) and the proposition. □

The multi-valued map $\mathscr{F}_T : X \to 2^X$ can be extended to a multi-valued map on bounded sets as follows.

The sets $\mathscr{F}_T(B)$. For a bounded set $B \subset X$ define $\mathscr{F}_T(B)$ as the set of all possible limits of sequences of the form $S_{w_k[n_k]}^{h_k[n_k]}(x_k)$, where $x_k \in B$, $n_k \nearrow \infty$, $w_k \in \Sigma_{\mathscr{G}}$, and $h_k \in \Sigma_{\mathscr{G}}$ such that $\lim_k$ *max-mesh* $(h_k[n_k]) = 0$. For $T = 0$ define $\mathscr{F}_0(B) = \overline{B}$.

Here are a few basic properties of the map $B \mapsto \mathscr{F}_T(B)$.

Proposition 2 *(a) If* $A \subset B$*, then* $\mathscr{F}_T(A) \subset \mathscr{F}_T(B)$.
(b) $\mathscr{F}_T(\overline{B}) = \mathscr{F}_T(B)$.
(c) If A is closed and bounded and $y \in \mathscr{F}_T(A)$*, then there exists an* $x \in A$ *such that* $y \in \mathscr{F}_T(x)$*. As a corollary,* $\mathscr{F}_T(A) = \cup_{x\in A}\mathscr{F}_T(x)$.
(d) For every bounded B the set $\mathscr{F}_T(B)$ *is compact.*
(e) For any $T_1, T_2 \ge 0$ *we have* $\mathscr{F}_{T_1}\left(\mathscr{F}_{T_2}(B)\right) = \mathscr{F}_{T_1+T_2}(B)$.

Proof The monotonicity is obvious and so is the inclusion $\mathscr{F}_T(B) \subset \mathscr{F}_T(\overline{B})$. Suppose $x \in \mathscr{F}_T(\overline{B})$. This means

$$x = \lim_k S_{w_k[n_k]}^{h_k[n_k]}(x_k)$$

where $x_k \in \overline{B}$. If $x_k = \lim_\ell x_k^\ell$, where $x_k^\ell \in B$, then for every k there is an $\ell = \ell_k$ such that

$$d\left(S_{w_k[n_k]}^{h_k[n_k]}(x_k), S_{w_k[n_k]}^{h_k[n_k]}(x_k^\ell)\right) \le 2^{-k}.$$

This implies $x \in \mathscr{F}_T(B)$.

In the setting of part (c), write y as the limit $y = \lim_k S_{w_k[n_k]}^{h_k[n_k]}(x_k)$, where $x_k \in A$. Sparse the sequence x_k to make it convergent, and let $x = \lim_k x_k$. Invoking Assumption **A4** we see that $y = \lim_k S_{w_k[n_k]}^{h_k[n_k]}(x)$, and hence $y \in \mathscr{F}_T(x)$.

The set $\mathscr{F}_T(B)$ is bounded. To show this it suffices to show that for every point $x \in B$ there is a ball $\mathscr{O}_\delta(x)$ around it such that its trajectory is bounded (then we assume that B is closed and cover B by a finite number of such balls). We claim that for every $x \in X$ and for every $T > 0$, there exists a $\delta > 0$ and there exists a constant $\tilde{C} = \tilde{C}(x, \delta, T)$ such that for any integer $n > 0$, any $h \in \Sigma_{\mathscr{G}}$ such that $\|h[n]\| \le T$, and any $w \in \Sigma_{\mathscr{G}}$,

$$d(y, S_{w[k]}^{h[k]}(y)) \le \tilde{C}, \quad \text{for} \quad 0 \le k \le n,$$

for any y in the δ-neighborhood of x. If the claim is wrong, then there exist x and T such that for every integer $m > 0$ one can find n_m, h_m with $\|h_m[n_m]\| \leq T$, w_m, and $y_m \in \mathcal{O}_{1/m}(x)$ such that

$$d(y_m, S_{w_m[k_m]}^{h_m[k_m]}(y_m)) > m$$

for some $k_m \leq n_m$. However, this is impossible because one has

$$d(y_m, S_{w_m[k_m]}^{h_m[k_m]}(y_m)) \leq d(y_m, x) + d(x, S_{w_m[k_m]}^{h_m[k_m]}(x)) + d(S_{w_m[k_m]}^{h_m[k_m]}(x), S_{w_m[k_m]}^{h_m[k_m]}(y_m)),$$

where the first term goes to 0 as $m \to \infty$ because $y_m \to x$, the last term is small for large m due to **A4**, and the middle term is bounded due to **A2**.

The set $\mathcal{F}_T(B)$ is closed. This can be checked by constructing a representation similar to (9). This proves our assertion (d).

Next we show that $\mathcal{F}_{T_2}\left(\mathcal{F}_{T_1}(B)\right) \subset \mathcal{F}_{T_1+T_2}(B)$. Let $y \in \mathcal{F}_{T_2}\left(\mathcal{F}_{T_1}(B)\right)$. Then $y = \lim_\ell S_{u^\ell[m^\ell]}^{s^\ell[m^\ell]}(y^\ell)$ with $y^\ell \in \mathcal{F}_{T_1}(B)$, $\|s^\ell[m^\ell]\| = T_2$, and *max-mesh* $(s^\ell[m^\ell]) \to 0$. In turn, every y^ℓ is a limit of the form $y^\ell = \lim_k z_k^\ell$ where $z_k^\ell = S_{w_k^\ell[n_k^\ell]}^{h_k^\ell[n_k^\ell]}(y_k^\ell)$, $y_k^\ell \in B$ and $\|h_k^\ell[n_k^\ell]\| = T_1$. Given an $\varepsilon > 0$, use the triangle inequality

$$d(y, S_{u^\ell[m^\ell]}^{s^\ell[m^\ell]}(z_k^\ell)) \leq d(y, S_{u^\ell[m^\ell]}^{s^\ell[m^\ell]}(y^\ell)) + d\left(S_{u^\ell[m^\ell]}^{s^\ell[m^\ell]}(y^\ell), S_{u^\ell[m^\ell]}^{s^\ell[m^\ell]}(z_k^\ell)\right),$$

choose ℓ to make the first term less than $\varepsilon/2$, and then choose k to make the second term also less than $\varepsilon/2$. This proves that $y \in \mathcal{F}_{T_1+T_2}(B)$.

To finish the proof of part (e), pick $y \in \mathcal{F}_{T_1+T_2}(B)$. Writing y as a limit of the end-points of switched trajectories can be always done in the form

$$y = \lim_k S_{u_k[m_k]}^{s_k[m_k]} \circ S_{w_k[n_k]}^{h_k[n_k]}(x_k),$$

where $x_k \in B$, $\|h_k[n_k]\| = T_{1,k}$, $\|s_k[m_k]\| = T_{2,k}$, $T_{1,k} + T_{2,k} = T_1 + T_2$, and $\lim_k T_{1,k} = T_1$, $\lim_k T_{2,k} = T_2$. Also, *max-mesh* $(h_k[n_k]) \to 0$ and *max-mesh* $(s_k[m_k]) \to 0$. Denote $z_k = S_{w_k[n_k]}^{h_k[n_k]}(x_k)$. We may assume B is closed and the sequence x_k converges to some $x_* \in B$. By Assumption **A4**, $\lim_k d(z_k, S_{w_k[n_k]}^{h_k[n_k]}(x_*)) = 0$. In turn, we may pass to a subsequence and use Proposition 1 part (b) to conclude that the (sub)sequence $S_{w_k[n_k]}^{h_k[n_k]}(x_*)$ converges to some point $z_* \in \mathcal{F}_{T_1}(x_*) \subset \mathcal{F}_{T_1}(B)$. Thus, (on this subsequence) $z_k \to z_* \in \mathcal{F}_{T_1}(x_*)$. Now,

$$d(y, S_{u_k[m_k]}^{s_k[m_k]}(z_*)) \leq d(y, S_{u_k[m_k]}^{s_k[m_k]}(z_k)) + d(S_{u_k[m_k]}^{s_k[m_k]}(z_k), S_{u_k[m_k]}^{s_k[m_k]}(z_*)).$$

The first term on the right goes to 0 by definition of z_k, while the second term can be made arbitrarily small by **A4**. This completes the proof. □

3.2 Continuous Limit Dynamics

There are many ways one can think of continuous limit dynamics. For us, the CLD corresponding to the maps S_j^τ is generated by the multi-valued maps $\mathcal{F}_t$ on the space of closed bounded subsets of X.

Denote by $X^\sharp$ the space of compact (= closed and bounded, in our situation) subsets of X equipped with the Hausdorff metric. It is well know that $X^\sharp$ is a complete metric space, [8]. The maps $\mathcal{F}_t$ are defined on $X^\sharp$.

Lemma 4 *Each map $\mathcal{F}_T$ is continuous as a map from $X^\sharp$ to $X^\sharp$.*

Proof Let A_n be a convergent sequence in $X^\sharp$, and let $A = \lim_n A_n$. We will show that $\mathcal{F}_T(A_n) \to \mathcal{F}_T(A)$. First show $e(\mathcal{F}_T(A_n), \mathcal{F}_T(A)) \to 0$. For this we show that any sequence $y^\ell \in \mathcal{F}_T(A_\ell)$ has a subsequence convergent to a point in $\mathcal{F}_T(A)$. That y^ℓ has a convergent subsequence follows from the fact that the set $B_1 = \bigcup_\ell A_\ell$ is bounded, and hence, $\mathcal{F}_T(B_1)$ is compact by Proposition 2(d), and $\mathcal{F}_T(A_\ell) \subset \mathcal{F}_T(B_1)$. Let y be the limit of this (sub)sequence y^ℓ. As we know from Proposition 2(c), for each y^ℓ there exists an $x^\ell \in A_\ell$ such that $y^\ell \in \mathcal{F}_T(x^\ell)$. Because $A_\ell \to A$, there is a subsequence of x^ℓ converging to some $x \in A$. Each y^ℓ can be represented as a limit,

$$y^\ell = \lim_k S_{w_k^\ell[n_k^\ell]}^{h_k^\ell[n_k^\ell]}(x^\ell).$$

Choose $k = k_\ell$ so that $d(y^\ell, S_{w_k^\ell[n_k^\ell]}^{h_k^\ell[n_k^\ell]}(x^\ell)) \leq 2^{-\ell}$. Then

$$S_{w_k^\ell[n_k^\ell]}^{h_k^\ell[n_k^\ell]}(x^\ell)\big|_{k=k_\ell} \underset{\ell\to\infty}{\to} y.$$

Because $x^\ell \to x$, we have (see **A4**) $S_{w_k^\ell[n_k^\ell]}^{h_k^\ell[n_k^\ell]}(x)\big|_{k=k_\ell}$ arbitrarily close to y as well. This proves that $y \in \mathcal{F}_T(x) \subset \mathcal{F}_T(A)$.

Next, we show $e(\mathcal{F}_T(A), \mathcal{F}_T(A_n)) \to 0$. For this it will suffice to show that $d(z, \mathcal{F}_T(A_n))$ goes to 0 for any $z \in \mathcal{F}_T(A)$. There exists an $x \in A$ such that $z \in \mathcal{F}_T(x)$. Choose a sequence $x^\ell \in A_\ell$ converging to x. It will suffice to show that $\lim_\ell d(z, \mathcal{F}_T(x^\ell)) = 0$. Write z as a limit,

$$z = \lim_k S_{w_k[n_k]}^{h_k[n_k]}(x).$$

Successively for $\ell = 1, 2, \ldots$ sparse the sequence of k's to obtain convergent sequences $S_{w_k[n_k]}^{h_k[n_k]}(x^\ell) \to z^\ell$. The limits, z^ℓ, belong to the corresponding sets $\mathcal{F}_T(x^\ell)$. Choosing the diagonal (sub)sequence and using the Assumption **A4**, we see that $z^\ell \to z$, and we are done. □

Definition 3 The CLD corresponding to the maps S_j^τ is the continuous semi-dynamical system generated by the maps $\mathcal{F}_T$ on the hyperspace $X^\sharp$.

From a different angle, we may want to think of transition from variable time-step dynamics with choice to continuous limit dynamics as transition from switched trajectories to continuous "integral curves" where the switching is no longer so pronounced. We phrase this in the form of a generalization of the Filippov–Ważewski relaxation theorem.

Definition 4 For every $B \in X^{\sharp}$, define $\mathscr{C}\mathscr{L}\mathscr{D}_{[0,T]}(B)$ as the set of all continuous maps (integral curves) $\gamma : [0, T] \to X$ such that $\gamma(t) \in \mathscr{F}_t(B)$ for all $t \in [0, T]$ and

$$\gamma(t_1 + t_2) \in \mathscr{F}_{t_2}(\gamma(t_1)) \tag{15}$$

for all $t_1, t_2 \in [0, T]$ with $t_1 + t_2 \in [0, T]$. The set $\mathscr{C}\mathscr{L}\mathscr{D}_{[0,T]}(B)$ is a subset of $C([0, T], X)$.

For integral curves to make sense, we need additional assumptions on the space X. We require that any two points in X that are sufficiently close to each other could be connected by the unique "shortest curve". We do not have space to bring into discussion the metric length spaces and related notions. Here the "shortest curves" are (understood to be) assigned to pairs of points. If X is a (Montel) vector space, then the shortest curve is the line segment connecting the points. If X is a finite-dimensional Riemannian manifold, then the shortest curve is the geodesic (unique, because the points are close to each other). In any case, the metric, d, should satisfy the following condition.

AX For every compact set A the following is true: For every $\varepsilon > 0$ there exist $\delta > 0$ and $\rho > 0$ such that whenever x_0 and x_1 lie in A together with the shortest curve $\gamma : [0, 1] \to X$ connecting them and $d(x_0, x_1) \le \rho$, $d(\gamma(t), \gamma(s)) \le \varepsilon$ if $|t - s| \le \delta$.

In applications we have in mind, when X is a finite-dimensional Riemannian manifold or X is a Fréchet-Montel space, condition **AX** is satisfied.

Theorem 1 *Assume the condition* **AX** *is satisfied. Then the set* $\mathscr{C}\mathscr{L}\mathscr{D}_{[0,T]}(B)$ *is nonempty. It is a compact in* $C([0, T], X)$. *At the same time, for every* $t \in [0, T]$,

$$\{x \in X : x = \gamma(t) \text{ for some } \gamma \in \mathscr{C}\mathscr{L}\mathscr{D}_{[0,T]}(B)\} = \mathscr{F}_t(B). \tag{16}$$

Proof The integral curves in $\mathscr{C}\mathscr{L}\mathscr{D}_{[0,T]}(B)$ will be constructed as limits of "broken trajectories" or, as we call them, approximation curves. Consider a sequence p^N of partitions of the interval $[0, T]$ with the property that *max-mesh*$(p^N) \searrow 0$. Pick an $x_0 \in B$, define $x_0^N = x_0$ and then successively pick the points $x_{n+1}^N \in \mathscr{F}_{t_{n+1}^N - t_n^N}(x_n^N)$. Connect x_0^N with x_1^N, x_1^N with x_2^N, etc., by the shortest curves. This yields a (continuous) curve $\gamma^N(t)$. We call the points x_n^N the nodes and the corresponding γ^N the approximation curve. We claim that the sequence of the approximation curves γ^N has a convergent subsequence in the space of X-valued continuous functions on $[0, T]$ and that its limit belongs to $\mathscr{C}\mathscr{L}\mathscr{D}_{[0,T]}(B)$.

The sequence $\{\gamma^N\}_N$ is uniformly bounded in the sense that there is a constant $C > 0$ such that

$$\sup_{0 \le t \le T} d(x_0, \gamma^N(t)) \le C \tag{17}$$

for all N. This is because the set $\bigcup_t \mathscr{F}_t(x)$ is compact by Proposition 1(d), and all points $\gamma^N(t)$ lie in the convex hull of this set, which is bounded.

Next, we show that the family $\{\gamma^N\}_N$ is equicontinuous. Denote temporarily by A the closure of the convex hull (in the case X is a vector space) of $\bigcup_t \mathscr{F}_t(x_0)$,

$$A = \overline{co}\left(\bigcup_t \mathscr{F}_t(x_0)\right).$$

(If X is a manifold, and if r_0 is the number such that any two points of A_0 at distance $\le r_0$ are connected by a unique geodesic, then A denotes the closure of the union of the set $A_0 = \bigcup_t \mathscr{F}_t(x_0)$ with the shortest geodesics connecting points of A_0 which are at most at distance r_0 from each other.) This is a compact set. Our assumptions **A3** and **A4** imply the following: for every $x \in X$ and for every $\varepsilon > 0$ there exist $\delta > 0$ and $\rho > 0$ such that if $d(y, x) < \rho$ and $\|s*\| < \delta$, then $d(S^{s*}_{w*}(y), y) < \varepsilon$. Fix an $\varepsilon > 0$. Cover the set A by balls $\mathscr{O}_\rho(x)$ of radius $\rho = \rho(x, \varepsilon)$. Because A is compact, there is a finite subcover, and we choose the smallest δ of the corresponding $\delta(x, \varepsilon)$. For all large enough N, the consecutive nodes of the approximation curves γ^N will be at distance $\le \rho$ from each other, and we can use condition **AX** to assure that the curves γ^N are uniformly continuous. The set of values of the maps γ^N being a subset of A is totally bounded. Hence we can apply the Arzelà-Ascoli theorem, see, e.g., [13, Sect. 16.4] and obtain a subsequence γ^{N_k} converging in the space of continuous maps $C([0, T], X)$ to some γ. To show that $\gamma \in \mathscr{CLD}_{[0,T]}(B)$, we have to check (15).

For every t, $\gamma(t) \in \mathscr{F}_t(x_0)$. Indeed, consider the approximation curves γ^{N_k} (from the convergent subsequence) and let t_{j_k} be a point in the partition of $[0, T]$ such that $t_{j_k} \le t < t_{j_k+1}$. Both sequences, $\gamma^{N_k}(t_{j_k})$ and $\gamma^{N_k}(t_{j_k+1})$ converge to $\gamma(t)$. By the construction of approximation curves, and thanks to Proposition 2(e), $\gamma^{N_k}(t_{j_k}) \in \mathscr{F}_{t_{j_k}}(x)$. The points $\gamma^{N_k}(t_{j_k})$ can be approximated by sequences of the form $y_k^\ell = S^{h_k^\ell[n_k^\ell]}_{w_k^\ell[n_k^\ell]}(x)$ so that $\gamma^{N_k}(t_{j_k}) = \lim_\ell y_k^\ell$. Using diagonalization and Proposition 1(b), we see that $\gamma(t) \in \mathscr{F}_t(x_0)$. Now consider $\gamma(t + \tau)$. This point is a limit of the points of the form $\gamma^{N_k}(t_{j_k} + \tau_{j_k})$. By construction, $\gamma^{N_k}(t_{j_k} + \tau_{j_k}) \in \mathscr{F}_{\tau_{j_k}}(\gamma^{N_k}(t_{j_k}))$. We know that $\gamma^{N_k}(t_{j_k} + \tau_{j_k}) \to \gamma(t + \tau)$ and, by Proposition 1(d), $\mathscr{F}_{\tau_{j_k}}(\gamma^{N_k}(t_{j_k})) \to \mathscr{F}_{t+\tau}(x)$. This proves (15).

Next we show that every map γ in $\mathscr{CLD}_{[0,T]}(B)$ is a limit of approximation curves. If $\gamma \in \mathscr{CLD}_{[0,T]}(B)$ and p^N is a sequence of partitions of $[0, T]$ with decreasing max-mesh, we build the approximation curves γ^N by choosing the nodes $x_n^N = \gamma(t_n^N)$ and connecting them by the shortest curves. As before, $\gamma^N \to \gamma$ in $C([0, T], X)$.

The fact that $\mathscr{CLD}_{[0,T]}(B)$ is compact in $C([0,T],X)$ follows, because if γ_n is a sequence in $\mathscr{CLD}_{[0,T]}(B)$ we can build approximation curves γ_n^N so that $\sup_t d(\gamma_n^{N_n}(t), \gamma_n(t)) \to 0$, and then show as before that the Arzelà-Ascoli theorem is applicable to the family $\gamma_n^{N_n}$.

It remains to show that through every point $z \in \mathscr{F}_t(B)$ passes a whole curve $\gamma \in \mathscr{CLD}_{[0,T]}(B)$. Indeed, if $z \in \mathscr{F}_t(B)$, there exists an $x_0 \in B$ such that $z \in \mathscr{F}_t(x_0)$. As a result, z can be represented as a limit,

$$z = \lim_k S_{w_k[n_k]}^{h_k[n_k]}(x_0),$$

where we can assume that $\|h_k[N_k]\| = T$ for some N_k, and *max-mesh* $(h_k[N_k]) \to 0$ as $k \to \infty$. With k fixed, define the nodes $x_m^k = S_{w_k[m-1]}^{h_k[m-1]}(x_0)$ for $m = 1, \ldots, N_k$. Connect the nodes with the shortest curves. This will give us the approximation curve $\gamma_k(\cdot)$. As before, we can show that the Arzelà-Ascoli theorem is applicable to this family, and obtain a limiting integral curve γ. Since $\gamma_k(t) = z_k \to z$ by construction, we have $\gamma(t) = z$. The theorem is proved. □

4 Differential Inclusions and CLD

In this section we present an example of a Continuous Limit Dynamics for a switched system and identify it with the solution set of the relaxed differential inclusion.

4.1 *Switched Solutions*

Throughout this section $f_1, \ldots, f_N : \mathbb{R}^d \to \mathbb{R}^d$ will be an arbitrary but fixed set of Lipschitz continuous vector-fields. Let L be the maximum of their Lipschitz constants, i.e., for all $x, y \in \mathbb{R}^d$,

$$|f_j(x) - f_j(y)| \le L\,|x - y|, \quad j = 1, \ldots, N.$$

With each f_j we associate the system of ordinary differential equations

$$\dot{x}(t) = f_j(x(t)) \tag{18}$$

whose solutions define a flow on $\mathbb{R}^d$. We denote by S_j^τ the corresponding evolution map from $\mathbb{R}^d$ to $\mathbb{R}^d$, where $S_j^\tau(\xi) = x(\tau)$ is the value at time τ of the solution $x(t)$ of system (18) with the initial condition $x(0) = \xi$. The Lipschitz continuity of f_j guarantees that solutions of (18) are global in time and unique.

Now, let $\mathscr{J}$ be the finite set $\{1, 2, \ldots, N\}$ with discrete metric. This is the set of possible regimes. The basic set of possible dwell-times will be $\mathscr{T} = [0, 1]$. As before, we denote by $\Sigma_{\mathscr{I}}$ the space of regime switching functions, and denote by $\Sigma_{\mathscr{I}}$ the space of dwell time switching functions. For every choice of $w \in \Sigma_{\mathscr{I}}$ and $h \in \Sigma_{\mathscr{I}}$, and for every choice of initial condition ξ, define $x_{(w,h)}(t,\xi)$ as follows. Set $t_0 = 0$, $t_1 = h(0)$, $t_2 = t_1 + h(1), \ldots, t_{n+1} = t_n + h(n), \ldots$ Then, starting with $x_{(w,h)}(0,\xi) = \xi$ define successively

$$x_{(w,h)}(t,\xi) = S_{w(n)}^{t-t_n}(x_{(w,h)}(t_n)), \quad \text{if} \quad t_n \leq t \leq t_{n+1}.$$

Definition 5 A switched solution of a family of Eq. (18), $j = 1, \ldots, N$, (or, a switched trajectory of the flows $\{S_j\}$) is any vector-function of the form $x_{(w,h)}(t,\xi)$.

An equivalent definition: $x_{(w,h)}(t,\xi)$ is the unique continuous solution of the initial value problem

$$\dot{x}(t) = \sum_{j=1}^{N} \varphi_j(w,h;t)\, f_j(x(t)), \quad x(0) = \xi, \tag{19}$$

where

$$\varphi_j(w,h;t) = \begin{cases} 1, \text{ if } j = w(n) \text{ and } t_n \leq t < t_{n+1} \\ 0, \text{ otherwise} \end{cases} \tag{20}$$

The solution of (19) is understood as a continuous function $x(t)$ satisfying

$$x(t) = \xi + \int_0^t \sum_{j=1}^{N} \varphi_j(w,h;s)\, f_j(x(s))\, ds. \tag{21}$$

To assure that the solution is defined for all t we need to assume that

$$\|h\| = \sum_{n=0}^{\infty} h(n) = +\infty. \tag{22}$$

This representation is convenient for obtaining estimates on the switched solutions. In particular, since

$$x(t) - \xi = \int_0^t \sum_{j=1}^{N} \varphi_j(w,h;s)\, \big(f_j(x(s)) - f_j(\xi)\big)\, ds + \int_0^t \sum_{j=1}^{N} \varphi_j(w,h;s)\, f_j(\xi)\, ds,$$

we obtain

$$|x(t) - \xi| \leq t\, f(\xi) + L \int_0^t |x(s) - \xi|\, ds,$$

where

$$|f(\xi)| = \max_j |f_j(\xi)|. \tag{23}$$

This leads to the estimate

$$\int_0^t |x(s) - \xi|\, ds \le |f(\xi)|\, \frac{e^{Lt} - 1 - Lt}{L^2}$$

which implies

$$|x_{(w,h)}(t, \xi) - \xi| \le |f(\xi)|\, \frac{e^{Lt} - 1}{L} \tag{24}$$

and, hence,

$$|x_{(w,h)}(t, \xi)| \le |\xi| + |f(\xi)|\, \frac{e^{Lt} - 1}{L}. \tag{25}$$

If $x_1(t)$ and $x_2(t)$ are solutions corresponding to different initial conditions, ξ_1 and ξ_2, then

$$x_1(t) - x_2(t) = \xi_1 - \xi_2 + \int_0^t \sum_{j=1}^N \varphi_j(w, h; s)\, \big(f_j(x_1(s)) - f_j(x_2(s))\big)\, ds$$

which implies

$$|x_1(t) - x_2(t)| \le |\xi_1 - \xi_2| + L \int_0^t |x_1(s) - x_2(s)|\, ds$$

and we obtain

$$|x_{(w,h)}(t, \xi_1) - x_{(w,h)}(t, \xi_2)| \le |\xi_1 - \xi_2|\, \frac{e^{Lt} - 1}{L}. \tag{26}$$

Notation:

- $\mathcal{SW}$—the set of all switched solutions (associated with the maps $f_1, \ldots, f_N$),
- $\mathcal{SW}(\xi)$—the set of all switched solutions initiating at the point ξ,
- $\mathcal{SW}_{[0,T]}(\xi)$—restrictions to the time interval $[0, T]$ of all switched solutions initiating at the point ξ,
- $\mathcal{SW}(\xi; s)$—the set of values at $t = s$ of all switched solutions $x_{(w,h)}(t, \xi)$.

4.2 Continuous Limit Dynamics (CLD)

Starting with the maps S_j^τ we define the corresponding CLD following the recipe of the previous section. The maps S_j^τ obviously satisfy Assumption **A0**. Assumptions **A1**, **A2**, and **A3** are satisfied thanks to estimate (24). Assumption **A1** is satisfied because each $S_j^\tau(x)$ is continuous in τ and x, and the continuity is uniform when $\tau \in [0, 1]$ and x varies on a compact set. Assumptions **A2** and **A3** are satisfied because the right-hand side of (24) depends on ξ and t only. Assumption **A4** is a consequence of (26), and Assumption **AX** is valid since $X = \mathbb{R}^d$. Thus, Theorem 1 is applicable in our situation. Recall Eq. (9) of the multi-valued map $\mathscr{F}_t$. It is instructive to compare the set $\mathscr{F}_t(\xi)$ with the set $\mathscr{SW}(\xi; t)$. Note that $\mathscr{SW}(\xi; t) \subset \mathscr{F}_t(\xi)$ because one can chop any time interval $h(n)$ into arbitrarily small subintervals and use the semi-group property of the maps S_j^τ. In other words, $S_{w(n)}^{h(n)}(x) = S_{w(n)}^{h(n)/N} \circ S_{w(n)}^{h(n)/N} \circ \cdots \circ S_{w(n)}^{h(n)/N}(x)$, so we can replace w by a string in which every symbol of w is repeated N times, and replace h by a string where each $h(n)$ is replaced by N repetitions of $h(n)/N$. The end-point, $x_{(w,h)}(t, \xi)$ will not change. Now it is clear that the closure of $\mathscr{SW}(\xi; t)$ in $\mathbb{R}^d$ is equal to $\mathscr{F}_t(\xi)$,

$$\overline{\mathscr{SW}(\xi; t)} = \mathscr{F}_t(\xi). \tag{27}$$

In fact, the closure of the set $\mathscr{S}W_{[0,T]}(\xi)$ in the space $C([0, T], \mathbb{R}^d)$ yields exactly all integral curves in $\mathscr{CLD}_{[0,T]}(\xi)$,

$$\overline{\mathscr{S}W_{[0,T]}(\xi)} = \mathscr{CLD}_{[0,T]}(\xi). \tag{28}$$

This shows the connection between the switched solutions and the CLD as defined in the previous section.

4.3 *Differential Inclusions*

Now we turn to differential inclusion. Define the multi-valued map

$$F(x) = \{f_1(x), \ldots, f_N(x)\} \tag{29}$$

and consider the differential inclusion

$$\dot{x}(t) \in F(x(t)). \tag{30}$$

Recall that a solution of (30) with the initial condition $x(0) = \xi$ is an absolutely continuous (a.s. for short) function $x(t)$ such that $x(0) = \xi$ and $x(t) \in F(x(t))$ for a.e. t. See [1] for information on differential inclusions. For this particular map $F(x)$, the existence of solutions is not a problem because any solution of (18) is a solution of (30). We will be interested in the totality of solutions of (30), so, by analogy with the switched solutions, we define the sets $\mathscr{DI}$, $\mathscr{DI}(\xi)$, etc. In particular, $\mathscr{DI}_{[0,T]}(\xi)$ is the set of restrictions to the interval $[0, T]$ of all solutions to (30) with the initial condition $x(0) = \xi$. Along with (30), it is important to consider its relaxed version,

$$\dot{x}(t) \in \overline{co}\, F(x(t)) \tag{31}$$

where

$$\overline{co}\, F(x) = \overline{co}\,\{f_1(x), \ldots, f_N(x)\}, \tag{32}$$

and $\overline{co}$ stands for the closure of the convex hull. Again, any solution to (18) is a solution of (31). The totality of solutions of (31) is known to have much better properties than that of (30), see [1]. In particular, the set $\mathscr{RDI}_{[0,T]}(\xi)$ of restrictions to the interval $[0, T]$ of all solutions to (31) with the initial condition $x(0) = \xi$ is known to be compact in the space of continuous functions $C([0, T]; \mathbb{R}^d)$, [1, Theorem 2.2.1]. All solutions in $\mathscr{DI}$ and in $\mathscr{RDI}$ are bounded,

$$|x(t, \xi)| \leq |\xi| + |f(\xi)|\, \frac{e^{LT} - 1}{L}. \tag{33}$$

What is more important to us is that the set $\mathscr{DI}_{[0,T]}(\xi)$ is dense in $\mathscr{RDI}_{[0,T]}(\xi)$ by the Filippov–Ważewski theorem, [3].

It is clear that $\mathscr{SW}_{[0,T]}(\xi) \subset \mathscr{DI}_{[0,T]}(\xi)$. Next we show that the closure of $\mathscr{SW}_{[0,T]}(\xi)$ in the space of continuous functions $C([0, T], \mathbb{R}^d)$ is equal to $\mathscr{RDI}_{[0,T]}(\xi)$.

Theorem 2

$$\mathscr{CLD}_{[0,T]}(\xi) = \mathscr{RDI}_{[0,T]}(\xi). \tag{34}$$

Proof The plan is to show that every solution $\gamma(t)$ of the relaxed inclusion (31) with $\gamma(0) = \xi$ can be approximated uniformly on the finite interval $[0, T]$ by switched solutions. Theorem 1 then will complete the proof of (34).

Let $\gamma(t)$ be a solution of (31) with the initial condition $\gamma(0) = \xi$. Fix the interval $[0, T]$. It is known, see [1, 1.14 Corollary 1] that there exist bounded measurable functions α_j on $[0, T]$ such that

$$0 \leq \alpha_j(t) \leq 1, \quad j = 1, \ldots, N, \quad \text{and} \quad \sum_{j=1}^{N} \alpha_j(t) = 1 \quad \text{a.e. on } [0, T], \tag{35}$$

and

$$\gamma(t) = \xi + \int_0^t \sum_{j=1}^{N} \alpha_j(s)\, f_j(\gamma(s))\, ds, \quad \text{for all } t \in [0, T]. \tag{36}$$

Approximate each α_i by simple functions as follows. For each positive integer n define the averages

$$\beta_{i,n}^k = \frac{n}{T} \int_{k\frac{T}{n}}^{(k+1)\frac{T}{n}} \alpha_i(s)\, ds, \; k = 0, \ldots, n-1,$$

and define the simple functions

$$\beta_{i,n}(t) = \beta_{i,n}^k \quad \text{for} \quad \frac{kT}{n} \le t < \frac{(k+1)T}{n}. \tag{37}$$

Then $\sum_{i=1}^N \beta_{i,n}(t) = 1$ for all $t \in [0, T)$, and all n. Moreover, the functions $\beta_{i,n}$ converge to α_i in $L^1([0, T])$, i.e.,

$$\lim_{n\to\infty} \int_0^T |\beta_{i,n}(s) - \alpha_i(s)|\, ds = 0.$$

For each n denote by γ_n the unique solution of the corresponding equation

$$\gamma_n(t) = \xi + \int_0^t \sum_{j=1}^N \beta_{j,n}(s)\, f_j(\gamma_n(s))\, ds.$$

It is clear that $\gamma_n \in \mathscr{RDI}_{[0,T]}(\xi)$ and it is easy to show that $\gamma_n \to \gamma$ in $C([0, T], \mathbb{R}^d)$.

From now on we drop subscript n and work with $\gamma(t, \xi)$, the solution of the equation

$$\dot\gamma(t) = \sum_{i=1}^N \beta_i(t) f_i(\gamma(t)), \;\; x(0) = \xi, \tag{38}$$

where each β_i is a constant between 0 and 1 on the intervals $[kT/n, (k+1)T/n]$ and $\sum_i \beta_i(t) = 1$. We will approximate γ by switched solutions. To do this, we use Proposition 3 below which is motivated by the classical Lie product formula:

$$e^{t(A+B)} = \lim_{m\to\infty} \left(e^{\frac{t}{m}A}\, e^{\frac{t}{m}B}\right)^m, \tag{39}$$

where A and B are square matrices. This formula shows how to approximate the evolution function of the equation $\dot x = (A + B)x$ by the evolution functions of the equations $\dot x = Ax$ and $\dot x = Bx$. Proposition 3 is a similar in spirit result. We consider the equation

$$\dot x(t) = \sum_{i=1}^N b_i f_i(x), \tag{40}$$

where $b_1, b_2, \dots, b_N$ are positive real numbers such that $\sum_{i=1}^N b_i = 1$. This equation has the same form as Eq. (38) on each of the intervals $[kT/n, (k+1)T/n]$. To model this when dealing with (40), we consider a fixed interval $[0, \tau]$. Denote by $S^t_{\sum_{i=1}^N b_i f_i}$ the solution map for (40) and denote by S_i^t the solution maps of the individual equations $\dot x = f_i(x)$. The analogue of (39) is the following formula

$$S^t_{\sum_{i=1}^N b_i f_i} = \lim_{m\to\infty} \left(S_N^{b_N \frac{t}{m}} \circ S_{N-1}^{b_{N-1}\frac{t}{m}} \circ \cdots \circ S_1^{b_1 \frac{t}{m}}\right)^m, \tag{41}$$

where the equality is understood in the strong sense, i.e., on every $\xi \in \mathbb{R}^d$. This will be a corollary of a more general statement.

Given an integer $m > 0$, define $h_m \in \Sigma_{\mathcal{I}_1}$ as follows:

$$h_m(0) = b_1 \frac{\tau}{m}, \quad h_m(1) = b_2 \frac{\tau}{m}, \quad \ldots, h_m(N-1) = b_N \frac{\tau}{m}, \tag{42}$$

and, for $k \geq N$,

$$h_m(k) = h_m(k(N)), \tag{43}$$

where, to shorten the notation,

$$k(N) = k \mod N.$$

Also, define $w_m \in \Sigma_{\mathcal{I}}$ as follows:

$$w_m(k) = 1 + k(N).$$

Simplify the notation for the switched solution $x_{(w_m, h_m)}(t, \xi)$ to $x_m(t, \xi)$.

Proposition 3 *For any choice of $\xi, \eta \in \mathbb{R}^d$, there exists a constant $C(\xi, \eta)$ such that*

$$|\gamma(t, \xi) - x_m(t, \eta)| \leq e^{L\tau} |\xi - \eta| + \frac{C(\xi, \eta)}{m} \tag{44}$$

for all $t \in [0, \tau]$. In particular, the functions $x_m(\cdot, \xi)$ converge to $\gamma(\cdot, \xi)$ in $C([0, \tau], \mathbb{R}^d)$ as $m \to \infty$.

Proof Pick ξ and η in $\mathbb{R}^d$. We will compare $\gamma(t, \xi)$ with $x_m(t, \eta)$. We use the representations

$$\gamma(t, \xi) = \xi + \int_0^t \sum_{i=1}^N b_i f_i(\gamma(s, \xi))\, ds \tag{45}$$

and

$$x_m(t, \eta) = \eta + \int_0^t \sum_{i=1}^N \varphi_i(w_m, h_m, s)\, f_i(x_m(s, \eta))\, ds. \tag{46}$$

The bounds on these solutions are given in (25) and (33), and on the finite interval $[0, \tau]$ and with the fixed ξ, we get the bounds:

$$|\gamma(t, \xi)|, |x_m(t, \xi)| \leq |\xi| + |f(\xi)| \frac{e^{Lt} - 1}{L} \leq |\xi| + |f(\xi)| \frac{e^{L\tau} - 1}{L}, \quad 0 \leq t \leq \tau. \tag{47}$$

Also, using (24) and the Lipschitz property of f_j's, we obtain

$$\max_j |f_j(\gamma(t,\xi))|, |f_j(x_m(t,\xi))| \le |f(\xi)|\, e^{Lt} \le |f(\xi)|\, e^{L\tau}, \quad 0 \le t \le \tau. \tag{48}$$

We next estimate the difference $\gamma(t,\xi) - x_m(t,\eta)$. To handle the integral in (46), we use (42) and (43) to partition the interval $[0,\tau]$. Denote $n_m = \lfloor mt/\tau \rfloor$. Next, denote

$$\lambda_0 = 0, \quad \lambda_j = \sum_{i=1}^{j} b_i, \quad j = 1, \ldots, N.$$

Define

$$t_0 = 0, \quad t_{k+1} = t_k + h_m(k).$$

Using λ_j's we can re-write this as

$$t_k = \left(\lfloor \frac{k}{N} \rfloor + \lambda_{k(N)} \right) \frac{\tau}{m}, \quad k = 0, 1, \ldots. \tag{49}$$

The interval $[0,t]$ with $t \le \tau$ will contain $N\, n_m$ of whole subintervals $[t_k, t_{k+1})$ of the partition plus possibly a leftover interval, $[t_{L_m}, t]$. Re-write the integral in (46) using this partition and the fact that

$$\varphi_i(w_m, h_m, s) = \delta_{i,k(N)} \quad \text{for} \quad t_{k-1} \le s < t_k.$$

We have

$$\begin{aligned}
&\int_0^t \sum_{i=1}^{N} \varphi_i(w_m, h_m, s)\, f_i(x_m(s,\eta))\, ds \\
&\quad = \sum_{k=0}^{Nn_m - 1} \int_{t_k}^{t_{k+1}} \sum_{i=1}^{N} \varphi_i(w_m, h_m, s)\, f_i(x_m(s,\eta))\, ds \\
&\quad\quad + \int_{n_m\tau/m}^{t} \sum_{i=1}^{N} \varphi_i(w_m, h_m, s)\, f_i(x_m(s,\eta))\, ds
\end{aligned}$$

We use (48) and the fact that $t - n_m\tau/m \le \tau/m$ to estimate the last integral:

$$\left| \int_{n_m\tau/m}^{t} \sum_{i=1}^{N} \varphi_i(w_m, h_m, s)\, f_i(x_m(s,\eta))\, ds \right| \le |f(\eta)| e^{L\tau}\, (\tau/m). \tag{50}$$

Invoking (49) and change of variables, we obtain

$$\sum_{k=0}^{Nn_m-1}\int_{t_k}^{t_{k+1}}\sum_{i=1}^{N}\varphi_i(w_m,h_m,s)\,f_i(x_m(s))\,ds=\sum_{k=0}^{Nn_m-1}\int_{t_k}^{t_{k+1}}f_{k(N)}(x_m(s))\,ds$$
$$=\sum_{k=0}^{n_m-1}\sum_{i=1}^{N}\int_{(k+\lambda_{i-1})\tau/m}^{(k+\lambda_i)\tau/m}f_i(x_m(s))\,ds$$
$$=\sum_{k=0}^{n_m-1}\sum_{i=1}^{N}b_i\int_0^{\tau/m}f_i(x_m((k+\lambda_{i-1})\tau/m+b_is,\eta)\,ds.$$

Now turn to the integral in representation (45). Here we partition $[0,t]$ into the intervals of size τ/m plus what is left.

$$\int_0^t\sum_{i=1}^{N}b_i\,f_i(\gamma(s))\,ds$$
$$=\sum_{k=0}^{n_m-1}\sum_{i=1}^{N}b_i\int_0^{\tau/m}f_i(\gamma(k\tau/m+s))\,ds+\int_{n_m\tau/m}^{t}\sum_{i=1}^{N}b_i\ f_i(\gamma(s))\,ds.$$

Again,

$$|\int_{n_m\tau/m}^{t}\sum_{i=1}^{N}b_i\ f_i(\gamma(s,\xi))\,ds|\le|f(\xi)|e^{L\tau}\,(\tau/m).\tag{51}$$

Thus,

$$|\gamma(t,\xi)-x_m(t,\eta)|\le|\gamma(n_m\tau/m,\xi)-x_m(n_m\tau/m,\eta)|+(|f(\xi)|+|f(\eta)|)\,e^{L\tau}\,(\tau/m).$$

Now we estimate the difference $\gamma(n\tau/m,\xi)-x_m(n\tau/m,\eta)$, where n is any integer between 0 and m. Write the difference as follows

$$\sum_{k=0}^{n-1}\sum_{i=1}^{N}b_i\int_0^{\tau/m}f_i(\gamma(k\tau/m+s,\xi))\,ds-\sum_{k=0}^{Nn-1}\int_{t_k}^{t_{k+1}}\sum_{i=1}^{N}\varphi_i(w_m,h_m,s)\,f_i(x_m(s,\eta))\,ds$$

which can be re-written as

$$\sum_{k=0}^{n-1}\sum_{i=1}^{N}b_i\int_0^{\tau/m}(f_i(\gamma(k\tau/m+s,\xi))-f_i(x_m((k+\lambda_{i-1})\tau/m+b_is,\eta))\,ds.$$

Its absolute value is not greater than

$$L \sum_{k=0}^{n-1} \sum_{i=1}^{N} b_i \int_0^{\tau/m} |\gamma(k\tau/m + s, \xi) - x_m((k + \lambda_{i-1})\tau/m + b_i s, \eta)| \, ds. \tag{52}$$

We use the triangle inequality

$$\begin{aligned} &|\gamma(k\tau/m + s) - x_m((k + \lambda_{i-1})\tau/m + b_i s)| \leq |\gamma(k\tau/m, \xi) - x_m(k\tau/m, \eta)| \\ &\quad + |\gamma(k\tau/m + s) - \gamma(k\tau/m)| + |x_m((k + \lambda_{i-1})\tau/m + b_i s) - x_m((k + \lambda_{i-1})\tau/m)| \\ &\quad + |x_m((k + \lambda_{i-1})\tau/m) - x_m(k\tau/m)|. \end{aligned}$$

The contribution of each of the last three terms on the right when integrated over the interval $[0, \tau/m]$ can be estimated using a simple Lipschitz continuity of the solutions valid simultaneously for $\gamma(t, \xi)$ and $x_m(t, \eta)$:

$$|\gamma(t_2, \xi) - \gamma(t_1, \xi)|, \; |x_m(t_2, \xi) - x_m(t_1, \xi)| \leq |f(\xi)| \, e^{L\tau} \, (t_2 - t_1), \;\; 0 \leq t_1 \leq t_2 \leq \tau. \tag{53}$$

Thus, we obtain

$$\int_0^{\tau/m} |\gamma(k\tau/m + s) - \gamma(k\tau/m)| \, ds \leq \frac{1}{2} |f(\xi)| \, e^{L\tau} \, (\tau/m)^2,$$

$$\int_0^{\tau/m} |x_m((k + \lambda_{i-1})\tau/m + b_i s) - x_m((k + \lambda_{i-1})\tau/m)| \, ds \leq \frac{1}{2} |f(\eta)| \, e^{L\tau} \, (\tau/m)^2,$$

and

$$\int_0^{\tau/m} |x_m((k + \lambda_{i-1})\tau/m) - x_m(k\tau/m)| \, ds \leq |f(\eta)| \, e^{L\tau} \, \lambda_{i-1} \, (\tau/m)^2 \leq |f(\eta)| \, e^{L\tau} \, (\tau/m)^2.$$

Collecting these inequalities to bound (52) we find that $|\gamma(n\tau/m, \xi) - x_m(n\tau/m, \eta)|$ is bounded by $|\xi - \eta|$ plus

$$L \sum_{k=0}^{n-1} \left(|\gamma(k\tau/m, \xi) - x_m(k\tau/m, \eta)| \, (\tau/m) + \left(\frac{1}{2} |f(\xi)| + \frac{3}{2} |f(\eta)| \right) e^{L\tau} \, (\tau/m)^2 \right).$$

Denote temporarily $z_k = |\gamma(k\,\tau/m, \xi) - x_m(k\,\tau/m, \eta)|$. For any $n = 1, \ldots, m$ we have the following inequalities:

$$z_n \leq z_0 + L\,(\tau/m) \sum_{k=0}^{n-1} z_k + \frac{3}{2} L\,(|f(\xi)| + |f(\eta)|) \, e^{L\tau} \, \tau^2/m.$$

Solving the discrete inequality

$$z_n \le z_0 + L\,(\tau/m) \sum_{k=0}^{n-1} z_k + C/m$$

we obtain

$$z_n \le (1 + L\,\tau/m)^{n+1}\, z_0 + (1 + L\,\tau/m)^n\, C/m$$

We conclude that

$$z_n \le (1 + L\,\tau/m)\, e^{L\tau}\, z_0 + \frac{3}{2}\, L\,(|f(\xi)| + |f(\eta)|)\, e^{2\,L\tau}\, \tau^2/m.$$

This completes the proof of the proposition. □

Proof of Theorem 2 **continued**. Now it is clear how to approximate the solution of (38) by switched solutions. There is a finite number of intervals of length $\tau = T/n$. We approximate γ by x_m choosing appropriate m on consecutive intervals. If $|\gamma(t, \xi) - x_{m_1}(t, \xi)| \le \varepsilon$ on the first interval, then we use estimate (44) and choose m_2 so that $|\gamma(t, \xi) - x_{m_2}(t, \xi)| \le e^{L\tau}\varepsilon + \varepsilon = (e^{L\tau} + 1)\varepsilon$ on the second interval. For the third interval we choose m_3 to have $|\gamma(t, \xi) - x_{m_3}(t, \xi)| \le (e^{L\tau} + 1)^2\varepsilon$ there. This way we construct a switched solution which is at most $(e^{L\tau} + 1)^n\varepsilon$ away from $\gamma(t)$ on the whole interval $[0, T]$. The theorem is proved. □

In the assumptions of the preceding theorem, consider the variable time-step dynamics with choice associated with the maps $S_j^\tau(x) = x + \tau\, f_j(x)$ (S_j^τ is the evolution map of the Euler method for the Eq. (18)). The corresponding switched trajectories are given by $x_{n+1} = S_{j_n}^{\tau_n}(x_n)$. Using the fact that the Euler method approximates the solutions of the constituent differential equations, and using Theorem 2, we conclude that the CLD for this variable time-step dynamics with choice coincides with the dynamics of the relaxed differential inclusion (31) and that equality (34) holds.

Acknowledgments This work has been completed during the fall 2012 program "Hamiltonians in Magnetic Field" at Institut Mittag-Leffler. LK thanks the organizers of the program and the leadership and staff of Institut Mittag-Leffler for their kind support and hospitality.

References

1. Aubin, J.P., Cellina, A.: Differential inclusions. Set-valued maps and viability theory. Grundlehren der Mathematischen Wissenschaften [Fundamental Principles of Mathematical Sciences], 264. Springer, Berlin (1984)
2. Branicky, M.: In: Hristu-Varsakelis, D., Levine, W.S. (eds.) Introduction to Hybrid Systems. Handbook of Networked and Embedded Control Systems, pp. 91–116. Birkhauser (2005)
3. Filippov, A. F.: Differential equations with discontinuous righthand sides. Translated from the Russian. Mathematics and its Applications (Soviet Series), 18. Kluwer Academic Publishers Group, Dordrecht (1988)

4. Goebel, R., Sanfelice, R.G., Teel, A.R.: Hybrid dynamical systems. IEEE Control Syst. Mag. **29**(2), 28–93 (2009)
5. Kapitanski, L., Zivanovic, S.: Dynamics with choice. Nonlinearity **22**, 163–186 (2009)
6. Kapitanski, L., Zivanovic, S.: Dynamics with a range of choice. Reliab. Comput. **15**, 290–299 (2011)
7. Kapitanski, L., Živanović Gonzalez, S.: Variable time-step dynamics with choice, submitted
8. Kuratowski, K.: Topology, vol. I. Academic Press, Polish Scientific Publishers, New York - London - Warszawa (1966)
9. Liberzon, D.: Switching in Systems and Control. Birkhäuser, Boston (2003)
10. Shorten, R., Wirth, F., Mason, O., Wulff, K., King, C.: Stability criteria for switched and hybrid systems. SIAM Rev. **49**(4), 545–592 (2007)
11. Van der Schaft, A., Schumacher, H.: An Introduction to Hybrid Dynamical Systems. Lecture Notes in Control and Information Sciences, vol. 251. Springer, Berlin (2000)
12. Zivanovic, S.: Attractors in Dynamics with Choice. PhD Thesis, University of Miami, Coral Gables, Fl, USA, (2009) Open Access Dissertations. Paper 210. http://scholarlyrepository.miami.edu/oa_dissertations/210
13. Zorich, V. A.: Mathematical Analysis II. Translated from the 2002 fourth Russian edition by Roger Cooke. Universitext. Springer, Berlin (2004)

Answers to Some Problems on Self-similar Sets and the Open Set Condition

Jakub Šotola

Abstract In their paper (Feng and Zhou, Anal Theory Appl 25:5–15, 2007) [3], Z. Zhou and L. Feng proposed some open problems concerning, among others, self-similar sets. In this paper we give the answer to two of them—the first one concerns identity of fixed points of generating systems of similitudes fulfilling the open set condition and the second one relates to connectedness and convexness of the so called feasible sets (we mean the open set from the open set condition).

Keywords Self-similar set · Open set condition · Feasible set · Finite clustering property · Fixed point

1 Introduction and Preliminaries

Zuoling Zhou and Li Feng have proposed some open problems in [3]. We answer positively to the following one:

Problem 1 Let E be a self-similar set satisfying the open set condition. Does there exist a generating system of contracting similitudes $\{\varphi_i\}_{i=1}^m$ such that

$$x_i \neq x_j, \quad i \neq j, \tag{1}$$

where x_i denotes the unique fixed point of the similitude φ_i?

The second problem concerns so called feasible sets. We will recall an older counterexample to the existence of a convex feasible set in order to answer this problem.

Now let us introduce the terminology and notation. A self-similar set is, roughly speaking, the union of its own smaller copies. The number of those copies is restricted to be greater than one, but finite.

J. Šotola (✉)
Silesian University in Opava, Na Rybníčku 1, 746 01 Opava, Czech Republic
e-mail: Jakub.Sotola@math.slu.cz

L. Alsedà i Soler et al. (eds.), *Difference Equations, Discrete Dynamical Systems and Applications*, Springer Proceedings in Mathematics & Statistics 180, DOI 10.1007/978-3-662-52927-0_20

Definition 1 Let m be an integer greater than 1 and let $\Phi = \{\varphi_i\}_{i=1}^m$ be a system of contractive similitudes on $\mathbb{R}^n$. Then the (unique) nonempty compact set E satisfying

$$E = \varphi(E) := \bigcup_{i=1}^{m} \varphi_i(E) \tag{2}$$

is called the *self-similar set*. The system Φ is called *the generating system* of E.

We denote the smaller copies of some set A in the following way

$$A_{\mathbf{i}} = A_{i_1 i_2 \dots i_k} := \varphi_{i_1 i_2 \dots i_k}(A) := \varphi_{i_1}(\varphi_{i_2}(\dots \varphi_{i_k}(A)\dots)) \tag{3}$$

and we call them *branches* with an *address* $\mathbf{i}$.

Two addresses of branches are *incomparable* if anyone of them is not prefix of the other.

Now we can easily describe the connection between a self-similar set and the space $\{1 \dots m\}^{\mathbb{N}}$, which is called the address of a point of a self-similar set.

Definition 2 Let E be a self-similar set. We say that a point $x \in E$ has an *address* $\mathbf{i} = i_1 i_2 \dots$, if

$$\{x\} = \bigcap_{k=1}^{\infty} E_{i_1 i_2 \dots i_k}. \tag{4}$$

Every point of a self-similar set has at least one address. But there are some points with multiple addresses, unless the branches are disjoint.

For example, the Sierpiński triangle generated by the similitudes

$$\psi_1(x,y) = \left(\frac{x}{2}, \frac{y}{2}\right),\ \psi_2(x,y) = \left(\frac{x}{2} + \frac{1}{2}, \frac{y}{2}\right),\ \psi_3(x,y) = \left(\frac{x}{2} + \frac{1}{4}, \frac{y}{2} + \frac{\sqrt{3}}{4}\right), \tag{5}$$

has points with two addresses in the intersection of any two smaller copies (branches) of itself (see Fig. 1).

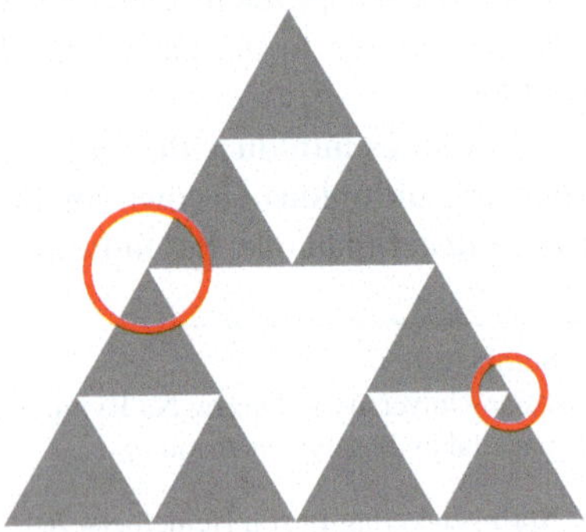

Fig. 1 Points with two addresses in Sierpiński triangle

The most of the known self-similar sets are those fulfilling the so called open set condition.

Definition 3 Let Φ be a generating system of some self-similar set E. We say that Φ satisfies the *open set condition* (OSC, for short), if there exists a nonempty bounded open set G such that

$$\varphi(G) \subset G \text{ and } \varphi_i(G) \cap \varphi_j(G) = \emptyset, \text{ for } i \neq j. \tag{6}$$

Such a set G is called a *feasible set.*

We say that a self-similar set E satisfies the OSC, if there exists a generating system of E satisfying the OSC.

2 Main Theorem

The following Theorem 1 solves the Problem 1. Its proof is based on a result by Ch. Bandt and H. Rao (see [2], Theorem 1(a)) concerning points with recurrent addresses.

Theorem 1 *Let $\Phi = \{\varphi_i\}_{i=1}^m$ be a generating system of a self-similar set E. Let x_i denote the unique fixed point of the similitude φ_i. If Φ satisfies the OSC, then the fixed points are pairwise distinct ($x_i \neq x_j$ for $i \neq j$).*

In the proof of this result we will use one of the conditions equivalent to the OSC (for a summary of more equivalents see [2], for the proof of Lemma 1 see [4], Theorem 2.2).

Lemma 1 (Finite clustering property) *Let Φ be a generating system of a self-similar set E. Then Φ satisfies the OSC if and only if there exists a positive integer N such that for every branch $E_{\mathbf{i}}$ of diameter ε there exist at most N branches $E_{\mathbf{j}}$ with incomparable addresses, diameter grater than ε and with distance from $E_{\mathbf{i}}$ smaller than ε.*

Proof (Proof of Theorem 1) Obviously, the fixed point x_i has address $iii\ldots$, because it has to lie in all branches $E_i,\ E_{ii},\ E_{iii},\ldots$.

Let us continue by contradiction. Let the point $x = x_i = x_j,\ i \neq j$ be the fixed point of two different generating similitudes. We will prove that the system Φ does not have the finite clustering property and hence, by Lemma 1, does not satisfy the OSC.

The point x has both addresses $iii\ldots$ and $jjj\ldots$. Let us choose a fixed positive integer N and let us denote by $\mathbf{i} := \underbrace{i\ldots i}_{a},\ a > N$ the prefix of the first address of the point x. Moreover let us denote by $\mathbf{i}_k := \underbrace{i\ldots i}_{a-k},\ k = 1,\ldots,N$, the prefix of $\mathbf{i}$.

To conclude the notions let

$$\mathbf{j}_k := \mathbf{i}_k \circ j \dots j, \tag{7}$$

where the number of j's is the biggest one such that the diameter of $E_{\mathbf{j}_k}$ is still greater or equal to the diameter of $E_{\mathbf{i}}$. Let us also suppose that the number a is big enough so that the number of j's is ever nonzero.

It is obvious that the point $\varphi_{\mathbf{i}_k}(x) = x$ lies in the intersection $E_{\mathbf{i}} \cap E_{\mathbf{j}_k}$ for every k. The addresses $\mathbf{j}_k$ are incomparable with the address $\mathbf{i}$, the branches $E_{\mathbf{j}_k}$ have diameter greater than the diameter of $E_{\mathbf{i}}$, but despite this they do intersect $E_{\mathbf{i}}$. So the proof is complete. □

Moreover, if we investigate the proof thoroughly, we can see that not only the identity of two fixed points contradicts the OSC, but also the identity of a fixed point with an address, say, $iii\dots$ and a point with an address $\mathbf{a}$ starting with a symbol different from i does. If you replace the address $jjj\dots$ with such an address, the proof will remain valid. The condition on the first symbol of $\mathbf{a}$ is essential. Since at least $\mathbf{j}_1$ would be equal to $\mathbf{i}$, if the condition was not fulfilled.

We may ask whether the inverse implication to the one in Theorem 1 is true. It is not hard to see, that it is not.

Let us take the following generating system: $\varphi_1(x) = x/3,\ \varphi_2(x) = x/3 + 2/3,\ \varphi_3(x) = 1 - x/3$. This system generates the Cantor middle-third set. Its fixed points are $x_1 = 0,\ x_2 = 1,\ x_3 = 3/4$, but the system does not satisfy the OSC.

3 A Counterexample to the Convexity of a Feasible Set

In [3], the authors stated that it would be useful to know whether the feasible set can be always chosen to be convex and they also ask whether there always exist a connected one. The answer is already known and is negative to both parts - there are generating systems with no connected (thus also not convex) feasible set.

The following example was presented by Ch. Bandt (see [1], Sect. 5), hence I call it the Bandt triangle.

Example 1 Let $E_B \subset \mathbb{R}^2$ be a self-similar set generated by the following similitudes

$$\psi_1(x,y) = \left(\frac{x}{2}, \frac{y}{2}\right),\ \psi_2(x,y) = \left(\frac{x}{2} + \frac{1}{2}, \frac{y}{2}\right), \psi_3(x,y) = \left(\frac{x}{2} + \frac{1}{4}, \frac{y}{2} + \frac{\sqrt{3}}{4}\right),$$
$$\psi_4(x,y) = \left(\frac{x}{2} + \frac{1}{4}, \frac{y}{2} + \frac{\sqrt{3}}{12}\right). \tag{8}$$

The only possible feasible set of this generating system was the interior of E_B minus some nowhere-dense set. From the construction demonstrated in the Fig. 2 it is quite obvious that such a feasible set cannot be even connected.

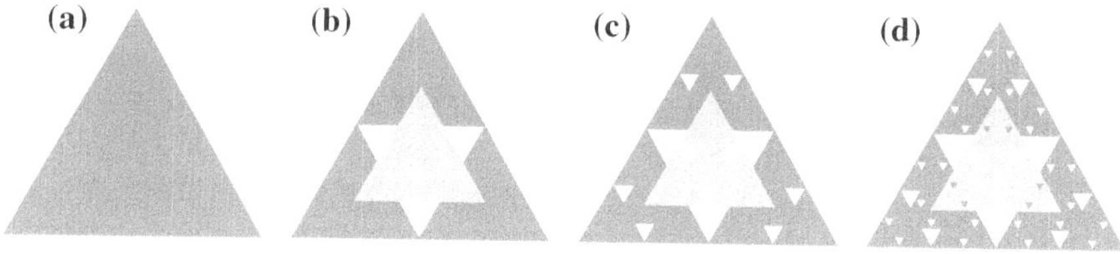

Fig. 2 The first four steps in construction of the Bandt triangle. **a** F. **b** $\psi(F)$. **c** $\psi^2(F)$. **d** $\psi^3(F)$

References

1. Bandt, Ch.: Self-similar sets 5. Integer matrices and fractal tilings of $\mathbb{R}^n$. Proc. Amer. Math. Soc. **112**, 549–562 (1991)
2. Bandt, Ch., Rao, H.: Topology and separation of self-similar fractals in the plane. Nonlinearity **20**, 1465–1474 (2007)
3. Feng, L., Zhou, Z.: Some problems on fractal geometry and topological dynamical systems. Anal. Theory Appl. **25**, 5–15 (2007)
4. Schief, A.: Separation properties of self-similar sets. Proc. Amer. Math. Soc. **122**, 111–115 (1994)

Taylor Domination, Difference Equations, and Bautin Ideals

Dmitry Batenkov and Yosef Yomdin

Abstract We compare three approaches to studying the behavior of an analytic function $f(z) = \sum_{k=0}^{\infty} a_k z^k$ from its Taylor coefficients. The first is "Taylor domination" property for $f(z)$ in the complex disk D_R, which is an inequality of the form

$$|a_k| R^k \leq C \max_{i=0,\dots,N} |a_i| R^i, \; k \geq N+1.$$

The second approach is based on a possibility to generate a_k via recurrence relations. Specifically, we consider linear non-stationary recurrences of the form

$$a_k = \sum_{j=1}^{d} c_j(k) \cdot a_{k-j}, \; k = d, d+1, \dots,$$

with uniformly bounded coefficients. In the third approach we assume that $a_k = a_k(\lambda)$ are polynomials in a finite-dimensional parameter $\lambda \in \mathbb{C}^n$. We study "Bautin ideals" I_k generated by $a_1(\lambda), \dots, a_k(\lambda)$ in the ring $\mathbb{C}[\lambda]$ of polynomials in λ. These three approaches turn out to be closely related. We present some results and questions in this direction.

Keywords Recurrence relations · Bautin ideals · Domination of initial Taylor coefficients

---

This author is supported by ISF, Grants No. 639/09 and 779/13, and by the Minerva foundation.

D. Batenkov
Department of Computer Science, Technion - Israel Institute of Technology,
32000 Haifa, Israel
e-mail: batenkov@cs.technion.ac.il

Y. Yomdin (✉)
Department of Mathematics, Weizmann Institute of Science,
76100 Rehovot, Israel
e-mail: yosef.yomdin@weizmann.ac.il

L. Alsedà i Soler et al. (eds.), *Difference Equations, Discrete Dynamical Systems and Applications*, Springer Proceedings in Mathematics & Statistics 180, DOI 10.1007/978-3-662-52927-0_21

1 Introduction

Let $f(z)$ be an analytic function represented in a disk D_R of radius R, centered at the origin by a convergent Taylor series $f(z) = \sum_{k=0}^{\infty} a_k z^k$. We assume that the Taylor coefficients a_k are explicitly known, or, at least, can be recovered by a certain explicit procedure. In many cases this is the only analytic information we possess on $f(z)$. A notorious example is the "Poincaré first return mapping" of a non-linear ordinary differential equation (see Sect. 4.2). We would like to investigate the behavior of $f(z)$ on the base of what we know, i.e. properties of the sequence a_k. In particular, we would like to bound from above the possible number of the zeroes of $f(z)$. In Sect. 2 below we provide an explicit connection between the two.

The main goal of this paper is to compare three approaches to the above question: the first is "Taylor domination", which is a bound on all the Taylor coefficients a_k of f through the first few of them. The second is a possibility to generate a_k via Difference Equations, specifically, by linear non-stationary homogeneous recurrence relations of a fixed length, with uniformly bounded coefficients. In the third approach we assume that $a_k = a_k(\lambda)$ are polynomials in a finite-dimensional parameter λ, and study "Bautin ideals" I_k generated by $a_1(\lambda), \ldots, a_k(\lambda)$ in the ring of polynomials in λ.

The main facts which we present are the following:

1. A sequence $a_0, a_1, \ldots$ can be obtained as a solution of a non-stationary linear homogeneous recurrence relations of a fixed length, with uniformly bounded coefficients, if and only if each its subsequence $a_m, a_{m+1}, \ldots$ possesses an appropriate Taylor domination property.
2. A sequence $a_1(\lambda), a_2(\lambda) \ldots,$ of polynomials in λ, under some natural assumptions, possesses a uniform in λ Taylor domination, with the parameters determined through the algebra of the Bautin ideals.
3. If a sequence $a_1(\lambda), a_2(\lambda) \ldots,$ of polynomials in λ is produced by *an algebraic* recurrence relation, its Bautin ideals can be computed explicitly. We discuss briefly the difficulties which arise for *differential-algebraic* recurrences (like in case of Poincaré mapping of Abel differential equation).

In our discussion we present some new specific results, and many known ones (some of them very recent, some pretty old). We believe that a general picture of interconnections between Taylor domination, recurrence relations, and Bautin ideals, given in this paper is new and may be instructive in further developments.

2 Taylor Domination and Counting Zeroes

"Taylor domination" for an analytic function $f(z) = \sum_{k=0}^{\infty} a_k z^k$ is an explicit bound of all its Taylor coefficients a_k through the first few of them. This property was classically studied, in particular, in relation with the Bieberbach conjecture, which asserts that for univalent f it always holds that $|a_k| \le k|a_1|$. See [7, 8, 18] and references therein. To give an accurate definition, let us assume the radius of convergence of the Taylor series for f is $\hat{R}$, $0 < \hat{R} \leqslant +\infty$.

Definition 2.1 Let a positive *finite* $R \le \hat{R}$, a natural N, and a positive sequence $S(k)$ of a subexponential growth be fixed. The function f is said to possess an $(N, R, S(k))$-Taylor domination property if for each $k \ge N + 1$ we have

$$|a_k|R^k \leqslant S(k) \max_{i=0,\dots,N} |a_i|R^i.$$

If $S(k) \equiv C$ we call this property (N, R, C)-Taylor domination.

The parameters $(N, R, S(k))$ of Taylor domination are not defined uniquely. In fact, the following easy result of [4] shows that each nonzero analytic function f possesses this property:

Proposition 2.1 (Proposition 1.1, [4]) *If* $0 < \hat{R} \leqslant +\infty$ *is the radius of convergence of* $f(z) = \sum_{k=0}^{\infty} a_k z^k$, *with* $f \not\equiv 0$, *then for each finite and positive* $0 < R \leqslant \hat{R}$, f *satisfies the* $(N, R, S(k))$*-Taylor domination property with* N *being the index of its first nonzero Taylor coefficient, and* $S(k) = R^k |a_k| (|a_N| R^N)^{-1}$, *for* $k > N$.

Consequently, the Taylor domination property becomes really interesting only for those *families* of analytic functions f where we can specify the parameters N, R, $S(k)$ in an explicit and uniform way. We concentrate on this problem below. Now we recall some well-known features of Taylor domination. Basically, it allows us to compare the behavior of $f(z)$ with the behavior of the polynomial $P_N(z) = \sum_{k=0}^{N} a_k z^k$. In particular, the number of zeroes of f can be easily bounded in this way. In one direction the bound is provided by the classical result of [8]. To formulate it, we need the following definition (see [18] and references therein):

Definition 2.2 A function f regular in a domain $\Omega \subset \mathbb{C}$ is called p-valent there, if for any $c \in \mathbb{C}$ the number of solutions in Ω of the equation $f(z) = c$ does not exceed p.

Theorem 2.1 (Biernacki, 1936, [8]) *If* f *is* p*-valent in the disk* D_R *of radius* R *centered at* $0 \in \mathbb{C}$ *then for each* $k \ge p + 1$

$$|a_k|R^k \le (A(p)k/p)^{2p} \max_{i=1,\dots,p} |a_i|R^i,$$

where $A(p)$ *is a constant depending only on* p.

In our notations, Theorem 2.1 claims that a function f which is p-valent in D_R, possesses a $(p, R, (Ak/p)^{2p})$-Taylor domination property.

For univalent functions, i.e. for $p = 1$, $R = 1$, Theorem 2.1 gives $|a_k| \le A(1)^2 k^2 |a_1|$ for each k, while the sharp bound of the Bieberbach conjecture is $|a_k| \le k|a_1|$.

Various forms of inverse results to Theorem 2.1 are known. In particular, an explicit bound for the number of zeroes of f possessing Taylor domination can be obtained by combining Proposition 2.1 and Lemma 2.2 from [26]:

Theorem 2.2 *Let the function f possess an $(N, R, S(k))$-Taylor domination property. Then for each $R' < R$, f has at most $M(N, \frac{R'}{R}, S(k))$ zeros in $D_{R'}$, where $M(N, \frac{R'}{R}, S(k))$ is a function depending only on N, $\frac{R'}{R}$ and on the sequence $S(k)$, satisfying $\lim_{\frac{R'}{R}\to 1} M = \infty$ and $M(N, \frac{R'}{R}, S) = N$ for $\frac{R'}{R}$ sufficiently small.*

We can replace the bound on the number of zeroes of f by the bound on its valency, if we exclude a_0 in the definition of Taylor domination (or, alternatively, if we consider the derivative f' instead of f).

3 Taylor Domination and Recurrence Relations

We start with a very general result obtained in [26]. Assume that a sequence Φ of mappings $\phi_k : \mathbb{C}^{d+1} \to \mathbb{C}$ is given, $k = d+1, d+2, \cdots$. For any $w = (w_0, \cdots, w_d) \in \mathbb{C}^{d+1}$ construct a sequence $a_k(w)$ as follows: $a_i = w_i$, $i = 0, \cdots, d$, and $a_j = \phi_j(w)$ for $j > d$. We also consider a power series $f_w(z) = \sum_{k=0}^{\infty} a_k(w) z^k$. Of course, any recurrence relation produces such a sequence Φ by iteration.

Assume also that each ϕ_k is a Lipschitzian mapping, satisfying

$$|\phi_k(w)| \le C^k |w| \tag{1}$$

for any $k \ge d+1$ and any $w \in B(0, \delta)$, with some given $\delta > 0$ and $C > 0$. This is the case in most of natural examples.

Theorem 3.1 (Theorem 4.1, [26]) *For C and δ as above and for any $w \in B(0, \delta)$, the series $f_w(x)$ converges on D_R, $R = 1/C$ and possesses there (d, R, K)-Taylor domination, with $K = [\max(1, C)]^d$.*

Next we restrict ourselves to linear recurrences, and produce more explicit bounds. We consider the class $\mathcal{S}$ of linear non-stationary homogeneous recurrence relations $\mathcal{R}$ of a fixed length, with uniformly bounded coefficients:

$$a_k = \sum_{j=1}^{d} c_j(k) \cdot a_{k-j}, \quad k = d, d+1, \ldots, \tag{2}$$

If for $j = 1, \ldots, d$ the coefficients $c_j(k)$ have a form $c_j(k) = c_j + \psi_j(k)$, with fixed c_j and with $\lim_{k\to\infty} \psi_j(k) = 0$, then recurrence relation (2) is said to be a *linear recurrence relation of Poincaré type* (see [22, 24]). We denote the class of such recurrences $\mathcal{S}_{\mathcal{P}}$.

We would like to write the bounds on $c_j(k)$ in a form

$$\left|c_j(k)\right| \leqslant K\rho^j, \qquad j = 1, \ldots, d, \; k = d, d+1, \ldots ,$$

for certain positive constants K, ρ. So for each $\mathcal{R} \in \mathcal{S}$ we define $K(\mathcal{R})$ and $\rho(\mathcal{R})$ to be the pair of constants providing the required bounds on $c_j(k)$, for which the product $\nu(\mathcal{R}) = (2K(\mathcal{R}) + 2) \cdot \rho(\mathcal{R})$ is minimal possible. We put $R(\mathcal{R}) \stackrel{\text{def}}{=} \nu(\mathcal{R})^{-1}$.

Theorem 3.2 (Theorem 3.1, [4]) *Let $\{a_k\}_{k=0}^{\infty}$ be a solution of the recurrence relation $\mathcal{R} \in \mathcal{S}$. Put $K = K(\mathcal{R})$, $\rho = \rho(\mathcal{R})$, $R = R(\mathcal{R})$. Then the series $f(z) = \sum_{k=0}^{\infty} a_k z^k$ converges in the open disk D_R and possesses there $(d-1, R, (2K+2)^{d-1})$ Taylor domination.*

By a proper rescaling, Theorem 3.2 can be easily extended to non-stationary linear recurrences with a subexponential (or exponential) growth of the coefficients $c_j(k)$. Consequently, generating functions of such recurrences allow for explicit bounds on their valency. On the other hand, a drawback of this approach is that in the case of linear recurrences with constant coefficients (and for Poincaré-type recurrences—see below) the disk D_R where the uniform Taylor domination is guaranteed, is much smaller than the true disk of convergence.

An important feature of Theorem 3.2 is that it allows us to provide an essentially complete characterisation of solutions of recurrence relations $\mathcal{R} \in \mathcal{S}$ through Taylor domination. The following result is new, although it follows closely the lines of Theorem 2.3 of [17]. Accordingly, we give only a sketch of the proof, referring the reader to [17] for details.

Theorem 3.3 *A sequence $\{a_k\}_{k=0}^{\infty}$ is a solution of the recurrence relation $\mathcal{R} \in \mathcal{S}$ of length d if and only if for each m its subsequence $\{a_{m+k}\}_{k=0}^{\infty}$ possesses a $(d-1, R, C)$ Taylor domination for some positive R and C.*

Proof In one direction the result follows directly from Theorem 3.2. Conversely, if for each m the subsequence $\{a_{m+k}\}_{k=0}^{\infty}$ possesses a $(d-1, R, C)$ Taylor domination, we use the corresponding bound on $|a_{m+d}|$ to construct step by step the coefficients $c_j(k)$, $j = 1, \ldots, d$ in $\mathcal{R}$, in such a way that they remain uniformly bounded in k. □

Let us now consider Poincaré-type recurrences. Characterization of their solutions looks a much more challenging problem than the one settled in Theorem 3.3. Still, one can expect deep connections with Taylor domination. One result in this direction (which is a sharpened version of Theorem 3.2), was obtained in [4]. For $\mathcal{R} \in \mathcal{S_P}$ the characteristic polynomial $\sigma^d - \sum_{j=1}^{d} c_j \sigma^{d-j} = 0$ and the characteristic roots $\sigma_1, \ldots, \sigma_d$ of $\mathcal{R}$ are those of its constant part. We put $\rho(\mathcal{R}) = \max_j |\sigma_j|$.

Theorem 3.4 (Theorem 5.3, [4]) *Let* $\{a_k\}_{k=0}^{\infty}$ *satisfy a fixed recurrence* $\mathcal{R} \in \mathcal{S}_{\mathcal{P}}$. *Put* $\rho \stackrel{def}{=} \rho(\mathcal{R}),\ R = 2^{-(d+3)}\rho^{-1}$. *Let* $\hat{N}$ *be the minimal of the numbers n such that for all* $k > n$ *we have* $|\psi_j(k)| \le 2^d \rho^j,\ j = 1, \dots, d$. *We put* $N = \hat{N} + d$, *and* $C = 2^{(d+3)N}$.

Then $\{a_k\}_{k=0}^{\infty}$ *possesses* (N, R, C)*-Taylor domination property.*

For Poincaré-type recurrences one can ask for Taylor domination in the maximal disk of convergence, which is typically ρ^{-1}. We discuss this problem in the next section.

3.1 Turán's Lemma, and a Possibility of Its Extension

It is well known that the Taylor coefficients a_k of a rational function $R(z) = \frac{P(z)}{Q(z)}$ of degree d satisfy a linear recurrence relation with constant coefficients

$$a_k = \sum_{j=1}^{d} c_j a_{k-j}, \quad k = d, d+1, \dots, \tag{3}$$

where c_j are the coefficients of the denominator $Q(z)$ of $R(z)$. Conversely, for any initial terms $a_0, \dots, a_{d-1}$ the solution sequence of (3) forms a sequence of the Taylor coefficients a_k of a rational function $R(z)$ as above. Let $\sigma_1, \dots, \sigma_d$ be the characteristic roots of (3), i e. the roots of its characteristic equation $\sigma^d - \sum_{j=1}^{d} c_j \sigma^{d-j} = 0$.

Taylor domination property for rational functions is provided by the following theorem, obtained in [4], which is, essentially, equivalent to the "first Turán lemma" ([20, 27, 28], see also [21]):

Theorem 3.5 (Theorem 3.1, [4]) *Let* $\{a_j\}_{j=1}^{\infty}$ *satisfy recurrence relation* (3) *and let* $\sigma_1, \dots, \sigma_d$ *be its characteristic roots. Put* $R \stackrel{def}{=} \min_{i=1,\dots d} \left|\sigma_i^{-1}\right|$. *Then for each* $k \ge d$

$$|a_k|\, R^k \le\ Q(k,d) \max_{i=0,\dots,d-1} |a_i| R^i, \tag{4}$$

where $Q(k,d) = [2e(\frac{k}{d}+1)]^d$.

Theorem 3.5 provides a uniform Taylor domination for rational functions in their maximal disk of convergence D_R, in the strongest possible sense. Indeed, after rescaling to the unit disk D_1 the parameters of (4) depend only on the degree of the function, but not on its specific coefficients.

We consider a direct connection of Turán's lemma to Taylor domination, provided by Theorem 3.5 as an important and promising fact. Indeed, Turán's lemma has numerous applications and connections, many of them provided already in the first Turan's book [27]. It can be considered as a result on exponential polynomials, and in this form it was a starting point for many deep investigations in Harmonic Analysis, Uncertainty Principle, Analytic continuation, Number Theory (see [20, 21, 27, 28] and references therein). Recently some applications in Algebraic Sampling were obtained, in particular, estimates of robustness of non-uniform sampling of "spike-train" signals ([3, 16]). One can hope that apparently new connections of Turán's lemma with Taylor domination, presented in [4] and in the present paper, can be further developed.

A natural open problem, motivated by Theorem 3.5, is a possibility to extend uniform Taylor domination in the maximal disk of convergence D_R, as provided by Theorem 3.5 for rational functions, to wider classes of generating functions of Poincaré type recurrence relations. Indeed, for such functions the radius of convergence of the Taylor series is, essentially, the same as for the constant-coefficients recurrences—it is the inverse of one of the characteristic roots: for some σ_j

$$\limsup_{k\to\infty} \sqrt[k]{|a_k|} = |\sigma_j| .$$

O.Perron proved in [22] that this relation holds for a general recurrence of Poincaré type, but with an additional condition that $c_d + \psi_d(k) \neq 0$ for all $k \in \mathbb{N}$. In [23] M.Pituk removed this restriction, and proved the following result.

Theorem 3.6 (Pituk's extension of Perron's Second Theorem, [23]). *Let $\{a_k\}_{k=0}^{\infty}$ be any solution to a recurrence relation $\mathcal{R}$ of Poincaré class $\mathcal{S}_{\mathcal{P}}$. Then either $a_k = 0$ for $k \gg 1$ or*

$$\limsup_{k\to\infty} \sqrt[k]{|a_k|} = |\sigma_j| ,$$

where σ_j is one of the characteristic roots of $\mathcal{R}$.

This result implies the following:

Theorem 3.7 (Theorem 5.2, [4]) *Let $\{a_k\}_{k=0}^{\infty}$ be any nonzero solution to a recurrence relation $\mathcal{R}$ of Poincaré class $\mathcal{S}_{\mathcal{P}}$ with initial data $\bar{a}$, and let R be the radius of convergence of the generating function $f(z)$. Then necessarily $R > 0$, and in fact $R = |\sigma|^{-1}$ where σ is some (depending on $\bar{a}$) characteristic root of $\mathcal{R}$. Consequently, f satisfies $(d-1, R, S(k))$-Taylor domination with $S(k)$ as defined in Proposition 2.1.*

Taylor domination in the maximal disk of convergence provided by Theorem 3.7, is only partially effective. Indeed, the number $d-1$ and the radius R are as prescribed by the constant part of the recurrence. However, Proposition 2.1 only guarantees that the sequence $S(k) = R^k|a_k| \cdot (\max_{i=0}^{d-1} R^i|a_i|)^{-1}$ is of subexponential growth but gives no further information on it. We can pose a natural question in this direction. For a sequence $\Delta = \{\delta_k\}$ of positive numbers tending to zero, consider a subclass $\mathcal{S}_{\mathcal{P},\Delta}$ of $\mathcal{S}_{\mathcal{P}}$, consisting of $\mathcal{R} \in \mathcal{S}_{\mathcal{P}}$ with $|\psi_j(k)| \le \delta_k \cdot \rho(\mathcal{R})^j$, $j = 1, \ldots, d$, $k = d, d+1, \ldots$

Problem 1 *Do solutions of recurrence relations $\mathcal{R} \in \mathcal{S}_{\mathcal{P},\Delta}$ possess $(N, R, S(k))$-Taylor domination in the maximal disk of convergence D_R, with $S(k)$ depending only on d and Δ? Is this true for specific Δ, in particular, for $\Delta = \{1, \frac{1}{2}, \frac{1}{3}, ..., \}$, as it occurs in most of examples (solutions of linear ODE's, etc.)?*

Taking into account well known difficulties in the analysis of Poincaré-type recurrences, this question may be tricky. Presumably, it can be easier for Δ with $\sum_{k=1}^{\infty} \delta_k < \infty$.

Some initial examples in this direction were provided in [32], via techniques of Bautin ideals. In Sect. 4 below we indicate an interrelation of the Poincaré-type recurrences with the Bautin ideals techniques. Another possible approach may be via an inequality for consecutive moments of linear combinations of δ-functions, provided by Theorem 3.3 and Corollary 3.4 of [31]. This inequality is closely related to Turán's lemma (and to the "Turán third theorem" of [20, 27, 28]). It was obtained via techniques of finite differences which, presumably, can be extended to Stieltjes transforms of much wider natural classes of functions, like piecewise-algebraic ones.

One can consider some other possible approaches to the "Turán-like" extension of Taylor domination to the full disk of convergence D_R for the Poincaré-type recurrences. First, asymptotic expressions in [10, 23] may be accurate enough to provide an inequality of the desired form. If this is the case, it remains to get explicit bounds in these asymptotic expressions.

Second, one can use a "dynamical approach" to recurrence relation (2) (see [11, 14, 19, 25, 29] and references therein). We consider (2) as a non-autonomous linear dynamical system T. A "non-autonomous diagonalization" of T is a sequence $\mathcal{L}$ of linear changes of variables, bringing this system to its "constant model" T_0, provided by the limit recurrence relation (3).

If we could obtain a non-autonomous diagonalization $\mathcal{L}$ of T with an explicit bound on the size of the linear changes of variables in it, we could get the desired inequality as a pull-back, via $\mathcal{L}$, of the Turán inequality for T_0. There are indications that this approach may work in the classes $\mathcal{S}_{\mathcal{P},\Delta}$ with $\sum_{k=1}^{\infty} \delta_k < \infty$.

The following partial result in the direction of Problem 1 above was obtained in [4]. It provides Taylor domination *in a smaller disk*, but with explicit parameters, expressed in a transparent way through the constant part of $\mathcal{R}$, and through the size of the perturbations.

Theorem 3.8 (Corollary 5.1, [4]) *Let $\Delta = \{\delta_k\}$ be a sequence of positive numbers tending to zero. Define $\hat{N}(\Delta)$ as a minimal number n such that for $k > n$ we have $\delta_k \leq 2^d$. Then for each $\mathcal{R} \in \mathcal{S}_{\mathcal{P},\Delta}$, the solution sequences of $\mathcal{R}$ possess (N, R, C)-Taylor domination, where $N = \hat{N}(\Delta) + d$, $R = 2^{-(d+3)} \rho(\mathcal{R})^{-1}$, $C = 2^{(d+3)N}$.*

3.2 An Example: D-Finite Functions

In this section we briefly summarize results of [4], concerning a certain class of power series, defined by the Stieltjes integral transforms

$$f(z) = S_g(z) = \int_a^b \frac{g(x)\,dx}{1 - zx}, \tag{5}$$

where $g(x)$ belongs to the class $\mathcal{PD}$ of the so-called *piecewise D-finite functions* [1], which are solutions of linear ODEs with polynomial coefficients, possessing a finite number of discontinuities of the first kind.

Using the expansion $(1 - zx)^{-1} = \sum_{k=0}^{\infty} (zx)^k$ for $|z| < \frac{1}{|x|}$, we obtain the following useful representation of $S_g(z)$ through the moments of g:

$$S_g(z) = \sum_{k=0}^{\infty} m_k z^k, \quad \text{where } m_k \stackrel{\text{def}}{=} \int_a^b x^k g(x)\,dx.$$

Obtaining uniform Taylor domination for S_g, where g belongs to particular subclasses of $\mathcal{PD}$ (in particular, g being piecewise algebraic), is an important problem with direct applications in Qualitative Theory of ODEs (see [9, 12] and references therein).

Definition 3.1 A real-valued bounded integrable function $g : [a, b] \to \mathbb{R}$ is said to belong to the class $\mathcal{PD}(\mathfrak{D}, p)$ if it has $0 \leqslant p < \infty$ discontinuities (not including the endpoints a, b) of the first kind, and between the discontinuities it satisfies a linear homogeneous ODE with polynomial coefficients $\mathfrak{D}\, g = 0$, where

$$\mathfrak{D} = \sum_{j=0}^{n} p_j(x) \left(\frac{d}{dx}\right)^j, \quad p_j(x) = \sum_{i=0}^{d_j} a_{i,j} x^i.$$

Let $g \in \mathcal{PD}(\mathfrak{D}, p)$, with $\mathfrak{D}$ as above. Denote the discontinuities of g by $a = x_0 < x_1 < \cdots < x_p < x_{p+1} = b$. In what follows, we shall use some additional notation. Denote for each $j = 0, \ldots, n$, $\alpha_j \stackrel{\text{def}}{=} d_j - j$. Let $\alpha \stackrel{\text{def}}{=} \max_j \alpha_j$.

Our approach in [4] is based on the following result:

Theorem 3.9 ([1]) *Let $g \in \mathcal{PD}(\mathfrak{D}, p)$. Then the moments $m_k = \int_a^b g(x)\,dx$ satisfy the recurrence relation*

$$\sum_{\ell=-n}^{\alpha} q_\ell(k)\, m_{k+\ell} = \varepsilon_k, \quad k = 0, 1, \ldots, \tag{6}$$

where $q_\ell(k)$ are polynomials in k expressed through the coefficients of $\mathfrak{D}$, while ε_k in the right hand side are expressed through the values of the coefficients of $\mathfrak{D}$, and through the jumps of g, at the points $x_0 < x_1 < \cdots < x_p < x_{p+1}$.

The recurrence (6) is inhomogeneous, and the coefficient of the highest moment may vanish for some k. Accordingly, we first transform (6) into a homogeneous matrix recurrence.

Definition 3.2 The vector function $\mathbf{y}(k) : \mathbb{N} \to \mathbb{C}^n$ is said to satisfy a linear system of Poincaré type, if

$$\mathbf{y}(k+1) = (A + B(k))\,\mathbf{y}(k), \tag{7}$$

where A is a constant $n \times n$ matrix and $B(k) : \mathbb{N} \to \mathbb{C}^{n \times n}$ is a matrix function satisfying $\lim_{k\to\infty} \|B(k)\| = 0$.

Put $\tau \stackrel{\text{def}}{=} n(p+2)$. Now define the vector function $\mathbf{w}(k) : \mathbb{N} \to \mathbb{C}^{\alpha+n+\tau}$ as

$$\mathbf{w}(k) \stackrel{\text{def}}{=} \begin{bmatrix} m_{k-n} \\ \vdots \\ m_{k+\alpha-1} \\ \varepsilon_k \\ \vdots \\ \varepsilon_{k+\tau-1} \end{bmatrix}.$$

Proposition 3.1 *The vector function* $\mathbf{w}(k)$ *satisfies a linear system of the form* (7). *This system is of Poincaré type, if and only if*

$$\alpha_n \geqslant \alpha_j \qquad j = 0, 1, \dots, n.. \tag{8}$$

This last condition is equivalent to the operator $\mathfrak{D}$ *having at most a regular singularity at* $z = \infty$. *The set* Z_A *of the eigenvalues of the matrix* A *is precisely the union of the roots of* $p_n(x)$ *(i.e. the singular points of the operator* $\mathfrak{D}$*) and the jump points* $\{x_i\}_{i=0}^{p+1}$.

Now we establish in [4] Taylor domination for the Stieltjes transform $S_g(z)$, combining two additional results: the first is the system version of Theorem 3.6.

Theorem 3.10 ([23]) *Let the vector* $\mathbf{y}(k)$ *satisfy the perturbed linear system of Poincaré type* (7). *Then either* $\mathbf{y}(k) = \mathbf{0} \in \mathbb{C}^n$ *for* $k \gg 1$ *or*

$$\lim_{k\to\infty} \sqrt[k]{\|\mathbf{y}(k)\|}$$

exists and is equal to the modulus of one of the eigenvalues of the matrix A.

Next main problem is: *how many first moments* $\{m_k\}$ *can vanish for a nonzero* $g \in \mathcal{PD}(\mathfrak{D}, p)$?. N in Taylor domination cannot be smaller than this number. In [2] we study this question, proving the following result.

Theorem 3.11 ([2]) *Let the operator $\mathfrak{D}$ be of Fuchsian type (i.e. having only regular singular points, possibly including ∞). In particular, $\mathfrak{D}$ satisfies the condition* (8). *Let $g \in \mathcal{PD}(\mathfrak{D}, p)$.*

1. *If there is at least one discontinuity point $\xi \in [a, b]$ of g at which the operator $\mathfrak{D}$ is nonsingular, i.e. $p_n(\xi) \neq 0$, then vanishing of the first $\tau + d_n - n$ moments $\{m_k\}_{k=0}^{\tau+d_n-n-1}$ of g implies $g \equiv 0$.*
2. *Otherwise, let $\Lambda(\mathfrak{D})$ denote the largest positive integer characteristic exponent of $\mathfrak{D}$ at the point ∞. In fact, the indicial equation of $\mathfrak{D}$ at ∞ is $q_\alpha(k) = 0$. Then the vanishing of the first $\Lambda(\mathfrak{D}) + 1 + d_n - n$ moments of g implies $g \equiv 0$.*

With these theorems in place, the following result is obtained in [4]:

Theorem 3.12 *Let $g \in \mathcal{PD}(\mathfrak{D}, p)$ be a not identically zero function, with $\mathfrak{D}$ of Fuchsian type. Then the Stieltjes transform $S_g(z)$ is analytic at the origin, and the series*

$$S_g(z) = \sum_{k=0}^{\infty} m_k z^k$$

converges in a disk of radius R which satisfies

$$R \geqslant R^* \stackrel{def}{=} \min\left\{\xi^{-1} : \xi \in Z_A\right\},$$

where Z_A is given by Proposition 3.1. Furthermore, for every

$$N \geqslant \max\{\tau - 1, \Lambda(\mathfrak{D})\} + d_n - n,$$

S_g satisfies $(N, R, S(k))$ Taylor domination, where $S(k)$ is given by Proposition 2.1.

4 Bautin Ideals

In this section we consider families

$$f_\lambda(z) = \sum_{k=0}^{\infty} a_k(\lambda) z^k, \tag{9}$$

where $\lambda = (\lambda_1, \ldots, \lambda_n) \in \mathbb{C}^n$. We assume that each $a_k(\lambda)$ is a polynomial in λ. It was a remarkable discovery of N. Bautin ([5, 6]) that in this situation the behavior of the Taylor coefficients of f_λ, and consequently, of its zeroes, can be understood in terms of the ideals $I_k = \{a_0(\lambda), \ldots, a_k(\lambda)\}$ generated by the subsequent Taylor coefficients $a_j(\lambda)$ in the polynomial ring $\mathbb{C}[\lambda]$. Explicit computation of I_k in specific examples may be very difficult. In particular, Bautin himself computed I_k for the Poincaré first return mapping of the plane polynomial vector field of degree 2, producing in this way one of the strongest achievements in the Hilbert 16th problem up to this day: at most three limit cycles can bifurcate from an isolated equilibrium point in vector field of degree 2.

Bautin's approach can be extended to a wide classes of families of the form (9). In particular, the following class was initially defined and investigated in [13, 15, 26], and further studied in [30, 32]:

Definition 4.1 Let $f_\lambda(z) = \sum_{k=0}^{\infty} a_k(\lambda) z^k$. f is called an A_0-series if the following condition is satisfied:

$$\deg a_k(\lambda) \leq K_1 k + K_2, \text{ and } ||a_k(\lambda)|| \leq K_3 \cdot K_4^k, \; k = 0, 1, \ldots,$$

for some positive K_1, K_2, K_3, K_4.

The ideal $I = \{a_0(\lambda), \ldots, a_k(\lambda), \ldots, \}$ generated by all the subsequent Taylor coefficients $a_j(\lambda)$ in the polynomial ring $\mathbb{C}[\lambda]$ is called the Bautin ideal of f, and the minimal d such that $I = I_d = \{a_0(\lambda), \ldots, a_d(\lambda)\}$ is called the Bautin index of f.

Such a finite d exists by the Noetherian property of the ring $\mathbb{C}[\lambda]$.

The following result of [15] connects the algebra of the Bautin ideal of f with Taylor domination for this series:

Theorem 4.1 *Let $f_\lambda(z) = \sum_{k=0}^{\infty} a_k(\lambda) z^k$ be an A_0-series, and let I and d be the Bautin ideal and the Bautin index of f. Then f possesses a (d, R, C)-Taylor domination for some positive R, C depending on K_1, K_2, K_3, K_4, and on the basis $a_0(\lambda), \ldots, a_d(\lambda)$ of the ideal I.*

Proof The idea of the proof is very simple: if for the A_0-series $f_\lambda(z) = \sum_{k=0}^{\infty} a_k(\lambda) z^k$ the first d Taylor coefficients $a_0(\lambda), \ldots, a_d(\lambda)$ generate the ideal $I = \{a_0(\lambda), \ldots, a_k(\lambda), \ldots, \}$, then for any $k > d$ we have

$$a_k(\lambda) = \sum_{i=0}^{d} \psi_i^k(\lambda) a_i(\lambda),$$

with $\psi_i^k(\lambda)$ - certain polynomials in λ. The classical Hironaka's division theorem provides a bound on the degree and the size of $\psi_i^k(\lambda)$, and hence we get an explicit bound on $|a_k(\lambda)|$ through $\max\{|a_0(\lambda)|, \ldots, |a_d(\lambda)|\}$ for each λ. □

Also here to get an explicit and uniform in λ Taylor domination in the full disk of convergence is a difficult problem. Some general additional conditions on A_0-series, providing such result, were given in [30]. These conditions are technically rather involved. A simpler special case was treated in [32].

4.1 Recurrence Relations and Bautin Ideals

A rather detailed investigation of A_0-series produced by recurrence relations was provided in [13]. It was done in several situations, including algebraic functions, linear and nonlinear differential equations with polynomial coefficients, and the Poincaré first return mapping for the Abel differential equation. In all these cases, except the last one, it was shown that the resulting series are A_0-ones, and their Bautin ideal was computed. In the present paper we generalize the result of [13], showing how to compute the Bautin ideal for general non-stationary polynomial recurrences. Let, as above, $\lambda = (\lambda_1, \dots, \lambda_n) \in \mathbb{C}^n$, and put $u = (u_1, \dots, u_d) \in \mathbb{C}^d$. Consider a sequence of polynomials

$$P_k(u_1, \dots, u_d) = \sum_{|\alpha| \le d_k} A_{k,\alpha}(\lambda) u^{\alpha},$$

where

$$A_{k,\alpha}(\lambda) = \sum_{|\beta| \le d_{k,\alpha}} A_{k,\alpha,\beta} \lambda^{\beta},$$

are, in their turn, polynomials in λ of degrees $d_{k,\alpha}$. Here $\alpha = (\alpha_1, \dots, \alpha_d) \in \mathbb{N}^d$, and $\beta = (\beta_1, \dots, \beta_n) \in \mathbb{N}^n$ are multi-indices, $|\alpha| = |\alpha_1| + \dots + |\alpha_d|$, $|\beta| = |\beta_1| + \dots + |\beta_n|$, $u^{\alpha} = u_1^{\alpha_1} \cdot \dots \cdot u_d^{\alpha_d}$, and $\lambda^{\beta} = \lambda_1^{\beta_1} \cdot \dots \cdot \lambda_n^{\beta_n}$. Finally, $A_{k,\alpha,\beta}$ are complex constants.

Definition 4.2 A recurrence of the form

$$a_k(\lambda) = P_k(a_{k-1}(\lambda), \dots, a_{k-d}(\lambda)), \ k = d, d+1, \dots \tag{10}$$

is called a non-stationary polynomial recurrence relation of length d.

The following result is purely algebraic, so we put no restrictions on the degrees and the size of the polynomials P_k.

Theorem 4.2 *Let polynomials $a_0(\lambda), \dots, a_{d-1}(\lambda)$ in λ be given, and let $a_k(\lambda)$, $k = d, d+1, \dots$ be produced by recurrence* (10). *Then the Bautin ideal I of the formal series $f_\lambda(z) = \sum_{k=0}^{\infty} a_k(\lambda) z^k$ is generated by $a_0(\lambda), \dots, a_{d-1}(\lambda)$.*

Proof Equation (10) applied with $k = d$ shows that $a_d(\lambda)$ belongs to the ideal I_{d-1} generated by $a_0(\lambda), \dots, a_{d-1}(\lambda)$. Applying this equation step by step, we see that all $a_k(\lambda)$, $k = d, d+1, \dots$ belong to I_{d-1}. Therefore $I = I_{d-1}$. □

We expect that if we assume the degrees d_k and $d_{k,\alpha}$, as well as the coefficients $A_{k,\alpha,\beta}$ to be uniformly bounded, then $f_\lambda(z)$ is, in fact, an A_0-series. We expect that this fact can be shown by the methods of [13], properly extended to non-stationary polynomial recurrence relation. Let us consider an example in this direction.

Let

$$P_k(u_1, \ldots, u_d) = \sum_{j=1}^{d} A_{k,j}(\lambda)u_j, \quad A_{k,j}(\lambda) = \sum_{i=1}^{n} A_{k,j,i}\lambda_i, \quad A_{k,j,i} \in \mathbb{C}.$$

We consider linear recurrence

$$a_k(\lambda) = \sum_{j=1}^{d} A_{k,j}(\lambda)a_{k-j}(\lambda). \tag{11}$$

Proposition 4.1 *Let polynomials* $a_0(\lambda), \ldots, a_{d-1}(\lambda)$ *of degrees* $0, 1, \ldots, d-1$ *in* λ *be given, and let* $a_k(\lambda)$, $k = d, d+1, \ldots$ *be produced by recurrence* (11), *with* $A_{k,j,i}$ *uniformly bounded in k. Then the formal series* $f_\lambda(z) = \sum_{k=0}^{\infty} a_k(\lambda)z^k$ *is, in fact, an* A_0 *series. In particular, the degree of* $a_k(\lambda)$ *is at most k, and the Bautin ideal I is generated by* $a_0(\lambda), \ldots, a_{d-1}(\lambda)$.

Proof Equation (11) shows that the degree in λ of a_k is by one higher than the maximum of the degrees of $a_{k-1}, \ldots, a_{k-d}$. Applying induction we see that the degree of $a_k(\lambda)$ is at most k. The last statement of Proposition 4.1 follows directly from Theorem 4.2. Now, write

$$a_k(\lambda) = \sum_{|\beta| \le k} a_{k,\alpha}\lambda^{\beta}.$$

Then Eq. (11) shows that $a_{k,\beta}$ satisfy a matrix recurrence relation, which in coordinates takes a form

$$a_{k,\beta} = \sum_{j=1}^{d}\sum_{i=1}^{n} A_{k,j,i}a_{k-j,\beta[i]}. \tag{12}$$

Here for $\beta = (\beta_1, \ldots, \beta_n)$ we define $\beta[i]$ as

$$\beta[i] = (\beta_1, \ldots, \beta_{i-1}, \beta_i - 1, \beta_{i+1}, \ldots, \beta_n).$$

So in the right hand side of (12) appear all the $a_{k',\beta'}$ with k' between $k-1$ and $k-d$, and β' smaller than β by one in exactly one coordinate.

Applying to the matrix recurrence (12) straightforward estimates, we obtain exponential in k upper bounds on the coefficients $a_{k,\beta}$, so $f_\lambda(z) = \sum_{k=0}^{\infty} a_k(\lambda)z^k$ is indeed an A_0 series. □

We expect that the Taylor domination bounds of Sect. 3 above, as applied to matrix recurrence (12) can be combined with algebraic consideration of the Bautin ideals. In this way one can hope to provide much more accurate parameters in Taylor domination than the general ones of Theorem 4.1.

4.2 *Poincaré Coefficients of Abel Equation*

Consider Abel differential equation

$$y' = p(x)y^2 + q(x)y^3 \tag{13}$$

with polynomial coefficients p, q on the interval $[a, b]$. A solution $y(x)$ of (13) is called "closed" if $y(a) = y(b)$. The Smale–Pugh problem, which is a version of the (second part of) Hilbert's 16th problem, is to bound the number of isolated closed solutions of (13) in terms of the degrees of p and q.

This problem can be naturally expressed in terms of the Poincaré "first return" mapping $y_b = G(y_a)$ along $[a, b]$. Let $y(x, y_a)$ denote the solution $y(x)$ of (13) satisfying $y(a) = y_a$. The Poincaré mapping G associates to each initial value y_a at a the value y_b at b of the solution $y(x, y_a)$ analytically continued along $[a, b]$. Closed solutions correspond to the fixed points of G. So the problem is reduced to bounding the number of the fixed points of G, or of zeroes of $G(y) - y$. Historically, one of the most successful directions in the study of the Poincaré mapping G was the direction initiated by Bautin in [5, 6]: to derive the analytic properties of G, in particular, the number of its fixed points, from the structure of its Taylor coefficients.

It is well known that $G(y)$ for small y is given by a convergent power series

$$G(y) = y + \sum_{k=2}^{\infty} v_k(p, q, a, b)y^k. \tag{14}$$

The "Poincaré coefficients" $v_k(p, q, x, a)$ of the Poincaré mapping from x to a satisfy the following differential-recurrence relation (see, for example, [12]):

$$\frac{dv_k}{dx} = -(k-1)pv_{k-1} - (k-2)qv_{k-2}, \;\; v_0 \equiv 0, \;\; v_1 \equiv 1, \;\; v_k(0) = 0, \; k \geq 2. \tag{15}$$

This recurrence is apparently not of the form considered above, i.e. it is not "polynomial recurrence". Still, it is easy to see from (15) that the Poincaré coefficients $v_k(p, q, b, a)$ are polynomials with rational coefficients in the parameters of the problem (i.e. in the coefficients of p and q). They can be explicitly computed for as large indices k as necessary. However, because of the derivative in the left hand side, (15) does not preserve ideals. So Theorem 4.2 is not applicable to it, and computing Bautin ideals for the Poincaré mapping $G(y)$ is in general a very difficult problem. Also Taylor domination for $G(y)$ is not well understood, besides some special examples. The only general result concerning the Bautin ideals of $G(y)$ we are aware of was obtained in [9] using an approximation of the "Poincaré coefficients" $v_k(p, q, b, a)$ with certain moment-like expressions of the form $m_k(p, q, b, a) = \int_a^b P^k(x)q(x)dx$, $P = \int p$. We would like to pose the investigation of the recurrence relation (15) in the lines of the present paper as an important open problem.

References

1. Batenkov, D.: Moment inversion problem for piecewise D-finite functions. Inverse Problems **25**(10), 105001 (2009)
2. Batenkov, D., Binyamini, G.: Moment vanishing of piecewise solutions of linear ODE's. arXiv:1302.0991. (Submitted to this volume)
3. Batenkov, D., Sarig, N., Yomdin, Y.: Accuracy of Algebraic Fourier reconstruction for shifts of several signals. Sampl. Theory Signal Image Process. **13**, 151–173 (2014)
4. Batenkov, D., Yomdin, Y.: Taylor Domination, Turán lemma, and Poincaré-Perron Sequences. Contemp. Math. **659**, 1–15 (2016)
5. Bautin, N.: Du nombre de cycles limites naissant en cas de variation des coefficients d'un etat d'equilibre du type foyer ou centre. C. R. (Doklady) Acad. Sci. URSS (N. S.) **24**, 669–672 (1939)
6. Bautin, N.: On the number of limit cycles which appear with the variation of coefficients from an equilibrium position of focus or center type. Am. Math. Soc. Transl. **100**, 19 (1954)
7. Bieberbach, L.: Analytische Fortsetzung. Springer, Berlin (1955)
8. Biernacki, M.: Sur les fonctions multivalentes d'ordre p. CR Acad. Sci. Paris **203**, 449–451 (1936)
9. Blinov, M., Briskin, M., Yomdin, Y.: Local center conditions for Abel equation and cyclicity of its zero solution. Complex analysis and dynamical systems II, Contemp. Math. Amer. Math. Soc. Providence, RI, **382**, 65–82 (2005)
10. Bodine, S., Lutz, D.A.: Asymptotic solutions and error estimates for linear systems of difference and differential equations. J. Math. Anal. Appl. **290**(1), 343–362 (2004)
11. Borcea, J., Friedland, S., Shapiro, B.: Parametric Poincaré-Perron theorem with applications. J. d'Analyse Mathématique **113**(1), 197–225 (2011)
12. Briskin, M., Roytvarf, N., Yomdin, Y.: Center conditions at infinity for Abel differential equations. Ann. Math. **172**(1), 437–483 (2010)
13. Briskin, M., Yomdin, Y.: Algebraic families of analytic functions I. J. Differ. Equ. **136**(2), 248–267 (1997)
14. Coppel, W.: Dichotomies and stability theory. In: Proceedings of the Symposium on Differential Equations and Dynamical Systems, pp. 160–162. Springer, Heidelberg (1971)
15. Francoise, J.-P., Yomdin, Y.: Bernstein inequality and applications to analytic geometry and differential equations. J. Funct. Anal. **146**(1), 185–205 (1997)
16. Friedland, O., Yomdin, Y.: An observation on Turán-Nazarov inequality. Studia Math. **218**(1), 27–39 (2013)
17. Friedland, O., Yomdin, Y.: (s, p)-valent functions, to appear in GAFA seminar notes
18. Hayman, W.K.: Multivalent functions, vol. 110. Cambridge University Press, Cmabridge (1994)
19. Kloeden, P., Potzsche, C.: Non-autonomous difference equations and discrete dynamical systems. J. Differ. Equ. Appl. **17**(2), 129–130 (2011)
20. Mitrinovic, D.S., Pecaric, J., Fink, A.M.: Classical and New Inequalities in Analysis, Series: Mathematics and its Applications, vol. 61, XVIII, p. 740 (1993)
21. Nazarov, F.L.: Local estimates of exponential polynomials and their applications to inequalities of uncertainty principle type. St Petersb. Math. J. **5**(4), 663–718 (1994)
22. Perron, O.: Über summengleichungen und Poincarésche differenzengleichungen. Mathematische Annalen **84**(1), 1–15 (1921)
23. Pituk, M.: More on Poincaré's and Perron's Theorems for Difference Equations. J. Differ. Equ. Appl. **8**(3), 201–216 (2002)
24. Poincare, H.: Sur les équations linéaires aux différentielles ordinaires et aux différences finies. Am. J. Math. **7**(3), 203–258 (1885)
25. Pötzsche, C.: Geometric Theory of Discrete Nonautonomous Dynamical Systems. Lecture Notes in Mathematics. Springer, Berlin (2010)
26. Roytwarf, N., Yomdin, Y.: Bernstein classes. Annales de l'institut Fourier **47**, 825–858 (1997)

27. Turán, P.: *Eine neue Methode in der Analysis und deren Anwendungen*. Akadémiai Kiadó, 1953
28. Turán, P., Halász, G., Pintz, J.: On a new method of analysis and its applications. Wiley-Interscience, New York (1984)
29. Yomdin, Y.: Nonautonomous linearization. Dynamical Systems (College Park, MD). Lecture Notes in Mathematics, vol. 1342, pp. 718–726. Springer, Berlin (1988)
30. Yomdin, Y.: Global finiteness properties of analytic families and algebra of their Taylor coefficients. The Arnoldfest (Toronto, ON, 1997), Fields Inst. Commun., vol. 24. Amer. Math. Soc., Providence, RI, pp. 527–555 (1999)
31. Yomdin, Y.: Singularities in algebraic data acquisition. Real and Complex Singularities. London Math. Soc. Lecture Notes Series, vol. 380, pp. 378–396, Cambridge Univeristy Press, Cambridge (2010)
32. Yomdin, Y.: Bautin ideals and Taylor domination. Publ. Mat. **58**, 529–541 (2014)

On the Set of Periods of the 2-Periodic Lyness' Equation

Guy Bastien, Víctor Mañosa and Marc Rogalski

Abstract We study the periodic solutions of the *non-autonomous periodic Lyness' recurrence* $u_{n+2} = (a_n + u_{n+1})/u_n$, where $\{a_n\}_n$ is a cycle with positive values a,b and with positive initial conditions. Among other methodological issues we give an outline of the proof of the following results: (1) If $(a, b) \neq (1, 1)$, then there exists a value $p_0(a, b)$ such that for any $p > p_0(a, b)$ there exist continua of initial conditions giving rise to $2p$-periodic sequences. (2) The set of minimal periods arising when $(a, b) \in (0, \infty)^2$ and positive initial conditions are considered, contains all the even numbers except 4, 6, 8, 12 and 20. If $a \neq b$, then it does not appear any odd period, except 1.

Keywords Difference equations with periodic coefficients · Elliptic curves · Lyness' type equations · QRT maps · Rotation number · Periodic orbits

1 Introduction

1.1 The Set of Periods

In this note we give a description of the *set of periods* of the 2-periodic Lyness' equations

---

G. Bastien · M. Rogalski
CNRS, Institut Mathématique de Jussieu-Paris Rive Gauche, Université Pierre et Marie Curie, 4 pl. Jussieu, 75005 Paris, France
e-mail: guy.bastien@imj-prg.fr

V. Mañosa (✉)
Departament de Matemàtiques; Control, Dynamics and Applications Group, Universitat Politècnica de Catalunya, Barcelona, Spain
e-mail: victor.manosa@upc.edu

M. Rogalski
Laboratoire Paul Painlevé, CNRS, Université des Sciences et Technologies de Lille, 59655 Villeneuve-d'Ascq, France
e mail: marc.rogalski@imj-prg.fr

L. Alsedà i Soler et al. (eds.), *Difference Equations, Discrete Dynamical Systems and Applications*, Springer Proceedings in Mathematics & Statistics 180, DOI 10.1007/978-3-662-52927-0_22

$$u_{n+2} = \frac{a_n + u_{n+1}}{u_n}, \quad \text{with} \quad a_n = \begin{cases} a \text{ for } n = 2\ell + 1, \\ b \text{ for } n = 2\ell, \end{cases} \tag{1}$$

where $(u_1, u_2) \in \mathcal{Q}^+ := \{(x, y) \in \mathbb{R}^2 : x > 0, y > 0\}; \ell \in \mathbb{N}$ and $a > 0, b > 0$. We will only outline the main results and focus on some methodological issues. We refer the reader to [4], where a further analysis and the proofs of the main results can be found.

Theorem 1 ([4, Corollary 3]) *Consider the 2-periodic Lyness' recurrence (1) for $a > 0$, $b > 0$ and positive initial conditions u_1 and u_2.*

(i) If $(a, b) \neq (1, 1)$, then there exists a number $p_0(a, b) \in \mathbb{N}$ such that for any $p > p_0(a, b)$ there exist continua of initial conditions giving rise to $2p$-periodic sequences.
(ii) The set of minimal periods arising when $(a, b) \in (0, \infty)^2$ and positive initial conditions are considered contains all the even numbers except 4, 6, 8, 12 *and* 20. *If $a \neq b$, then it does not appear any odd period, except* 1.

As we will see, the number $p_0(a, b)$ is computable with our tools, in an open and dense set of the parameter space.

1.2 A Dynamical System Approach

To study the recurrence (1) we will adopt a dynamical systems approach, by using the associated *composition map:*

$$F_{b,a}(x, y) := (F_b \circ F_a)(x, y) = \left(\frac{a + y}{x}, \frac{a + bx + y}{xy}\right), \tag{2}$$

being F_a and F_b the Lyness maps $F_\alpha(x, y) = (y, (\alpha + y)/x)$, associated to the autonomous lyness recurrence $u_{n+2} = (\alpha + u_{n+1})/u_n$.

It is very easy to observe that the first component of the iterates of the composition map gives the odd terms of the recurrence (1), while the second one gives the even terms.

The maps $F_{b,a}$ are particular cases of the family of *QRT maps*, which is a big class of mappings introduced in [13, 14], that contains many families of discrete dynamical systems arising in mathematical physics. The QRT one constitutes a paradigmatic family of discrete integrable maps, and the reader is referred to exhaustive monograph [8] for a deeper approach.

The main feature of QRT maps is that they are defined by a geometrical construction, in such a way they preserve a *fibration* of the plane given by some algebraic curves of a prescribed form. In the case of the map $F_{b,a}$, this preserved fibration is given by the family of cubics:

$$\mathcal{C}_h := \{(bx + a)(ay + b)(ax + by + ab) - hxy = 0\}.$$

Of course, this is equivalent to the fact that the map possess a *first integral* (invariant function) given by

$$V_{b,a}(x,y) := \frac{(bx+a)(ay+b)(ax+by+ab)}{xy}. \tag{3}$$

Recall that V is a first integral of a map F if and only if $V(F) = V$, so that the orbits of the dynamical system described by F lie on the level sets of the function V (the *energy levels* from now on). The expression of $V_{b,a}$ was also found, in another context, in [10] by following a technique for finding invariants for recurrences with periodic coefficients introduced in [9].

Since our approach depends strongly on the fact that the maps $F_{b,a}$ are QRT, it is worth noticing that the only k-periodic Lyness' difference equations such that their associated composition maps are QRT for all values of the parameters are the cases $k \in \{1, 2, 3, 6\}$ (see [7, Corollary 4]).

It is known that if $a, b > 0$, then any map $F_{b,a}$ has a unique fixed point $(x_c, y_c) \in \mathcal{Q}^+$ given by the solution of the system

$$\{x^2 = a + y, \quad y^2 = b + x,$$

which corresponds to the unique global minimum of $V_{b,a}$ in $\mathcal{Q}^+$.

Setting $h_c := \{V_{b,a}(x_c, y_c)\}$, the level sets $\mathcal{C}_h^+ := \left\{\{V_{b,a} = h\} \cap \mathcal{Q}^+\right\}$ are given by the curves

$$\mathcal{C}_h^+ := \{(bx+a)(ay+b)(ax+by+ab) - hxy = 0\} \cap \mathcal{Q}^+ \quad \text{for} \quad h > h_c,$$

which are closed ones ([6, Theorem 2]; see also Fig. 1).

It is also known (see [6], and also [4]) that the dynamics of $F_{b,a}$ restricted on each connected component of the invariant curves $\mathcal{C}_h$ is *conjugate to a rotation on the unit circle*.

Coming back to the problem of determining the periodic orbits of Eq. (1), and since the first components of the iterates of the map $F_{b,a}$ give the odd terms of the recurrence and the second one gives the even terms, in fact, by previously proving that if $a \neq b$ there are no periodic orbits with odd period different from 1 ([4, Lemma 22]), we obtain that Theorem 1 follows as a corollary of the next result:

Theorem 2 ([4, Theorem 2]) *Consider the family of maps $F_{b,a}$ given in (2) for $a, b > 0$.*

(i) If $(a, b) \neq (1, 1)$, then there exists a value $p_0(a, b) \in \mathbb{N}$ such that for any $p > p_0(a, b)$ there exist at least a continuum of initial conditions in $\mathcal{Q}^+$ (an oval $\mathcal{C}_h^+$) giving rise to p-periodic orbits of $F_{b,a}$.

(ii) The set of periods arising in the family $\{F_{b,a},\ a > 0, b > 0\}$ restricted to $\mathcal{Q}^+$ contains all minimal periods except 2, 3, 4, 6 *and* 10.

It is interesting to notice that the minimal periods 2 and 3 do not appear for any a and b in the whole domain of definition of the dynamical system defined by $F_{b,a}$, but periods 4, 6, and 10 appear for some $a, b > 0$ and some initial conditions outside $\mathcal{Q}^+$, see the example of Sect. 3.1 for instance.

2 Searching for Periods

In this section we explain the strategy to compute the allowed periods for a particular map $F_{b,a}$, and finally for the whole family $\{F_{b,a},\ a, b > 0\}$.

2.1 Each Map $F_{b,a}$ Has an Infinite Number of Periods

In this section we give an sketch of the proof of Theorem 2(i). First observe that since $F_{b,a}$ is conjugated to a rotation on each set $\mathcal{C}_h^+$, we can consider the *rotation number function* parameterized by the energy level $h \geq h_c$: $\theta_{b,a}(h)$, see Fig. 1.

In [6] (and also in [4] using another approach) it is proved that $\theta_{b,a}(h)$ is a continuous function in $[h_c, \infty)$, in fact analytic in (h_c, ∞). Furthermore, we can compute $\theta_{b,a}(h_c) := \lim_{h\to h_c} \theta_{b,a}(h)$ and $\lim_{h\to+\infty} \theta_{b,a}(h)$, resulting that *generically* (that is, in an open and dense subset of the parameter space) $\theta_{b,a}(h_c) \neq \lim_{h\to+\infty} \theta_{b,a}(h)$. This means that generically there exists a non-degenerate interval containing some allowed rotation numbers $I(a, b) = \langle\theta_{b,a}(h_c), 2/5\rangle \subseteq \theta_{b,a}(h_c, +\infty)$ (see Corollary 1), which means that the problem of partially knowing the set of periods of the map is equivalent to the problem of determining the set of denominators of the irreducible fractions in the interval $I(a, b)$. In summary:

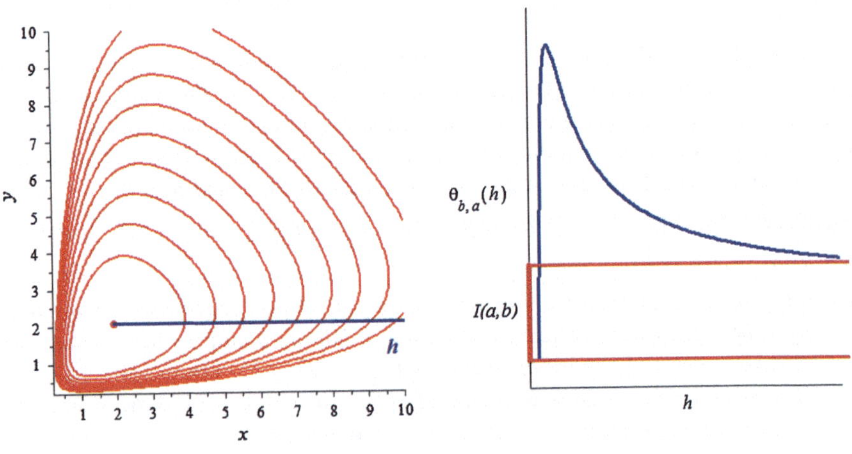

Fig. 1 The invariant curves $\mathcal{C}_h^+$, and the sketch of the graph a possible function $\theta_{b,a}(h)$ with the definition of the interval $I(a, b)$

Proposition 1 ([4, Proposition 13]) *The rotation number map $\theta_{b,a}(h)$ is continuous in $[h_c, +\infty)$. Furthermore*

$$\lim_{h\to h_c^+} \theta_{b,a}(h) = \sigma(a,b) := \frac{1}{2\pi}\arccos\left(\frac{1}{2}\left[-2+\frac{1}{x_c\, y_c}\right]\right) \quad \textit{and} \quad \lim_{h\to+\infty} \theta_{b,a}(h) = \frac{2}{5}.$$

As a consequence of the above result we have the following corollary:

Corollary 1 *Set $I(a,b) := \langle\sigma(a,b), \frac{2}{5}\rangle :=$* $(\min(\sigma(a,b), 2/5), \max(\sigma(a,b), 2/5))$.

(i) *If $\sigma(a,b) \neq 2/5$, then for all $\theta \in I(a,b)$, there exists an oval $\mathscr{C}_h^+$ such that $F_{b,a}(\mathscr{C}_h^+)$ is* conjugate to a rotation, *with a rotation number $\theta_{b,a}(h) = \theta$.*
(ii) *For all irreducible $q/p \in I(a,b)$, there exists* periodic orbits *of $F_{b,a}$ of minimal period p.*

The above result guarantees that if the parameters a and b are such that $\sigma(a,b) \neq 2/5$, then there exists a *known* non-degenerate interval containing the information of some allowed periods. The existence of this interval ensures the existence of a *computable* number $p_0(a,b) \in \mathbb{N}$, such that for all natural number $p > p_0(a,b)$ there exists an irreducible fraction $q/p \in I(a,b) \subseteq \theta_{b,a}(h_c, +\infty)$, and hence proving Theorem 2(i) for this case. The computability of $p_0(a,b)$ is explained in the next subsection, now we focus on the fact that the property $\sigma(a,b) \neq 2/5$ is a generic one. Indeed, consider the parameter space $\mathscr{P} := \{(a,b),\ a,b > 0\}$, then the subset $\{\sigma(a,b) = 2/5\} \subset \mathscr{P}$ for $a, b > 0$ is given by the curve

$$\Gamma := \{\sigma(a,b) = 2/5,\ a,b > 0\} = \left\{(a,b) = \left(\frac{t^3-\phi^2}{t}, \frac{\phi^4-t^3}{t^2}\right),\ t \in (\phi^{\frac{2}{3}}, \phi^{\frac{4}{3}})\right\},$$

where $\phi = (1+\sqrt{5})/2$ see [4, Corollary 19] and Fig. 2.

Of course $\mathscr{P}\backslash\Gamma$ is open and dense in $\mathscr{P}$, and therefore the fact that $I(a,b) \neq \emptyset$ and the computability of $p_0(a,b)$ are generic properties.

Suppose that $\sigma(a,b) = 2/5$. As mentioned before (see also [4, Lemma 22]) it is proved that if $a \neq b$ there are no period orbits of odd period, and that $(a,b) = (1,1)$ is the only point in the parameter space satisfying both that $\sigma(a,b) = 2/5$ and $a = b$. Then, if $(a,b) \neq (1,1)$ it is clear that $J(a,b) := \theta_{b,a}([h_c, +\infty])$ is a closed interval with nonempty interior (i.e. not the single value 2/5), since otherwise there will be globally 5-periodic maps $F_{b,a}$ with $a \neq b$, a contradiction. This last fact, implies the existence of the number $p_0(a,b) \in \mathbb{N}$, such that for all natural number $p > p_0(a,b)$ there exists an irreducible fraction $q/p \in \theta_{b,a}(h_c, +\infty)$. This number is non-computable with our method.

Collecting all the above considerations one gets the proof of Theorem 2(i), and hence Theorem 1(i).

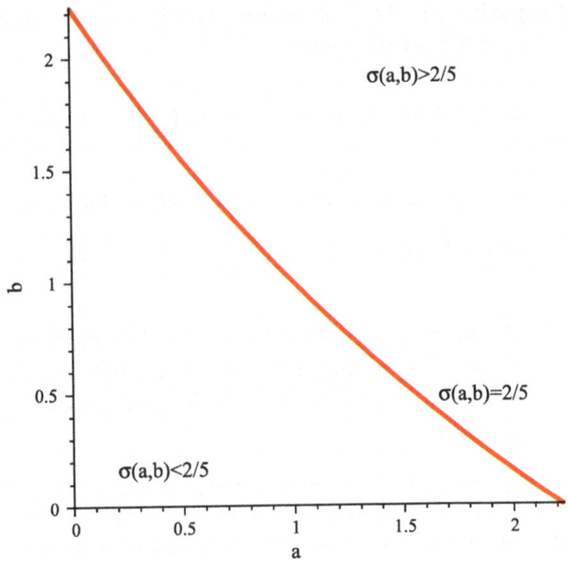

Fig. 2 The curve $\Gamma = \{\sigma(a, b) = 2/5$ for $a, b > 0\}$ in the parameter space $\mathscr{P}$

2.1.1 Determining the Particular Periods of a Given Map $F_{b,a}$

In this section we briefly outline a methodology for determining the set of periods of a map $F_{b,a}$, for some given fixed values $a, b > 0$, in the case $\sigma \neq 2/5$. In this case it is clear that to determine which are the periods of a particular $F_{b,a}$ is equivalent to determine which are the irreducible fractions in $I(a, b)$, and this can be done following the next steps:

1. If $\sigma(a, b) \neq 2/5$, since the interval $I(a, b)$ is known, it is possible to obtain *constructively* a value p such that for any $r > p$, there exists an irreducible fraction $q/r \in I(a, b)$. This can be done in many ways, for instance, by using the following result

Lemma 1 ([5, Theorem 25 and Corollary 26]) *Consider an open interval* (c, d) *with* $0 \leq c < d$; *denote by* $p_1 = 2, p_2 = 3, p_3, \ldots, p_n, \ldots$ *the set of all the prime numbers, ordered following the usual order. Also consider the following natural numbers:*

- *Let* p_{m+1} *be the smallest prime number satisfying that* $p_{m+1} > \max(3/(d - c), 2)$,
- *Given any prime number* p_n, $1 \leq n \leq m$, *let* s_n *be the smallest natural number such that* $p_n^{s_n} > 4/(d - c)$.
- *Set* $p := p_1^{s_1-1} p_2^{s_2-1} \cdots p_m^{s_m-1}$.

Then, for any $r > p$ *there exists an irreducible fraction* q/r *such that* $q/r \in (c, d)$.

2. Finally, a finite checking (that can be done with a computer using integer arithmetics, if the above value p is large) determines which values of $r \leq p$ are such that there exists an irreducible fraction $q/r \in I(a,b)$. The minimum value is then $p_0(a,b)$.

It is known that the rotation function $\theta_{b,a}(h)$ is not always monotonic, hence $\overline{I(a,b)} \subset \theta_{b,a}([h_c,+\infty])$. So still the finite set of forbidden periods have to be determined to know exactly the set of periods of the map $F_{b,a}$. To do this, it is useful the algebraic geometric approach described in Sect. 3.1.

2.2 The Periods of the Family $\{F_{b,a},\ a,b>0\}$

Now we give an sketch of the proof of Theorem 2(ii). By using the characterization of the interval $I(a,b)$ for the subfamily determined by $a=b^2$, it is straightforward to obtain that

$$\bigcup_{b>0} I(b^2,b) = \left(\frac{1}{3},\frac{1}{2}\right) \setminus \left\{\frac{2}{5}\right\},$$

see [4, Lemma 16 and Proposition 17]. Hence, since

$$\bigcup_{b>0} I(b^2,b) = \left(\frac{1}{3},\frac{1}{2}\right) \setminus \left\{\frac{2}{5}\right\} \subseteq \bigcup_{a>0,\,b>0} I(a,b) \subseteq \bigcup_{a>0,\,b>0} \theta_{b,a}\left(h_c,+\infty\right),$$

and the map $F_{1,1}$ is globally 5-periodic (with $\theta_{1,1}\equiv 2/5$), we have the following result:

Proposition 2 *For each θ in $(1/3,1/2)$ there exists $a,b>0$ and at least an oval $\mathscr{C}_h^+$, such that $F_{b,a}(\mathscr{C}_h^+)$ is conjugate to a rotation with rotation number $\theta_{b,a}(h)=\theta$. In particular, for all irreducible $q/p \in (1/3,1/2)$, there exists p-periodic orbits of $F_{b,a}$.*

We will be able, then, to characterize some periods of $\{F_{b,a},\ a,b>0\}$ (in fact all of them), if we are able to characterize the irreducible fractions in $(1/3,1/2)$. To this end we apply Lemma 1 to the interval $(1/3,1/2)$, obtaining that for all $r\in\mathbb{N}$, such that $r>p$ with $p:=2^4\cdot 3^3\cdot 5\cdot 7\cdot 11\cdot 13\cdot 17 = 12\,252\,240$, there exists an irreducible fraction $q/r\in(1/3,1/2)$. Again, a finite checking determines which values of $r\leq p$ are such that $q/r\in(1/3,1/2)$, resulting that there appear irreducible fractions with all the denominators except 2, 3, 4, 6 and 10. Hence from Proposition 2 we have that for all $p\notin\{2,3,4,6,10\}$ there exists $a,b>0$ and at least an oval $\mathscr{C}_h^+$ with associated rotation number $\theta_{b,a}(h)=q/p$, thus giving rise to a continuum (at least) of p-periodic orbits of $F_{b,a}$.

Notice that still it must be proved that 2, 3, 4, 6 and 10 are forbidden. The reader is referred to [4] where, to this end, it is used the algebraic geometric approach described in Sect. 3.1.

3 Asymptotic Behavior of $\theta_{b,a}(h)$

The key point in all the proofs of the above results is the characterization of the intervals $I(a,b)$, and in particular the asymptotic value of $\theta_{b,a}(h)$ at infinity given in Proposition 1 (the value at the fixed point can be computed straightforwardly using tools like [3, Proposition 8], for example). In this section we sketch the proof of this result, by following the methodology introduced in [1] to study the autonomous Lyness equation. This methodology has been applied also for other equations, see [2, 3], for instance.

3.1 The Map $F_{b,a}$ from an Algebraic Geometric Approach

In this section we will give an easy formulation of map $F_{b,a}(\mathscr{C}_h)$ in terms of an inner operation of the invariant curve. This description is useful to see that the action of $F_{b,a}$ on each connected component of $\mathscr{C}_h$ is conjugated to a rotation, and to study and locate the invariant curves filled with periodic orbits with a prescribed period, as well.

We start by extending the curves $\mathscr{C}_h$ and the map $F_{b,a}$ to the complex projective space. Indeed, The curves $\mathscr{C}_h$, in homogeneous coordinates $[x:y:t]\in\mathbb{C}P^2$, are

$$\widetilde{\mathscr{C}_h}=\{(bx+at)(ay+bt)(ax+by+abt)-hxyt=0\}.$$

Observe that there exists three points at the infinity line: $H=[1:0:0]$; $V=[0:1:0]$; and $D=[b:-a:0]$ which are common to all the curves (see Fig. 3).

The key point in our approach is the following result:

Proposition 3 ([4, Proposition 4]) *If $a>0$ and $b>0$, then for all $h>h_c$, the curves $\widetilde{\mathscr{C}_h}$ are elliptic.*

Recall that a *cubic* curve is *elliptic* if and only if it does not contain any singular point. Also recall that in any elliptic curve an inner sum can be defined, endowing it with a group structure. This inner operation is given by the *chord-tangent process*, described bellow (see [16], for instance, for further details):

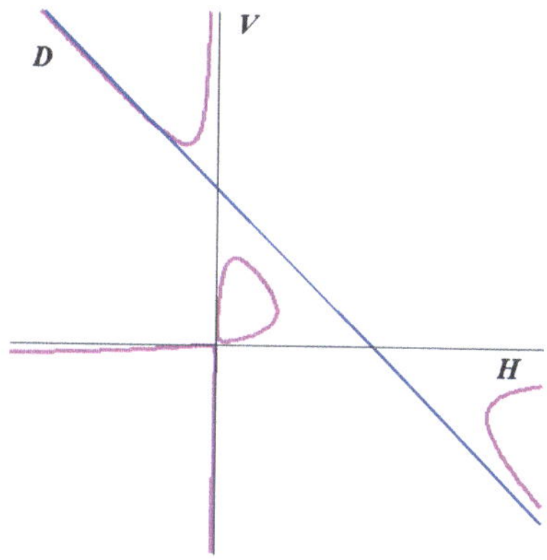

Fig. 3 A typical curve $\widetilde{\mathscr{C}}_h$ with $h > h_c$ in the real affine plane. The infinity points V, H, and D, are common to all curves. Observe that in the projective plane the three real unbounded components form a closed curve

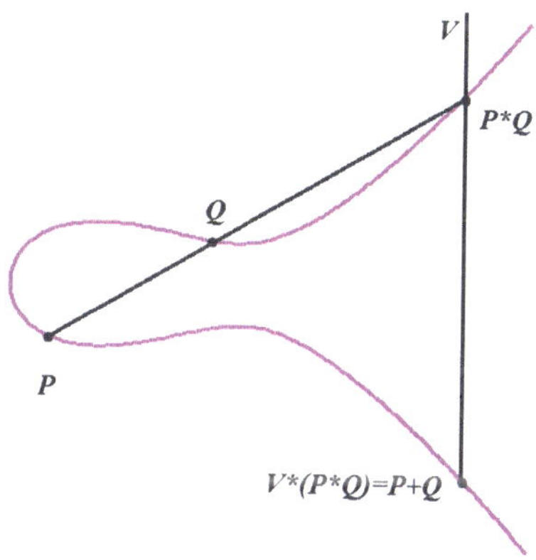

Fig. 4 The chord-tangent law defined on an elliptic curve, taking the infinite vertical point V as a neutral element. The chord passing through P and Q intersects the curve at a third point denoted by $P * Q$. The point $P + Q$ is then defined as $V * (P * Q)$

The chord-tangent law in a non-singular cubic. Consider a nonsingular cubic $\mathscr{C}$. Take two points P and $Q \in \mathscr{C}$.

1. Select a point $\mathscr{O}$ on the curve, to be the neutral element of the inner sum.
2. Take the chord passing through P and Q (the tangent line if $P = Q$). It will always intersect $\mathscr{C}$ at a third point denoted by $P * Q$.
3. The point $P + Q$ is then defined as $\mathscr{O} * (P * Q)$.

The curve endowed with this operation, denoted by $(E, +, \mathscr{O})$, is an abelian group (Fig. 4).

Any map $F_{b,a}$ extends to $\mathbb{C}P^2$ as the polynomial map: $\widetilde{F}_{b,a}([x : y : t]) = [ayt + y^2 : at^2 + bxt + yt : xy]$. As for any QRT map, in fact for any birational map preserving an elliptic curve, there is a relation between the dynamics of the map and the group structure of the curve, see [11, Theorem 3]. In our case it is very easy to verify the following result (see Fig. 5):

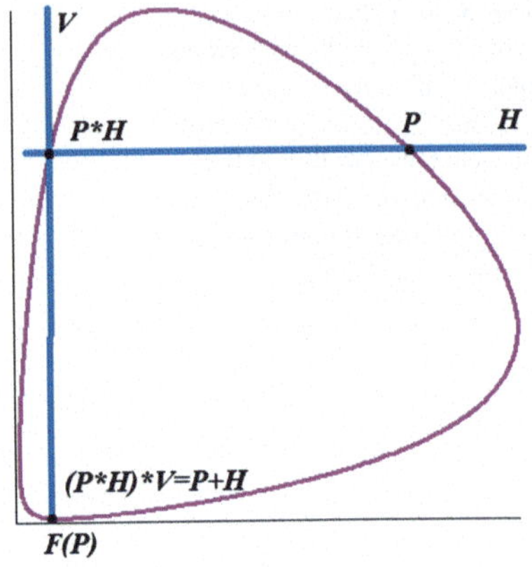

Fig. 5 The relation between the dynamics of $\widetilde{F}_{b,a}$ and the *group structure* of $\widetilde{\mathscr{C}}_h$, which also follows from the fact that $F_{b,a}$ is a QRT map

Proposition 4 *For each value h such that $\widetilde{\mathscr{C}}_h$ is elliptic,*

$$\widetilde{F}_{b,a|\widetilde{\mathscr{C}}_h}(P) = P + H,$$

where + is the addition of the group law of $\widetilde{\mathscr{C}}_h$ taking the infinite point V as the neutral element.

As a corollary of this result the action of $F_{b,a}$ on each connected component of $\mathscr{C}_h^+$ is conjugated to a rotation.

Observe that, as a consequence of the above result

$$\widetilde{F}^n_{b,a}(P) = P + n\,H.$$

Hence, the set $\widetilde{\mathscr{C}}_h$ is full of p-periodic orbits if and only if the following equation is satisfied

$$pH = V, \tag{4}$$

that is, if and only if H is a *torsion point* of $(\widetilde{\mathscr{C}}_h, +, V)$. This equation is very useful to characterize the level sets where the periodic points are located, as well to characterize the forbidden periods of a particular $F_{b,a}$, or the whole family as well, as it is shown in the following example.

Example. Here we will characterize which are the level sets $\widetilde{\mathscr{C}}_h$, where the 4-periodic orbits of $\widetilde{F}_{b,a}$ are located, by using Eq. (4).

First, we notice that since the neutral element V is not an inflection point of $\widetilde{\mathscr{C}}_h$, then the usual collinearity relation writes $A + B + C = V * V$, instead of $A + B + C = V$. In consequence for any given a point P on $\widetilde{\mathscr{C}}_h$, we have $-P = P * (V * V) = P * [-a/b : 0 : 1]$, see [4, Remark 7].

Now, observe that condition $4H = V$ is equivalent to $2H = -2H$, and then $2H = 2H * [-a/b : 0 : 1]$. An straightforward computation gives that $2H = [0 : -a : 1]$ and

$$2H * [-a/b : 0 : 1] = \left[-\frac{a^3 - a^2b^2 - ab - h + b^3}{ab\left(a - b^2\right)}, -\frac{ab + h - b^3}{a\left(a - b^2\right)} : 1 \right],$$

hence there will be 4-periodic orbits if and only if $a = \left(ab + h - b^3\right) / \left(a\left(a - b^2\right)\right)$, and then the energy level where the 4-periodic orbits are located is given by: $h_4 := (a - b^2)(a^2 - b)$. In [4, Proposition 21], it is proved that when $a, b > 0$, then $h_4 < h_c$, and therefore there are not 4 periodic orbits in $\mathcal{Q}^+$.

3.2 The Weierstrass Form of the Curves $\widetilde{\mathscr{C}}_h$ and the Conjugated Dynamics on Them

To compute the asymptotic behavior of $\theta_{b,a}(h)$ at infinity, we follow the scheme introduced in [1]. According to this, instead of looking for a normal form of the map $\widetilde{F}_{b,a}$ we look for a normal form of the invariant curve $\widetilde{\mathscr{C}}_h$. The idea is to find a birational change of variables preserving the group structure (a group's isomorphism) sending each curve $(\widetilde{\mathscr{C}}_h, +, V)$ to the corresponding Weierstrass Normal Form $(\widehat{\mathscr{E}}_L, +, V)$, where

$$\widehat{\mathscr{E}}_L := \{\, y^2 t = 4\,x^3 - g_2\, x t^2 - g_3\, t^3 \},$$

being $g_i := g_i(a, b, h)$, some functions whose expression can be found in [4, Proposition 11]; and where the group law is the chord-tangent one with the vertical point V as a neutral element. This isomorphism, can be found in [4, Sect. 2.2.1],[1] and it gives a conjugation between the action of $\widetilde{F}_{b,a}(\widetilde{\mathscr{C}}_h)$ and the action of certain birational map $\widehat{G}(\mathscr{E}_L)$:

$$\begin{array}{rcl} i : (\widetilde{\mathscr{C}}_h, +, V) & \stackrel{\cong}{\rightarrow} & (\widehat{\mathscr{E}}_L, +, V) \\ \widetilde{F}_{b,a}|_{\widetilde{\mathscr{C}}_h} : P \mapsto P + H & \rightarrow & \widehat{G}|_{\mathscr{E}_L} : P \mapsto P + \widehat{H} \end{array}$$

where $\widehat{H} = i(H)$, (see [4, Proposition 11(iii)]). The purpose of working with the Weierstrass normal form is the following:

1. We can parameterize the Weierstrass curve by using the Weierstrass $\wp$ function.
2. This parametrization allows us to obtain the *integral expression* (8) for the rotation number function.
3. The *asymptotic* behavior of the integral expression (8) can be studied, using the tools in [1], giving the desired result.

By Proposition 3, the cubic curves $\widetilde{\mathscr{C}}_h$ with $h > h_c$ are elliptic (thus non-singular), and therefore the curves $\widehat{\mathscr{E}}_L$ are also elliptic. In this case, the *degree-genus formula* for complex algebraic curves [12],

---

[1] The Weierstrass Normal Form can be computed using computer algebra systems like Magma or Sage, see the Appendix. In [4] the isomorphism is explicitly constructed in order to use them in some steps of the proof of Proposition 11.

$$g = \frac{1}{2}(d-1)(d-2),$$

(where d and g stand for the *degree* and *genus* of the curve, respectively), gives $g = 1$, and hence each curve $\widehat{\mathscr{E}_L}$ is (holomorphic) homeomorphic to a two dimensional torus $\mathbb{T}^2$. It is well known (the reader is addressed to any standard book of Algebraic Geometry like [12] or [15] for instance) that there is a suitable parametrization of the curve $\widehat{\mathscr{E}_L}$ obtained in the following way:

The standard parametrization of the Weierstrass Cubic. For a given Weierstrass cubic $\widehat{\mathscr{E}_L}$, there exists a lattice

$$\Lambda = \{2n\omega_1 + 2m\, i\omega_2 \text{ such that } (n,m) \in \mathbb{Z}^2\} \subset \mathbb{C},$$

such that the map

$$\begin{aligned} \phi : \mathbb{T}^2 \cong \mathbb{C}/\Lambda &\longrightarrow \widehat{\mathscr{E}_L} \\ z &\longrightarrow \begin{cases} [\wp(z) : \wp'(z) : 1] \text{ if } z \notin \Lambda, \\ V = [0:1:0] \text{ if } z \in \Lambda, \end{cases} \end{aligned}$$

where $\wp(z)$ is the *Weierstrass $\wp$ function relative to Λ*

$$\wp(z) = \frac{1}{z^2} + \sum_{\lambda \in \Lambda \setminus \{0\}} \left[\frac{1}{(z-\lambda)^2} - \frac{1}{\lambda^2} \right],$$

is an holomorphic homeomorphism, and it gives a parametrization of $\widehat{\mathscr{E}_L}$. Observe that, in consequence, the function $\wp$ satisfies the differential equation

$$\wp'(z)^2 = 4\wp(z)^3 - g_2\wp(z) - g_3. \tag{5}$$

In our case ω_1 and ω_2 depend on a, b and h, and the lattice Λ can be chosen in such a way that the associated parametrization posses the following properties that are interesting for our purposes, see Fig. 6:

1. The circle given by $[0, 2\omega_1]/\Lambda$ is projected bijectively into the *real* unbounded branch of the Weierstrass curve $\widehat{\mathscr{E}_L}$, namely $\mathscr{E}_L = \{y^2 = 4x^3 - g_2 x - g_3\}$, (from now on we drop the hat when referring to the curve in the real plane).
2. The interval $[0, \omega_1] \in \Lambda$ is projected bijectively into the real unbounded half branch of $\mathscr{E}_L$ with *negative coordinates*.
3. The point $\omega_1 \in \mathbb{C}/\Lambda$ is projected to the real affine point $(e_1, 0)$ which is the intersection of the real unbounded branch of $\mathscr{E}_L$, with the real abscissa axis (see Fig. 6) so

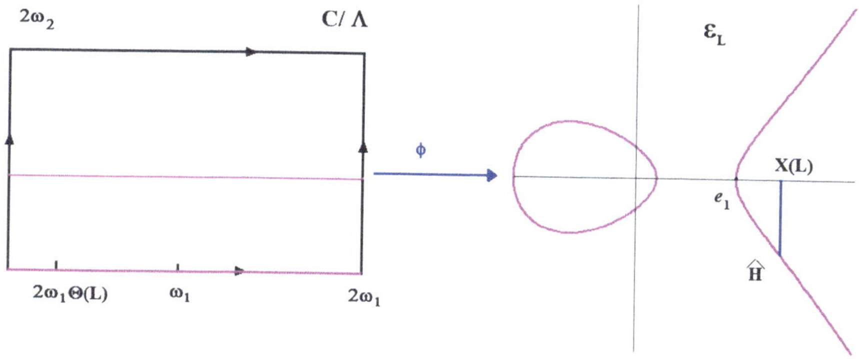

Fig. 6 Relationship between the geometry of the Weierstrass elliptic curve $\mathscr{E}_L$ and the action of $\widehat{G}$

$$e_1 = \wp(\omega_1) \tag{6}$$

With all the above geometric consideration, by direct integration of the differential equation (5) on $[0, u)$ with $u \le \omega_1$, we have that in real variables

$$u = \int_{\wp(u)}^{+\infty} \frac{\mathrm{d}s}{\sqrt{4s^3 - g_2 s - g_3}}. \tag{7}$$

On the other hand the map $\widehat{G}_{|\mathscr{E}_L}$ has the following properties:

1. The map $\widehat{G}_{|\mathscr{E}_L}$ is conjugated to *rotation* with rotation number $\Theta(L) \in \left[0, \frac{1}{2}\right)$.
2. Since V is the neutral element of $(\mathscr{E}_L, +, V)$, then

$$\widehat{G}_{|\mathscr{E}_L}(V) = V + \widehat{H} = \widehat{H}$$

3. The point $\widehat{H} = i(H)$, is located on the real unbounded half branch of $\mathscr{E}_L$ with *negative coordinates*, [4, Proposition 11(iii)].

In summary, $\widehat{G}$ maps the point V, with associated parameter $u = 0$ in the interval $[0, \omega_1]/\Lambda$, to the point $\widehat{H}$ with some parameter $u = 2\omega_1\Theta(L) \in [0, \omega_1]$. Hence, the abscissa of $\widehat{H}$, say $X(L)$, must satisfy $X(L) = \wp(2\omega_1\Theta(L))$. Using the integral expression (7) we have that

$$2\omega_1\Theta(L) = \int_{X(L)}^{+\infty} \frac{\mathrm{d}s}{\sqrt{4s^3 - g_2\, s - g_3}}.$$

Now, using Eq. (6) and again (7), we get the desired integral expression for the rotation number

$$\Theta(L) = \frac{1}{2}\frac{\displaystyle\int_{X(L)}^{+\infty}\frac{\mathrm{d}s}{\sqrt{4s^3-g_2\,s-g_3}}}{\displaystyle\int_{e_1}^{+\infty}\frac{\mathrm{d}s}{\sqrt{4s^3-g_2\,s-g_3}}}. \tag{8}$$

The asymptotic analysis of this equation can be done using the tools developed in [1], resulting that $\Theta(L) \sim 2/5$. To this end, the asymptotic expansions of some of the elements of the curve and of $X(L)$ must to be known, and this is one of the main computational obstruction in the proof, see [4] for further details.

Acknowledgments The second author is partially supported by Spain's Ministry of Economy and Competitiveness (MEC) through grants DPI2011-25822 and DPI2016-77407-P. CoDALab group is supported by Generalitat de Catalunya through the SGR program.

Appendix

The following, is a Magma code for computing the Weierstrass Normal form $\widehat{\mathscr{E}}_L$ of the curves $\widetilde{\mathscr{C}}_h$.

```
K<a,b,h>:=FunctionField(Rationals(),3);
A<x,y>:=AffineSpace(K,2);
C:=Curve(A,(b*x+a)*(a*y+b)*(a*x+b*y+a*b)-h*x*y);
CP:=ProjectiveClosure(C);
P:=CP![0,1,0];
E:=EllipticCurve(CP,P);
WeierstrassModel(E);
```

It can be processed in the Magma's site http://magma.maths.usyd.edu.au/calc/. Using Magma V2.18-8, it gives the expression of $\widehat{\mathscr{E}}_L$ in 0.220 s with a total memory usage of 12.06 MB.

References

1. Bastien, G., Rogalski, M.: Global behavior of the solutions of Lyness' difference equation $u_{n+2}u_n = u_{n+1} + a$. J. Differ. Equ. Appl. **10**, 977–1003 (2004)
2. Bastien, G., Rogalski, M.: On the Algebraic Difference Equation $u_{n+2}u_n = \psi(u_{n+1})$ in $\mathbb{R}_*^+$ Related to a Family of Elliptic Quartics in the Plane. Adv. Differ. Equ. **3**, 227–261 (2005)
3. Bastien, G., Rogalski, M.: On algebraic difference equations $u_{n+2} + u_n = \psi(u_{n+1})$ in $\mathbf{R}$ related to a family of elliptic quartics in the plane. J. Math. Anal. Appl. **326**, 822–844 (2007)
4. Bastien G., Mañosa V., Rogalski M.: On periodic solutions of 2-periodic Lyness difference equations, Int. J. Bifurc. Chaos **23**, 1350071, 18p (2013)
5. Cima, A., Gasull, A., Mañosa, V.: Dynamics of the third order Lyness difference equation. J. Differ. Equ. Appl. **13**, 855–884 (2007)

6. Cima, A., Gasull, A., Mañosa, V.: On 2- and 3-periodic Lyness difference equations. J. Differ. Equ. Appl. **18**, 849–864 (2012)
7. Cima, A., Gasull, A., Mañosa, V.: Integrability and non-integrability of periodic non-autonomous Lyness recurrences. Dyn. Syst. **28**, 518–538 (2013)
8. Duistermaat, J.J.: Discrete Integrable Systems: QRT Maps and Elliptic Surfaces. Springer, New York (2010)
9. Feuer, J., Janowski, E.J., Ladas, G.: Invariants for some rational recursive sequence with periodic coefficients. J. Differ. Equ. Appl. **2**, 167–174 (1996)
10. Janowski, E.J., Kulenović, M.R.S., Nurkanović, Z.: Stability of the kth order Lyness' equation with period-k coefficient. Int. J. Bifurc. Chaos **17**, 143–152 (2007)
11. Jogia, D., Roberts, J.A.G., Vivaldi, F.: An algebraic geometric approach to integrable maps of the plane. J. Phys. A: Math. Gen. **39**, 1133–1149 (2006)
12. Kirwan, F.: Complex Algebraic Curves. Cambridge University Press, Cambridge (1992)
13. Quispel, G.R.W., Roberts, J.A.G., Thompson, C.J.: Integrable mappings and soliton equations. Phys. Lett. A **126**, 419–421 (1988)
14. Quispel, G.R.W., Roberts, J.A.G., Thompson, C.J.: Integrable mappings and soliton equations II. Phys. D **34**, 183–192 (1989)
15. Silverman, J.: The Arithmetic of Elliptic Curves. Springer, New York (2009)
16. Silverman, J., Tate, J.: Rational Points on Elliptic Curves. Springer, New York (1992)

The manufacturer's authorised representative in the EU is Springer Nature Customer Service Centre GmbH, Europaplatz 3, 69115 Heidelberg, Germany. If you have any concerns regarding our products, please contact ProductSafety@springernature.com

Printed and bound by CPI Group (UK) Ltd, Croydon, CR0 4YY
22/09/2026
02221062-0018